Around the World in 84 Days

To Grayson & Toby ~
Remember to keep
reaching for the sky.
My best regards.

Jerry Carr
CDR/SKYLAB4

Around the World

in 84 Days

The Authorized Biography of Skylab
Astronaut Jerry Carr

by

Dave Shayler

An Apogee Books Publication

We acknowledge the financial support of the Government of Canada through the Book Publishing Industry Development Program for our publishing activities.

Published by Apogee Books an imprint of Collector's Guide Publishing Inc., Box 62034, Burlington, Ontario, Canada, L7R 4K2, http://www.cgpublishing.com

Printed and bound in Canada

Around the World in 84 Days (First Edition) by Dave Shayler
ISBN 9781-894959-40-7 - ISSN 1496-6921

Contents

Acknowledgements

This has been a long and involved project that has been assisted by a large number of contacts and correspondents over many years.

Of course this book would not have been possible without the immense help and support of Jerry Carr, who opened his family archives, and searched his memory over a period of almost 20 years since the project was first suggested. His assistance with editing and proofreading the manuscript was invaluable.

The support and cooperation of Jerry's immediate family was also critical to the development and detailing of this project and sincere thanks go to:

Pat Musick (Carr) for spending time detailing her own life and career, reflecting on Jerry and Pat's 29 year marriage, and help editing and proofreading the manuscript; JoAnn Carr for very personal memories of her marriage to Jerry and her own 'mission' on Earth while Jerry was in Space; Jerry's brother Ronald Carr for expanding the history of the Carr dynasty; Jerry's six children: Jennifer, Jeff, Jamee, John, Jessica and Josh who all provided details of Jerry's life and family activities as well as their own careers since leaving high school. The Carr family also provided unprecedented information on what it was like to have Dad as an astronaut. They also revealed the family life before, during and after spaceflight, and the triumphs and setbacks that entails.

Jerry's own recollections and extensive archive of his years both in the US Marine corps and at NASA have been supplemented by interviews and research at NASA facilities and with former colleagues. Most notable of these were his fellow Skylab 4 crewmates Bill Pogue and Ed Gibson, together with their Skylab colleagues, all of whom helped supplement the story of Skylab told in the companion volume *Skylab: America's Space Station* [Springer-Praxis 2001].

Staff at the NASA JSC Public Affairs Office Space Center in Houston, the former JSC History Office; Rice University, NARA Fort Worth, PAO at Kennedy Space Center in Florida provided background material and access. Special thanks are due to: former and current PAO members: Iva 'Scotty' Scott; Barbara Schwartz, Eileen Hawley, James Hartsfield and Jerry's son Jeff Carr; all of whom helped in arranging visits, tours and interviews at the Houston space centre, at contractors and support sites during the years 1988-2002.

JSC History office contacts (1988-1994) were: Janet Kovacevich, Joey Pellerin (from 1994) Dave Portree and more recently Glen Swanson. At Rice University's Fondren Library Joan Ferry provided access to the Skylab files when they were located there (1988-1994), then at University of Clear Lake and at NARA facilities located in Fort Worth.

At NASA HQ History Office, former archivist Lee Saegesser, and former NASA historian Roger Launius provided further background details over many years, on Jerry's career, the Skylab program, Jerry's fellow astronauts and human spaceflight in general.

At KSC NASA, Historian and Archivist Ken Nail supplied information on Skylab launch processing.

Several former astronauts have provided extensive details on Skylab issues and missions over the years. These included: Walter Cunningham, Owen Garriott, Fred Haise, Bill Lenoir, Don Lind, Jack Lousma, Story Musgrave, Bill Thornton, and Paul Weitz.

The work of Michael Cassutt and his two books on Deke Slayton and Tom Stafford were of great help in detailing the history, organization and workings of the Astronaut Office and astronaut selection system.

Thanks to my family for supporting this long project and giving the encouragement to see it through. Especially to my brother Mike who spent hours editing and preparing the manuscript and illustrations, corresponding between myself and Jerry on numerous occasions over several years. To Ruth Shayler, Mike's wife, who painstakingly transcribed numerous family interviews for the book.

Special thanks to Rob Godwin and the staff of Apogee Books who gave the project a home, and despite extended delays and hurdles, continued to support and believe in the project to its publication.

Special thanks to Rex Hall, past President of the British Interplanetary Society, whose staff allowed me access to their impressive library and archive.

Finally, love and admiration to my partner Bel Edge, who not only allowed me to spend hours at the computer turning notes and scribbles into formal chapters, but also helped on the transcribing of interviews and scanning of original documents for use in the text.

To one and all a very large "Thank You."

Dave Shayler FBIS

Introduction

As far as I am concerned, writing a book ranks just below going to the dentist on my list of things I would like to do. For years, friends, family and acquaintances have badgered me to write my story, and I have steadfastly resisted. I greatly admired the effort that Bill Pogue and Ed Gibson, my soul mates on Skylab 4, had put into book writing. I read and enjoyed their literary pieces, but wanted no part of writing one of my own.

Then one day I was being interviewed by David Shayler who subsequently authored the book *SKYLAB: America's Space Station*, published by Springer-Praxis in 2001. In the course of the interview I mentioned that I had written a diary while on orbit. Over the next year or two Pat and I developed a close friendship with Dave. One evening, while cooking dinner together, Dave, in his delightful British Birmingham accent, said, "You have a good story to tell. Whoy don't you wroit a book aboot it?" I responded, as usual, "I'd rather go to the dentist." Dave quickly replied that he would write it for me, but I would have to feed him the information, and by all means, we would want to include entries from that diary. He would want to interview all of the members of my family and quote them in the book as well.

Pat and I were intrigued by Dave's concept for the book. No one, to my knowledge, has done a biography of an astronaut that included his personal diary. We decided that we could trust him to be a faithful biographer and complete what, for me, was an intimidating task.

So, I spent hours babbling into a tape recorder and made all of my mother's scrapbooks, my education, military personnel, NASA office and correspondence files available to him. He dug like a terrier and found things I had completely forgotten – some embarrassing and most not. One thing is for sure. Dave is a thorough researcher, and I wish him well on a job I wouldn't touch with a ten foot pole.

Jerry Carr 2007

Foreword

Skylab in Context

Jerry and I met in the fall of 1965 while Brooks Air Force Station medical personnel in San Antonio were screening us for astronaut selection. We made it through that evaluation and the interviews the following spring and in May of 1966 we reported in to the Manned Spacecraft Center to begin our training. The classroom instruction was ho-hum but we really had a blast during our training on geology field trips.

In late 1966 we were assigned to support crews and eventually, in 1970, Jerry, Ed Gibson and I were selected to fly on Skylab. The Skylab missions taught us a lot, especially about the interaction between Mission Control and the flight crews. The medical and physiological experiments we performed are still considered the most rigorous of all medical/physiological studies related to long-duration space flight.

In 1984 Jerry and I were approached by the Boeing Company to support its effort to win a contract for Space Station Freedom. We settled on an arrangement whereby Jerry's company, CAMUS, Inc. would have a contract with Boeing and I would work for Jerry as a subcontractor. It was a great relationship and we worked together for 14 years. Jerry has a very stable personality, is of unimpeachable character and it was a pleasure to work with him. I know you'll enjoy reading about a great American, Gerald P. Carr.

WILLIAM R. (Bill) Pogue
Former NASA astronaut, Pilot Skylab 4, Colonel USAF (Retired)

Author's Preface

During Christmas 1968, I avidly followed the mission of Apollo 8 to the moon. As a young teenager I was fascinated by the prospect of men walking on the moon and devoured anything relating not only to the mission, but to space exploration in general. This passion had developed over several years to a point where, by December 1968, it had become a daily part of my life.

It was from reports and accounts of Apollo 8 that I first heard the name 'Jerry Carr', who was one of several astronauts back on Earth supporting the mission. Jerry was, as I learned, a 'CAPCOM', a fellow astronaut who talked to the crew in space. Little did I expect four decades later to be writing a book on the life and career of this 'CAPCOM', a project that ended up taking almost 20 years to complete.

In those early years, I began to collect information on each manned spaceflight (including the crews) and became fascinated by the careers of the men who became astronauts and cosmonauts. In particular, I began to collect a wealth of information on each man's career before, between and following spaceflights. This included those selected in 1966, and of course Jerry Carr. The Apollo lunar program was one of those seminal moments in history for which I could proudly say, "I was there to witness it," on TV if not in person. Inspired by Apollo 8, I avidly sat through the excitement of the Apollo 11 mission as it happened, experienced the frustration of losing the TV pictures from the Apollo 12 moonwalks, feverishly willed the Apollo 13 crew home safely, and finally felt the disappointment of the end of the lunar program with Apollo 17. But there was still the prospect of a new vehicle called Space Shuttle and, especially, an American space station based on unused Apollo hardware, called Skylab.

I followed the Skylab program with the same interest as the moon landings, even if the general media did not. In 1973, while serving in HM Royal Marines in the south west of England, I saw the Skylab space station fly over as a bright dot towards the south. I had seen the Apollo 10 capsule tour England some time before, but this was Skylab – and it was still in space. I was hooked. Skylab became a passion of mine, and has remained so for over 30 years.

My reward for this devotion was the publication my book on the program in 2001. In that research, many former Skylab astronauts offered their valuable time and assistance in detailing aspects of their involvement. That strengthened my resolve to press on with the other project I was working on at that time – the biography of Skylab 4 astronaut Jerry Carr.

I first contacted Jerry by old fashioned snail mail in 1982, conducting research for a book on the Group 5 astronauts which I published myself through Astro Info Service the following year. Jerry was one of the former Group 5 astronauts who had supplied personal information to supplement the research I had completed. By 1988, I was ready to embark on my own maiden 'space mission' – a visit to NASA JSC in Houston Texas. A meeting with Jerry was arranged with the help of his son Jeff, who was working at JSC Public Affairs at that time.

In August 1988, Jerry and I met face to face for the first time. During the interview, we spoke about Jerry's work in pressure suits and on EVA (for a different project of mine). In 1989 we met again at JSC and discussed Jerry's flying career (again for a separate project). The interest in Skylab missions, astronaut careers and cancelled Apollo missions led to my interest in Jerry's brief assignment to what would have become Apollo 19 and in his record-breaking mission on Skylab. I was sure there was a story to tell here of a military pilot, who worked on Apollo, and lost the moon, but gained an important mission in the history of manned space exploration.

Over the next few years, Jerry and I communicated over Skylab and the details of his career, but it was not until 1994 that the idea of a book on Jerry was finally agreed and work could begin.

At the outset we decided to tell the story of Skylab 4 not from the daily activities reports, but from Jerry's point of view, using the diary he had written in orbit twenty years before and never published. Right from the start I suggested it was important to include accounts from his family, as no one at that time had, as far as I was aware, ever included details of what the family on the ground had to go through while an astronaut was in space. For Skylab 4, 'dad' would be away for three months, an unprecedented duration in space.

From 1995 we commenced working on the layout and draft of Jerry's book. At that time, the Praxis Skylab book was still six years in the future, so part of the joint project with Jerry would include background on Skylab. When the Praxis book was published, this alleviated the need to record operations and hardware on Skylab in great depth, and left room to expand on Jerry's personal story in more detail. The title of the book came easily.

What took longer to decide upon was the way the story was to be told. Jerry did not want to write the book in first hand, or use me as a ghost writer. This allowed me to expand *my account* of his life beyond what Jerry would have done himself. It was clear in discussions with Jerry, and in learning about the man behind 'the astronaut', that this would not be a 'tell-all' account of the highs and lows at NASA during 1966-1977. Neither would we be revealing skeletons in the closets of the Astronaut Office. It was not in the nature and character of the man to do so.

What we did decide was to look at Jerry's career in detail. Jerry's mother had collected a scrap book of his career, from his early years with cuttings of his schooling and scouting records, to his wedding, his Marine years and his astronaut career. Jerry himself had kept the official Cruise Record books of his service in the Marines, including *every* movement order Jerry had received between 1954 and 1977. Adding this to his spaceflight dairy and contributions from his family, a very personal story would unfold.

Research unearthed details of the training and early assignments of the 1966 astronaut class, and what might have occurred if Apollo 19 had indeed flown to the moon with Jerry as Lunar module Pilot. He would probably have become one of the few humans to have stepped onto the lunar surface thus far.

The book has also included life after NASA and spaceflight, a period some astronauts have struggled with. The openness and friendship of Jerry's family has provided me with opportunities I could only have dreamed of all those years ago, when I first learned of a NASA astronaut called Jerry Carr.

Fond memories of staying with Pat and Jerry at their 'ranch' in Arkansas, of many meals and conversations over the years 1997-2002, will forever remain vivid supplements to these words. Conversations with Jerry's first wife JoAnn and their six children provide some of the highlights of this research, along with accompanying Pat Musick and Jerry as they installed artworks, and working at 'the ranch'.

In compiling these pages, Jerry and Pat amended and expanded the original copy many times as new thoughts surfaced from memories of a past long forgotten. This greatly expanded and improved the account. So, although Jerry is not writing the text first hand, it is my interpretation of those first hand interviews and records, family accounts and Jerry's personal explanations that make this book as close to his autobiography as we could get.

Dave Shayler FBIS
Director AIS.
www.astroinfoservice.co.uk

West Midlands, England
July 2007.

Chapter 1

One of Nineteen

On April 1, 1966, almost five years after Soviet cosmonaut Yuri Gagarin had become the first man in space, and US President John F. Kennedy had committed America to reach the Moon by the end of the decade, US Marine Corps aviator Major Gerald ('Jerry') P. Carr, was conducting an engineering design review meeting at Litton Industries Data Systems Division in Van Nuys, California. This was part of a Marine evaluation project to develop an airborne tactical data link system. He was in a conference room with about 15 persons in attendance when a secretary came in and approached him, stating: "Sir, there is a Captain Shepard on the phone who'd like to speak with you, and he says that it is urgent."

At that time, Jerry was working with a young Marine Corps captain named Shepard at the Marine Corps Air Facility in Santa Ana, California, and he wondered what was so important that he needed to interrupt Jerry at this meeting. Then he reasoned that if it was *that* important, he had better take the call from his colleague. There happened to be a phone on the conference table and Jerry calmly lifted the receiver, expecting to speak to his young fellow officer.

"This is Major Carr."

"Jerry, this is Alan Shepard down at the Manned Spacecraft Center in Houston." Jerry replied "Oh…Yes," suddenly remembering his recent application to join the astronaut program. Glancing at all the people in the room, who were looking at him, but were unaware of who he was speaking to on the phone, Jerry tried to remain calm and matter of fact as Shepard indicated that he had been selected as an astronaut by NASA, and that if he was interested, he was to report to Houston on May 1. Jerry replied with a surprised, "Oh yes… Yes Sir."

He was told not to tell anyone, apart from his wife, until the official announcement was made in a couple of days. They ended the brief conversation and Jerry expected most of the people in the room to wonder what was going on as he was sure he must be floating three feet off the chair. Wanting to tell the whole room, but unable to utter a word about the phone call that would change his life, Jerry claims to have no recollection of what happened for the rest of the conference. He could not wait to tell his wife, JoAnn, who replied upon hearing the news, "You're kidding, it's April Fools Day!" A few days later, the press releases came out and local papers picked up the story of his selection and pending move to Houston. On April 5, 1966, Major Gerald P. Carr was named one of America's latest nineteen astronauts, in the fifth selection by NASA since April 1959.

In front of him were months, even years, of training and preparation that might lead to an assignment on a spaceflight crew to orbit. In 1966, his prospects, on the surface, looked good for more than one flight into space, and perhaps even to the Moon. Indeed, the reason for the selection of the nineteen new astronauts was to support the initial testing of hardware in a program designed to land men on the Moon – the Apollo program. After initial training, Carr's group was expected to support the first few landings on the Moon and then crew the follow-up missions under what was then called the Apollo Applications Program (AAP). This program was expected to utilize Apollo hardware for other missions once the first lunar landing had been achieved. This included extended duration lunar surface and lunar orbital exploration missions, Earth orbital science missions, and the creation of America's pioneering space station, known as the Saturn Orbital Workshop. All of this was in the future, and nothing was certain in the space program, but in April 1966 it was all the motivation needed for Jerry and his family to undertake the move from California to Houston. With six children, this would be a challenge in itself.

Steps to the moon

In the first two NASA astronaut selections in 1959 and 1962, as well as the initial Soviet cosmonaut selections, the emphasis had been on flying experience. In the United States, test pilot training and extensive flight experience were deemed essential. Though selection was initially restricted to military pilots, this requirement was relaxed for the second selection to include civilian test pilots. This would give the agency a broader group from which to select candidates. As the pioneering manned space programs were exploring unknown environments, the belief that a test pilot would be best suited to coping with the stresses, strains and unexpected incidents of such an unknown was proven to be valid. The first few manned flights into space were successful, but it soon became evident that the physical and psychological stresses were more benign than first thought. Bodily functions such as swallowing, urination and defecation, as well as senses of sight, smell and taste, appeared to be minimally affected.

By late 1963, a third group of pilot astronauts had been selected. This time, the test pilot requirement had been dropped. This selection brought the total number of NASA astronauts to thirty, sufficient to provide crews for the one-man Mercury and two-man Gemini programs. Of these thirty, six had already flown Mercury missions. With that program now completed, the group was focusing on Gemini and the initial test flights of Apollo hardware in Earth orbit. It was expected that all of the flight crews for the Gemini and early Apollo missions (probably leading up to the first landing) would come from these thirty men. However, as 1964 dawned, plans were being developed for an ambitious program of missions, both to the Moon and in Earth orbit after the initial lunar landings. The plan was to utilize Apollo-type hardware, along with new vehicles under development. To crew these missions, which would begin to rely on science as well as engineering, NASA began looking at adding scientists to the astronaut team and bringing in more pilots to command and fly these more scientifically demanding missions.

There were intended to be three groups, or 'blocks', of Apollo manned missions, each preceded by a series of unmanned qualification and engineering flights. Block I Apollo missions featured spacecraft capable of supporting a crew in Earth orbit, but without docking equipment or the ability to fly to the Moon. These would be engineering test flights of the main spacecraft, known as the Command and Service Module (CSM), prior to committing crews to fly the more advanced Block II. The CSM actually consisted of two modules – the Command Module (CM) housed the crew compartment, and the Service Module (SM) provided service support functions (oxygen, water, electricity, propulsion). The Block II spacecraft were the lunar mission vessels, featuring docking and internal crew transfer facilities with the lunar landing module. This vehicle was called the Lunar Excursion Module (LEM), later simplified to Lunar Module (LM). The proposed Block III CSMs were envisioned as advanced main Apollo spacecraft with scientific experiment packages installed in them. They were expected to serve as transportation to the first American space stations, which were to be developed using as many other elements of Apollo hardware as possible.

As the plans for the initial manned lunar landings were beginning to unfold, ideas about what should follow were beginning to be discussed. In 1964, a NASA Future Programs Task Group had reported on the development of techniques and procedures which would support the initial lunar effort and potential future objectives. Manned spaceflight in Earth orbit for between one and two weeks would be accomplished by Gemini and Block I Apollo missions, while orbital maneuvering, rendezvous and docking would be achieved by Gemini and Block II Apollo missions. Flights of between one and two months in Earth orbit would be supported by Block III Apollo missions, in readiness for more advanced new spacecraft during the 1970s and early 1980s.

It was the third phase Apollo missions for which Jerry's group were selected. Indeed, the selection of new astronauts had been split into two groups. The first, chosen in June 1965, was the initial group of scientists selected for astronaut training. Those who had not previously qualified as jet pilots were required to undergo a year-long jet pilot training course at a USAF base before coming to NASA to start astronaut training. The second, Jerry's group, was the fourth 'pilot' selection, although some of the candidates, like Jerry, had delayed applying for test pilot training in order to take advanced academic courses to supplement their flying credentials.

These qualifications were exactly what NASA was looking for to supply crews for the Block III missions. This phase was eventually called the Apollo Applications Program (AAP). Previous titles had been Apollo A, Apollo X (experimental) and Apollo Extension System (AES). These missions would include pilot astronauts to fly the vehicle and utilize scientist astronauts to operate experiments in biosciences, physical sciences and astronomy/astrophysics. There would be missions in various orbits around the Earth – equatorial, polar, high inclination and synchronous. They would include rendezvous and docking for inspection, repair and rescue, and a program of Earth-orientated investigations in atmospheric sciences, technology, communications, Earth resources and mapping. AAP would help to develop new operational techniques and evaluate systems for future spacecraft and programs, as well as conduct biomedical and behavioral investigations. All of this would lead, step by step, to the development of new manned spacecraft, and ultimately to flexible Earth-orbital operations that envisaged the creation of large, 50- to 100-person space laboratories.

Beyond the moon

The long-term vision was not confined to Earth orbit. Apollo Applications planning also encompassed lunar missions devoted to extensive lunar mapping and stays on the surface of up to fourteen days. This would lead to the creation of a large lunar research facility and extensive lunar roving operations, with some manned expeditions lasting up to two weeks from the main base. From this would emerge the creation of a manned lunar orbital space station, leading to the first expeditions to the planet Mars using upgraded Apollo hardware and nuclear-propelled Saturn rocket stages. That was the grand plan, but by 1966 it had already begun to unravel.

The main lunar landing program was envisaged as up to eighteen three-man Apollo missions, with AAP likely to be up to twenty three-man flights. The launch manifest of 1964/1965 for Apollo hardware therefore projected around 50 Gemini and Apollo/AAP missions between 1965 and 1975, with more hoped for during 1975-1980. With at least 140 crew seats to fill, the NASA astronaut corps of thirty would clearly be insufficient. Without taking into account natural attrition, and even giving each astronaut three or four missions, they would barely fill the flight crew requirements, let alone provide back up crews and the numerous support roles required for each flight.

NASA finally decided that some of the unmanned and manned test flights of Apollo were unnecessary and deleted them, essentially eliminating hardware that would not be used for the lunar missions. Former Mercury astronaut Donald 'Deke' Slayton, the Assistant Director of Flight Crew Operations, who was in charge of crew selection and flight operations along with Chief Astronaut Alan B. Shepard, knew that NASA had enough astronauts to complete the Gemini program, as well as up to eight crews for the Apollo flights leading up to the first landing. They preferred not to take on any more astronauts, especially with the group of scientists coming on-board at the same time from flight school. However, looking further ahead it was obvious the same 30 astronauts would neither want nor be able to remain at the forefront of crew selection for the next 10-15 years. As it turned out, many of them flew only one or two missions before either returning to their parent service, retiring, moving into civilian careers, or staying at NASA in managerial roles.

The nation's first scientist astronauts had already been selected to support the AAP program, and Slayton realized that NASA would need to select more pilot astronauts to replace those who would start retiring from the program after their one or two missions. It was also clear that although ambitious plans were on the drawing boards, no extra funding was forthcoming. In fact, there were indications that the Saturn V production line would be closed after building only fifteen vehicles. This would be a significant shortfall compared to the earlier ambitious planning for AAP. It was beginning to become apparent that the Vietnam War would take its toll on the national budget, and the space program would feel the pinch.

On January 15, 1965, Robert Gilruth, Director of the Manned Spacecraft Center in Houston, sent Deke Slayton to NASA Headquarters in Washington to put the case for more pilot astronauts to the NASA Associate Administrator for Manned Spaceflight, George Mueller. Slayton received a lukewarm response to the idea of more pilot astronauts, but it was conceded that possibly a group could be brought into the program in the Fall of 1965. The decision would depend upon the evaluation of techniques such as rendezvous and docking, long duration spaceflight and the first American Extra Vehicular Activity (EVA, or spacewalking) demonstration, which would take place during the Gemini program. These evaluations were important for determining how realistic President Kennedy's target of a Moon landing by the end of the decade would actually be.

A case for more pilots

On September 10, 1965, NASA issued a statement indicating that it would accept new applications for astronaut candidates until December 1, 1965, to 'ensure adequate numbers for the first Apollo crews and future manned missions.' To qualify, prospective candidates had to fulfill several basic criteria:

1) Be a citizen of the United States of America, born after December 1, 1929. He could be no taller than 6 feet.
2) Recipient of a Bachelor of Science degree in engineering, the physical sciences or the biological sciences.
3) An experienced jet pilot with 1000 hours flying time, or a graduate of an armed forces test pilot school.

Applications would be accepted from either military or civilian candidates. The height restriction was necessary for any astronaut to be able to fit inside the Command and Lunar Modules comfortably, and to access the vehicles through the hatches while wearing a pressurized suit. The size of the spacecraft was governed by the dimensions and capabilities of the Saturn launch vehicles designed to launch them. Jerry's fellow Marine candidate Jack Lousma claimed he stood at "five feet thirteen" inches and had to do some real scrunching down when they measured him.

The age limit was raised from 34 to 36 years of age. This meant that after training, and a potentially long wait for their first mission, applicants could well be in their fifties by the time they were expected to fly, in the mid 1980s (which was the case). When considered with the other criteria, it was fair to assume that no pilot younger than 26 would have gained the required academic credentials, flying experience or maturity – thus there was an implied ten-year window for suitable applicants. It was not well highlighted at the time, but the offer was also open to women and minority applications.

In the late 1950s Jerry, as an all-weather jet pilot with the US Marines, had given the rapid developments of the space age no more than passing interest. In April 1959, Marine pilot John Glenn became one of America's first astronauts. Less than three years later, in February 1962, he became a household name as the first American to orbit the Earth. Jerry's immediate goals, however, were not as lofty as his Marine colleague. He was busy trying to gain a Masters degree and then graduate from test pilot school, giving him the chance to fly the top planes available. The prospect of becoming an astronaut was not yet on his horizon.

In October 1963, NASA announced the selection of a third group of astronauts for the forthcoming Gemini and Apollo programs. Of the fourteen selected, one was a Marine pilot named Clifton C. (C.C.) Williams, whom Jerry had known as a contemporary at MCAS Cherry Point. This gave the astronaut corps a more personal connection. For the first time, the activities of the NASA program received more than a cursory interest from Jerry.

The Application

In October 1965, while reading an 'All Marines' bulletin (which is circulated to every Marine Corps unit), Jerry learned that NASA was about to select another new group of pilot astronauts and that anybody with the listed qualifications could apply. Though still planning to attend test pilot school, Jerry read the announcement, realized that he met NASA's qualifications, and decided that he could apply for consideration. Since it would mean a major move to Houston and another hazardous assignment, Jerry talked it over with his wife, JoAnn. They agreed that Jerry was certainly qualified and should apply. His letter, together with two letters of recommendation from former commanding officers, went to Headquarters, Marine Corps in Washington DC. As a serving military officer, Jerry's application first had to go through a USMC astronaut selection panel prior to consideration by NASA. On 27 October 1965, Jerry Carr formally applied for astronaut training.

JC: "I thought I'd just try and see how I measured up, knowing that C.C. (Williams) had made it, and I was as surprised as anyone to receive a letter back from NASA saying I'd made the first cut."

As a pilot's wife, JoAnn had learned to live "with the odds." As Jerry later commented at the time of his selection, "We don't feel that there is any more danger involved as an astronaut than as a pilot flying in Vietnam." At 33, Jerry was below the age limit, and at 5 feet 9.5 inches tall and 148 lbs, he was well within the requirements set out by NASA. He held a Bachelor of Engineering (Mechanical) degree earned from the University of Southern California in 1954, a Bachelor of Science degree (Aeronautical Engineering) from the USN Postgraduate school, Monterey, California in 1961, and a Master of Science degree (Aeronautical Engineering) from Princeton University in 1962. All his college transcripts were on file at USMC Headquarters for examination. His pilot experience by October 1965 was 1,903 total hours, of which 1,369 were in jets. In his application letter, Jerry categorized this experience by aircraft type and model, with appropriate dates.

Letters of recommendation were forwarded from current and former commanding officers, and were quite complimentary. The first, from Jerry's immediate commanding officer, Lt. Colonel Edwin A. Burns of Marine Air Control Squadron-3, stated: "Carr is a very intelligent, highly motivated Marine Corps officer with a quick grasp of situations and a practical solution to problems. His demonstrated initiative, devotion to duty and his intense interest in aviation and related fields qualify him for greater responsibilities… His technical background and operational flight experience, primarily in high performance jet aircraft, makes him an ideal selection for the astronaut training program."

The second letter of recommendation came from Lt. Colonel Dale L Ward, who was Jerry's commanding officer at Marine All Weather Fighter Squadron 122 from December 1962 through February 1964. At that time Jerry, "by virtue of his educational background," was assigned collateral duties in the Aircraft Maintenance Department. His duty assignment was as Assistant Maintenance Officer for a period of eight months. Then he became the Aircraft Maintenance Officer during his squadron's deployment in the Far East. Ward wrote that, professionally, Jerry "was a competent, meticulous worker whose concern for detail is supported by the fact that not one maintenance error accident occurred during my tenure as Squadron Commander. This fact also attests to his superior leadership capabilities. During the many inspections to which his section was subjected, no discrepancies of a great magnitude were discovered, and very few of even a minor nature were noted." According to Ward, Jerry possessed, "the rare faculties of both high intelligence and common sense. His pleasant personality belies a firm self-discipline and determination to achieve. These fine attributes enabled Jerry to create a relaxed working environment, with the right amount of discipline being used at the proper time." Jerry had often displayed, "the highest set of moral standards, and with the support of his wife and family, enjoys a happy family life." Ward also mentioned Jerry's courageous nature, pointing out that despite his extensive and demanding duties as a maintenance officer, he still ensured that he was available for flight time when he could easily and legitimately have flown fewer hours than the average pilot. Summarizing his recommendation, Lt. Colonel Ward stated, "In my estimation, [Jerry] possesses in great abundance all those qualities necessary to excel in the astronaut training program. Of all the Marine aviators with whom I have been associated [Carr] is the best qualified for the program he has requested. I recommend him, without reservation, for astronaut training."

Jerry thought that his experience and professionalism would stand him in good stead for an astronaut career, but did not initially entertain the thought that he would actually make it to selection. On November 16, the first batch of official forms were dispatched to Jerry from USMC HQ. These had to be returned by November 25, along with a copy of his last flight physical examination, for transmission to NASA by December 1. These forms included a multi-section Statement of Personal History [DD398, Standard form 86], an Application for Federal Employment [Standard form 57] and an Aviation Training Summary.

In a letter dated December 17, 1965, Jerry was informed of his recommendation for the NASA Astronaut Training Program by USMC Deputy Chief of Staff (Air), L.B. Robertshaw. The letter explained that the USMC Astronaut Selection-board established at Headquarters Marine Corps had reviewed Jerry's application and official records and had commented: "Your superior performance of duty throughout your Marine Corps career, your motivation, aviator skills, and other qualities, are such as to warrant your selection as a Marine Corps nominee to NASA for astronaut training." The letter ended with congratulations "on the enviable record which merited your nomination by the Marine Corps," and a message of good luck for the ensuing selection process.

Recommendation for MOL

The letter also indicated that Jerry had been recommended by the USMC selection-board to the Chief of Staff of the USAF 'for consideration by the Manned Orbiting Laboratory (MOL) Selection-board , although Jerry was unaware of this at the time. A second pilot from MACS Three, Captain Stanley P. Lewis, was nominated for MOL along with Jerry.

The USAF MOL program had been initiated on December 10, 1963. It was conceived to fly pairs of military astronauts in a pressurized laboratory attached to the rear of a modified Gemini spacecraft for up to 30 days. Launched by a Titan III, each mission would feature classified military experiments and research, Earth observations and biomedical studies, and evaluations of new technologies. Five manned missions were initially planned, which would be launched from the Vandenberg AFB in California. Despite continual difficulties in justifying the MOL program in the face of the developing NASA Orbital Workshop program, advancements in unmanned surveillance satellites and the escalating South-East Asia war, the program survived until 1969, when it was terminated without a single manned flight taking place.

To be considered for assignment to MOL, a candidate had to be a qualified military pilot, a graduate of the USAF Aerospace Research Pilot School (ARPS) at Edwards AFB in California, a serving military officer recommended by a commanding officer, and a US citizen by birth. Though at least 20 candidates were wanted, a total of 17 were chosen in the three groups selected on November 12, 1965 (8 astronauts), June 30, 1966 (5 astronauts) and June 30, 1967 (4 astronauts). A fourth selection was in progress when MOL was cancelled in 1969. Jerry (along with fellow Marines Jack R. Lousma and Robert F. Overmyer) was nominated for the second MOL selection.

Over 500 applications were screened by the USAF during November 16-19, 1965 (a few days after the first MOL astronaut group was named). From these, 25 were identified for MOL consideration. Jerry did not make the 25, probably because he had not attended the ARPS school at Edwards. The letter declining his application did commend him for his "initiative and ambition in applying for this duty," however. Jack Lousma also never made it to MOL, probably for the same reason as Jerry, but Bob Overmyer, who was a 1966 ARPS graduate, did make the selection and was named to MOL in the June 1966 announcement. When MOL was cancelled, Overmyer transferred to NASA in August 1969, along with six other former MOL astronauts, forming the agency's seventh astronaut group.

On December 8, 1965, Jack G. Cairl, the NASA Astronaut Selection Coordinator, wrote to Jerry to confirm receipt of all the documents supporting his application to join the NASA pilot astronaut program. Preliminary screening by NASA was scheduled for the middle of December. Five days before Christmas, Deke Slayton wrote to Jerry (and all the other successful NASA applicants) informing him of his participation in the final stages of the Manned Space Craft Center Pilot Astronaut selection program. Jerry was told to be prepared to spend a week in either January or February 1966 at Brooks AFB in Texas for physical evaluations, and one week in March at MSC in Houston for final evaluations. Along with the letter was a clutch of security investigation forms to be filled out and returned as soon as possible. As well, he was advised that additional medical forms were on the way to him, which would have to be filled out and collated with his formal USMC records for further evaluation. Since most of the remaining candidates under consideration would not be selected, Jerry was told to have "as little discussion as possible with regard to the program." Slayton continued: "We ask you to use discretion in discussing your own status and any instructions or information we may give you…best wishes for the holidays." It was a great Christmas present, but one which he could not fully 'unwrap' for some time, assuming he passed all the tests and investigations.

Attached to the letter from Slayton was another from Charles A. Buckley, the Chief Security Officer at NASA MSC, who informed Jerry that, as an astronaut applicant, it was necessary for a full background investigation on him to be completed by the US Civil Service Commission. This would be conducted regardless of any previous investigations conducted or security clearance granted, including those attained in the USMC. Standard Form 86 was the Security Investigation Data for Sensitive Positions, and Standard form 87 was a Fingerprint chart. Jerry dutifully filled in and returned the documents to NASA MSC. All he could do now was wait.

By the cut off date of December 1, 1965, NASA had received over 5,000 applications for the next group of astronaut trainees. Of these, only 351 met the selection requirements, including Jerry's. There had been six applications from female pilots and one from a young USN officer, Lt. Frank E. Ellis, who had lost both legs in a jet crash in July 1962. Ellis maintained that despite his handicap, his flying skills were unimpaired and that being unable to run and jump was irrelevant for an astronaut in 'zero-g'. Neither Ellis (who did go on to complete some special work for NASA) nor the female or minority candidates made it to the finals, but a total of 159 applications (100 from the military and 59 civilians) were considered for further evaluation for the expected 15 positions.

Ten days at Brooks

On January 6, 1966, Jerry received a package from Captain Henry C. Reinhard Jr. of the USAF School of Aerospace Medicine at Brooks AFB. The package contained forms, including an aero-medical survey, which had to be sent to Brooks (along with Jerry's medical and dental records) by January 10. The medical survey was to be reviewed by Jerry's current flight surgeon, whose opinions and findings were a pertinent part of the evaluation. Jerry was to report to Brooks AFB by January 19, as the medical evaluations would begin at 07:30 hrs on January 20 and would last for approximately ten days.

Among the instructions Jerry received was a dietary program that he would have to adhere to in preparation for the medical examinations. For three days prior to his arrival, his diet had to include at least two ounces of any type of meat; one three-ounce serving per day of any type of potatoes; two pieces of any bread up to three times a day; one cup of sugared coffee once or twice a day; and three candy bars each day. This was to standardize his carbohydrate intake, reducing the amount of time each candidate would need in dietary preparation and medical evaluation. He could eat any other foods he wished, but any prescribed medication had to be discontinued at least three weeks prior to the reporting date (which was only two weeks away when Jerry received the instructions!). The guidelines instructed the applicant to eat at least 3 to 5 candy bars per day, even if lunches were also taken. Alcohol was not allowed during the period of dieting and fasting, but smoking was, although this was not a problem for Jerry, who did not smoke.

On the night prior to the first appointment, the candidate could take nothing by mouth except water, and a urine sample had to be provided shortly after arrival, ideally the first specimen of that morning. Sugar tolerance tests would also be carried out just after arrival and the candidates were warned that compliance with the instructions was essential to ensure they passed the medical evaluation.

Jerry had been scheduled to travel to Litton Industries in Van Nuys, California on January 12 for a 12-day visit, dealing with technical data system matters in support of his Marine assignment. But he now had to complete his work there by January 13 in order to prepare for the week at Brooks. On January 19, 1966, Jerry took a ten-minute taxi ride from his home on Louise Street, Santa Ana, California to the local airport coach station, followed by a ninety-minute coach ride to Los Angeles Airport. Flying on Continental Airlines flight CO66, he arrived at San Antonio Airport, Texas, where a government car was waiting to take him to Brooks AFB. Arriving at the base at 21:25 hrs, Jerry checked in for the week of medical evaluations. The candidates would be billeted at the Visiting Officer Quarters (VOQ) on base for the week, both for easy access to the medical facilities and to preserve the security of their selection process.

The 10-day 'week' at Brooks has been described by Captain Lawrence J. Enders, Chief of the Flight Medical Evaluation Section, as "pleasant and enlightening, and even exciting at times… certainly the most thorough medical evaluation you will ever have." The week was also described in detail in Tom Wolfe's book, *THE RIGHT STUFF (1979)* and depicted further in the 1984 film based on the book. Wolfe's description of the evaluation was considerably more chaotic than that of Captain Enders, but Jerry agrees that it was indeed most thorough. He met many of the men with whom he would eventually be selected. The group went through a battery of medical tests, which had evolved from the strenuous program given to the original Mercury astronauts in 1958-1959. Nearly all of the tests were still demanding, however. One test included pouring warm water in the ear, followed immediately by ice water, which, as Jerry recalled, "threw the inner ear and sense of balance into a wild frenzy." Fellow astronaut applicant Charlie Duke was sure that these tests were designed to see if the applicants' eyeballs could be detached!

Duke later wrote about the week at Brooks in his 1990 biography, recalling: "The whole process featured being poked, prodded, tested and analyzed from every angle, in every conceivable position, and from every opening of our bodies. They did tests I couldn't pronounce, much less figure out what they were. Some were crazy things, and the physical also included psychological testing. It was the first time I'd ever seen inkblots, and we'd all decided ahead of time that when we looked at those blots we weren't going to see anything sexy or perverted."

In 1959, Charles 'Pete' Conrad (whom Jerry would work very closely with at NASA) was in line for selection to the Mercury program when he underwent the same Rorschach ink blot test. After five days of medical examinations, he'd had enough and had decided it was all about pain. For NASA, these seemingly irrelevant tests on a 'fighter-jock' like Conrad were essential in revealing a candidate's ability to cope with the stresses of spaceflight (although no one had flown in space yet) and other biomedical and environmental aspects of long periods in confined conditions. When Conrad was shown an ink blot on a card, he came up with a very colorful story about what he could see. When shown a blank card, he told the psychiatrist he couldn't reveal anything about the card, as it was upside down! Conrad was not selected for the Mercury program, having been deemed "unsuitable for long duration flight," a statement that often amused him. Conrad became more interested in the space program, and smarter in the selection process, and was selected in the second group of astronauts in 1962. He went on to make four spaceflights, two of them long-duration.

Jerry made sure he put all of his natural effort, professionalism and dedication into any test that was thrown at him, whether it was blots of ink on paper, IQ tests, or problem solving tasks, such as the speed test to match square or round pegs with appropriate holes (designed to test the candidate's ability to make accurate and rapid decisions while under pressure). In all, there were about 35 hours of medical evaluations and observations, performed on a very tight schedule. Dr Owen Garriott, who was selected in the first scientist astronaut group the year before Jerry, kept a diary of his week at Brooks which was very similar to that completed by Jerry and his colleagues for the fifth astronaut selection.

Garriott recorded that the first day was taken up with a battery of medical examinations and discussions of each man's medical history. There were blood sugar tests and dental X-rays, followed by abdominal examinations and lengthy electrocardiograms. A "cold processor test" involved placing a hand in ice water for a minute while doctors monitored the candidate's heart rate and blood pressure. The following day a series of cardiograms were completed, followed by a session on the tilt table to record the ability of the candidate's body to regulate blood pressure under simple stress situations. There were also runs (without pressure suits) in a centrifuge to around 4 or 5G. These runs were later increased to around 10.5 G to simulate an off-the-pad abort. In the afternoon, candidates completed 48 trips over a turnstile and a session on a treadmill at 3.5 mph, with the elevation gradually increased by one degree per minute for around 20 minutes on average.

Day three involved a "breakfast cocktail of tritium" and another blood sample to record total body water levels. Psychology tests followed, along with IQ examinations, these lasting about four hours. The day ended with a neurological check up and an EEG (with 15 quarter-inch needles placed just under the skin and taped down to make a good electrical contact. That night, the candidates took six tablets, to help in the gastrointestinal "barium milkshake" test conducted the following morning. Day four continued with a fluoroscope, a program of X-rays and respiratory tests. A 30-minute visit to a physicist completed the day. Day five included ocular tests, measurements of body density and fat percentage (by immersing the candidate in water) and a full dental check up. An evening cocktail party at the Officer's Club was mooted somewhat by the thought of visiting the altitude chamber the next day and a warning from the medical department to avoid a supper or breakfast of foods likely to produce gas in the intestine.

Day six featured the expected altitude chamber runs, initially up to 5,000 feet, then up to 43,000 feet at a rate of 7,000 feet per minute and back down to 25,000 feet. At this point, candidates were instructed to remove their oxygen masks to experience a hypoxic state. The chamber was then regulated to ground level, back up to 8,000 feet to simulate a rapid decompression experience, then up to 23,000 feet, and finally back to ground pressure levels. Day seven was usually a day off, where the candidates rested. Day eight, according to Garriott, was to feature written examinations and an exercise on an aircraft simulator, with further medical tests completed on Day nine.

Most of this was a blur for Jerry, having few memories of the process other than sitting in the VOQ in the evenings with Paul Weitz and Stan Lewis wondering what was coming next. Years later, when reading *The Right Stuff*, Jerry found that author Tom Wolfe had described the astronaut selection process fairly accurately, and humorously.

JC: "He certainly captured the essence of the physical exams, and I got a kick out of the antics of the actor playing the role of Pete Conrad, although Pete's name wasn't mentioned."

After completing these tests, Jerry returned home on January 28. The trip had cost him a grand total of $74.57, and it was one more step towards selection. During the week at Brooks, Jerry was suffering from hay fever and, thinking that NASA only selected 'perfect specimens', assumed that he would progress no further. However, a letter arrived a few days later indicating that he had reached the second cut of around 50 applicants. He was invited to an interview in Houston.

The Interview

On February 27, Jerry flew to Houston, where he was met by a representative of NASA who drove him to the Rice Hotel. He and the other candidates would stay here while undergoing the next stage of the selection process.

The following day, news came through that Gemini 9 astronauts Elliott See and Charles Bassett had been killed in an air crash of their T-38 jet at the McDonnell Plant in St Louis. This cast a shadow over events in the hotel, and Jerry recalls that people were nervous and distracted. The interviews went ahead, however. Having arrived at one of the ballrooms at the appointed time the next morning, Jerry recalls opening the door and seeing a long table with a dark green tablecloth. There were seven people seated behind it. They were Assistant Director for Flight Crew Operations Deke Slayton, Chief Astronaut Alan Shepard, Gemini 10 commander John Young (representing the USN), Gemini 10 pilot Mike Collins (representing the USAF), Gemini 11 and 12 back up pilot C.C. Williams (for the USMC), Mercury spacecraft designer Max Faget, and astronaut training officer Warren North. Jerry, noticed that the chair for the candidates was under a spotlight out in front of the table, and the thought crossed his mind that the scene looked more like a court martial than an interview.

An effort was made at the start of the interview to relax the applicants, not from a desire to catch anybody out, but more to exchange information on both sides and to discover the personality behind the reams of paper and medical information that had been submitted on each applicant. However, it was still very difficult for any of the applicants to feel at ease. The Pilot Selection Panel fired questions from all angles and subjects. Jerry was questioned on why he wanted to become an astronaut, and on his career and achievements, with the panel evaluating his responses and his ability to maintain his composure.

The week also featured more written tests, orientation briefings and tours of the Manned Spacecraft Center (MSC) near Clear Lake, about 30 miles south east of downtown Houston. At that time, the centre had only been operating for a few years and was still being developed, far removed from the sprawling metropolis of Houston. On March 4, Jerry returned to California and his USMC work at Litton Industries, still thinking he would not make the final cut. The phone call he received from Alan Shepard on April 1 would change his life and career path in a way he would never have dreamed.

Class of '66

When Deke Slayton was asked by the rest of the selection panel how many candidates he wanted from the 35 finalists, he replied "as many qualified guys as you can find," to fill out the expected Apollo and AAP flights planned at that time. The board came up with nineteen names that were announced to the press on April 4, 1966 as the 5th NASA Astronaut Class (4th pilot selection). The nineteen included one 33-year-old USMC Major, Gerald P. Carr, who had made it through the long and difficult selection process on his first attempt. Jerry's 18 classmates from the 1966 intake were:

Vance D. Brand, 34, civilian, former US Marine
John S. Bull, 30, Lieutenant, USN
Charles M. Duke Jr, 30, Captain USAF
Joe H. Engle, 33, Captain USAF
Ronald E. Evans, 32, Lt. Commander USN
Edward G. Givens Jr, 36, Major USAF
Fred W. Haise Jr. 32, civilian, former US Marine
James B. Irwin, 36, Major USAF
Don L. Lind, 35, civilian, former US Navy
Jack R. Lousma, 30, Captain USMC
Bruce McCandless II, 28, Lieutenant USN
Thomas K. Mattingly II, Lieutenant USN
Edgar D. Mitchell, 35, Commander USN
William R. Pogue, 36, Major USAF
Stuart A. Roosa, 32 Captain USAF
John L. Swigert Jr, 34, civilian, former USAF
Paul J. Weitz, 33, Lt. Commander USN
Alfred M. Worden, 34, Captain USAF

Formal NASA photo of the Group 5 astronaut selection (all in jackets), together with trainers and models of the Apollo CM, LM and Saturn V. Front row seated from left: Evans, Roosa, Duke, Givens, Engle, Lind, Lousma, Pogue and McCandless. Back row standing from left: Swigert, JERRY CARR, Weitz, Worden, Mitchell, Irwin, Bull, Haise, Brand and Mattingly.

Seventeen of the group were married men with families, the other two were bachelors. Among them, they held two PhDs and ten Masters degrees, and half of them had test pilot experience or training. Eight (Duke, Engle, Givens, Haise, Mattingly, Mitchell, Roosa and Worden) were former graduates of the USAF Aerospace Research Pilot School at Edwards, while Brand and Bull had graduated from the USN Test Pilot School, Patuxent River, Maryland. Others, like Jerry, were operational pilots, with Evans and Weitz having combat experience in Vietnam. Don Lind was a NASA physicist as well as a pilot. Joe Engle had also flown the X-15 rocket research aircraft eight times and already held USAF Astronaut Wings, with three of his flights having exceeded 50 miles in altitude. Ed Givens had worked on the USAF Astronaut Maneuvering Unit at MSC, while Ed Mitchell had completed assignments in support of the USAF Manned Orbiting Laboratory Program. Their average flight time was about 2,700 hours, with almost 2,000 hours of that in jets. The group had an average age of 32.8 years.

Their selection brought the number of active astronauts to fifty, and though a flight on a later Apollo mission was a possibility, it was by no means certain that all 19 would even be assigned to a mission, let alone get the chance to fly into space. The competition to be assigned to the already declining flight seats would be tough, but the first hurdle had been passed, and Jerry was now one of America's new astronauts. Though some define 'real astronauts' as those who have actually flown in space, it was still a major achievement to be selected to one of the most exclusive groups in history. In the early selections to NASA, astronaut candidates, as they were called in the Shuttle era, were not known as such. Instead, they were considered fully fledged (if unflown) astronauts from day one. It was an honor that would remain with Jerry for the rest of his life, and one of which he continues to be immensely proud.

Jerry received the letter confirming his selection from Deke Slayton, who informed him he was to report for duty on May 2, just over a month away. A letter of appreciation from Deputy Chief of Staff A.J. Armstrong of USMC Headquarters noted his achievement in the face of the stringent qualifications required and the strong competition. Armstrong also wrote: "I am confident that you will continue to serve with distinction in your coming assignment. The hope and pride of all Marines go with you on your new endeavor. Best wishes for your success in this most challenging assignment."

In the days immediately following the official announcement, the local papers in Santa Ana, where Jerry had lived since he was a young boy, carried headlines about the local resident Marine who was one of the nation's 19 new astronauts. Noting that he was the fourth astronaut selected from Orange County (the others being Scott Carpenter in 1959, Walter Cunningham in 1963 and Ed Gibson in 1965), the papers also reported that his selection was as part of the team that was working to place Americans on the Moon later in the decade. On the subject of making such a flight himself, Jerry commented that he would love to land on the Moon. He noted that he and members of his class would probably "get in on the tail end of the series of Apollo flights with a possible assignment to a later landing crew." But he added, "It's only a guess. I don't know what the plans are, except for a lot of training for all 19 of us."

The Carr's six children had been told in January that their dad was under consideration for astronaut training, but that it must be kept a big secret. The older ones, aged 10 and 7, were 'bubbling over' awaiting permission to tell their friends at school. When the all-clear came through, it was the Easter holidays, so they had to wait to spread the news. By then of course, everyone already knew. Summarizing his pride on selection, Jerry told reporters "This is a real turning point in my career and I'm looking forward to training for the Moon with great anticipation."

JoAnn had been confident that Jerry would be selected: "I don't think it was a big surprise. It would have been a surprise if he *hadn't* been selected. He had always been at the top of everything he did and I just knew he would make it."

Meanwhile, over in Lakewood, a suburb of Denver, Colorado, Jerry's' grandmother, Mrs. Ferdie Barker, was interviewed by the *Denver Post*. Beaming with pride that her grandson was one of the new astronauts, she did admit to being a little apprehensive about a potential flight to the Moon: "Anybody would be a little scared up there… but don't ask me to go along. He can go if he wants to, but I'd rather stay here. Heights bother me. I don't care to fly, but I never miss a blast off on television."

Going to Houston

The first challenge in his new assignment was to move from California to Houston within a month. Moving a household is a stressful time for any family, and to do so in just a few weeks particularly so. However, changing bases is part of the life of a serving military officer and his or her family. It was decided that Jerry would move to Houston first, with JoAnn and the children following when school was finished in June. Jerry decided to make the trip by driving to Houston in his favorite 1953 MG TD sports car. According to Jerry, this was an experience itself, taking the trip during the rainy season and encountering flash floods in the desert. The Permanent Change of Station order from the Marines came on 15 April, with his official new station listed as the NASA Manned Spacecraft Center, Houston, Texas, from May 2. Jerry had accumulated enough leave to allow him to help with the preparations for the move to Texas. He started his four day trip at 9:00 p.m. on April 28. He arrived at Ellington AFB near the Space Center at 2:30 a.m. on May 2. After a few hours sleep, he would report for duty at MSC later the same day. Jerry recorded that the journey had logged 1,583 miles on the car's odometer. It was a trip not without incident.

Jerry and Jamee washing the MG in preparation for Jerry's drive to Houston after being named one of NASA's new astronauts.

He left late in the day in order to cross the southern California desert in the cool of the night. His first stop was at the Naval Air Station, Yuma, Arizona. He arrived very early in the morning, checked into the bachelor officer's quarters and slept until the early afternoon. When he awoke, he felt hot and flushed, and to his surprise when he looked into the mirror, his face and neck were covered with a red rash. He rushed to the medical dispensary, and the flight surgeon took one look at him and burst into laughter. He said, "Major, you have a classic case of three-day measles!" When Jerry explained his need to get to Houston, the medic told him to continue traveling by night to avoid getting overheated, gave him some aspirin and wished him good luck. The motel night desk clerks gave him some very dubious stares as he registered at 2:00 and 3:00 in the morning during the remaining two days of the trip.

As Jerry drove in West Texas near Fort Stockton, he was carefully following another car at night during a heavy rainstorm, passing signs warning of flash floods on the road. Suddenly, the lights of the car in front of him disappeared. Jerry slowed down, and was startled to see a sudden wave of brown water coming over the front right side of the car. Jerry suddenly found himself sitting in a soaking wet car, shivering from the cold and icy water that was filling it.

Not wanting to stall in the water, he downshifted and gunned the engine to get through the pool he had driven into. As he slowly emerged, he saw the car he had been following sitting stationary in the middle of the pool, without lights. The wave the first car had created entering the flood ahead of Jerry, along with the force of the flood waters coming from the right, had actually washed Jerry's car to the left, which fortunately meant that he completely missed the stalled car ahead of him. The old MG kept on running. All of the electrics and ignition on the left side escaped the deluge of water from the right because it was shielded by the engine block. Pulling up on the other bank of the wash, he left the MG running and checked to make sure the occupants of the other car were okay. The occupants were safe but cold and wet and needed to be towed, so Jerry drove another four or five miles to a filling station, where he let out the water from inside by opening the doors. He told the owners about the stranded car, and they dispatched help to rescue them.

Finally, arriving in Clear Lake after midnight on May 2, Jerry stayed at the bachelor's quarters at Ellington AFB, prior to reporting for NASA duty later the same day and commencing his astronaut training program.

Back in California, JoAnn was preparing to move the household and six children to Houston. Jerry, meanwhile, began looking for a new house near the Manned Spacecraft Center. Several houses were available in Clear Lake, a couple of miles east of the MSC. They had been built by developers on speculation, so the new astronauts found some excellent deals. The Carr's new home in El Lago (Seabrook) had been a show house, so it needed little work done to it before they moved in. Jerry had time to take photographs to send to the family in Santa Ana so they could see what their new home looked like. JoAnn was pleased to be forewarned about the house, but the heat of Houston would be a bit of a shock that would take some getting used to.

The family prepares to make the move from Santa Ana to Houston. Jerry gets a little help from John (wearing his Batman outfit) while JoAnn helps Jennifer, Jeff and Jamee put the For Sale board up. To the right is Jessica.

JoAnn began making arrangements to drive the family station-wagon down to Texas with all six children. She enlisted the help of 19-year-old Becky Bachman, who lived next door and had just graduated from high school, to share the driving and to help settle the family into the new home while Jerry was immersed in the astronaut training program.

Finally, on Friday June 10, 1966, JoAnn and her gang moved out of their home on Louise Street in Santa Ana and prepared for the drive to Houston. For a few days prior to the long drive, JoAnn and the children would stay with her sister Adele Hasenyager, who also lived in Santa Ana. The furniture and bulky belongings were being shipped by a local moving firm. When the moving crew arrived to the semi-organized chaos of six children running around and JoAnn trying to maintain some sort of discipline, moving man Stanley Alexander commented dryly, "No wonder he wants to go to the Moon." JoAnn was sorry to be leaving Santa Ana but excited about the move to Houston. She was not expecting to spend more than ten years in Houston before the family moved on again, but she was still living in the same house four decades after first moving in.

Driving across Arizona was a challenge, especially when the motel they stopped at for the first night had enough room – just – but no cold water. As the drive progressed, JoAnn snatched some extra sleep in the back seat of the station wagon, but everyone was getting more tired and the inevitable arguments between the children added to the stress levels. For ten-year-old Jennifer, the Carr's eldest child, the move to Houston was just like any other military move, which she had experienced many times before. "When we arrived in Houston, we stayed at what used to be the Kings Inn. We spent two or three days there because the house wasn't ready for us yet, and then we moved in. But there were no houses either side of us then." JoAnn and the family settled in to their new home over the summer break before resumption of lessons at a brand new school, Jerry was about to start his own new learning process, as an astronaut trainee.

Though the astronaut selection process had seen the family undertake a long journey from Santa Ana to Houston, Jerry's personal road to space had begun many years before. He had grown up with a love for aviation, all things mechanical and a zest for life and adventure that one day would take him to heights much loftier than the fantastic airplanes he watched for hours as a child growing up in California during World War II.

Chapter Two

The Early Years (1932-1950)

Gerald Paul Carr's first solar orbit began at 9:30 a.m. on August 22, 1932, when he was born at St. Anthony Hospital, Denver, Colorado, USA. Despite the official name recorded on his birth certificate, he was known as 'Jerry' from the start. His parents, Thomas Ernest Carr (born October 8, 1909) and Freda Wright Carr (born April 12, 1911), lived in their own house near the intersection of W. 6th Avenue and, by coincidence, Carr Street (no connection) in Edgewater, a suburb of Denver. Jerry had been born at a time when such a thing as 'astronauts' or a 'space program' did not exist. Less than thirty years before Jerry was born, the Wright Brothers had completed the first powered flight. Just over thirty years after his birth, Jerry would be selected for one of America's early groups of astronauts for a potential journey to the Moon. The rapid pace of aviation in the 20th century was to capture the imagination of the young Jerry.

The family Carr

Thomas Carr had a brother, Robert F. Carr, and two half brothers, Victor and Jack Borcherdt. The year after Jerry's father had been delivered at 817 East 16th Avenue in Denver by Dr. A.H. Harris, his parents (Jerry's grandparents) had moved their young family to Lakewood, another suburb of Denver. Jerry's paternal grandfather was Theodore Ernest Carr, born on September 12, 1886 in Montgomery, Alabama, who had worked as a railway brakeman. His paternal grandmother was the former Mary Fredericka Kephart, housewife, born on June 6, 1888 in Frostburg, Maryland. Always known as 'Ferdie,' Jerry's grandmother outlived three husbands (Theodore Carr, Mr. Borcherdt and Mr. Barker) and later, in her seventies, would often speak with pride about the success of her grandson in becoming one of the early space pioneers. She recalled that she, too, was "a real pioneer" when, at the age of 21, her family had moved to Colorado by horse-drawn covered wagons, "stopping often to poke the mud from between the spokes of the wagon wheels."

Jerry's mother, Freda Leatha Carr (née Wright) had been born in Pryor, Oklahoma exactly 50 years to the day before Yuri Gagarin became the first person to fly in space. She had moved to Denver with her family in the late 1920s. She had a half-brother named William and a half-sister named Lois. Her mother, Jerry's grandmother, was Ethel Fay Wright (née Denton), who was born on October 21, 1894 in Crooker, Pulaski County. She preferred to be called, 'Fay' and was the daughter of Robert Lee Denton and Jenny Elizabeth Denton (née Carrell), together with her three sisters Jeannie (born and died September 8, 1893), Sylvia (born September 15, 1896) and Lena Cesal (born October 12, 1900). Jerry vaguely remembers hearing his mother and grandmother talking about Aunt Sylvia and Aunt Lena, the sisters of his grandmother, when he was a young boy, but cannot remember very much about them.

Jerry proudly holds his younger brother Ron (1936)

At the time of Jerry's birth, his father had been working for the Dixon Paper Products Company for five years, selling paper to a range of businesses in Denver and Cheyenne, Wyoming. His mother had been working as a clerk in a department store. In 1935, while the family was still in Denver, Jerry's younger brother, Ronald Ernest, was born.

Coincidentally, on November 8, 1936 in Buffalo, New York, another baby entered the world. Edward George Gibson would become a member of Jerry's Skylab crew some three decades later. The third member of that crew already had a head start on both of them. William Reid Pogue was born in Okemah, Oklahoma, on January 23, 1930.

From Denver to California

In September 1938, Jerry's education began at the Mountain View School in Denver. He attended 1st and most of 2nd grade there until June 1940 and gained mainly Grade

Bs in his studies. In 1940 his father moved the family to California, where he took up a new position with a different paper company. Jerry and his brother completed their education there. Jerry's mother took a position as a clerical secretary with an electrical supply company.

1940 Mountain View School photo -
Jerry is seated front row, sixth from the left

Jerry's family settled in Santa Ana, California when he was 8 and Jerry came to consider that this was his hometown. The move to California would have an important influence on Jerry's future career, because the area was a hotbed of aviation at that time. As a young boy during World War II, Jerry often saw some extremely unusual airplanes, then in development, flying over his house. He still remembers the sight of the Flying Wing passing overhead, being amazed that a plane could fly without a tail! His love of technology was later to be expanded considerably.

Jerry finished 2nd grade at the Chapman Avenue School in Gardena, California, in June 1941. When the family moved to Santa Ana, he attended Wilson Elementary School. Jerry was an avid reader. In the 4th Grade, he was awarded a Certificate of Perfect Attendance and in 1942 he received the Santa Ana Public Library Junior Department Vocational Reading Club Certificate for reading "10 good books during the summer vacation." He also attended the Young Men's Christian Association (YMCA) and was awarded a Certificate of Achievement for passing Junior Commando Tests at Santa Ana YMCA in September 1943, a reflection of his interest in youth and church activities. He remained at Wilson Elementary for three years, leaving in June 1944 for Junior High School.

In his reports for the 4th, 5th and 6th grade, Jerry's results were listed as 'satisfactory'. Considering the upheaval of the move from Denver, Jerry had settled in well, however, and soon showed a keen ability, making significant contributions to the class. His grade marks improved from mainly B (Good) to straight A (Excellent), with particular aptitude for spelling and reading.

His 4th grade teacher, Marian Valley, recorded: "Jerry seems very interested in his schoolwork and is doing well in all subjects." His 5th grade teacher, Janet Loupe, reported: "Jerry seems to take pride in doing well in his work," though she also noted that Jerry did not always follow directions very accurately, and she hoped he would improve. On occasion, she noted, Jerry seemed to be unduly argumentative, although he continued to hand in very good project papers. He was a recognized hard worker, with excellent spelling and good arithmetic skills. In his tests, Jerry scored above his grade level in most subjects, and it was noted that his attitude had begun to improve by then. In his 6th grade report, his teacher, Sophie Scott, recorded that Jerry was improving well in sportsmanship, but noted that although he was getting along well with his fellow classmates, "he should watch his choice of expression on the playground, where he uses slang unconsciously." Jerry's mother replied to this comment and recorded: "We will call his attention to his slang around the home and see what the outcome is." ('Slang' was their euphemism for profanity.)

Jerry attended Willard Junior High (now called Intermediate) School (Grades 7-9) from September 1944 to June 1947, and in 1945 he attained a second Santa Ana Library Reading Club certificate, this time for reading 25 good books during his summer vacation. His literary heroes, John Paul Jones and Horatio Hornblower, were the subject of much of his reading. As a result of this and also his Boy Scout activity during his junior high years, Jerry became very interested in the idea of a military career. Jerry's favorite subject at Willard was journalism, and in June 1946 he received a Certificate of Honor, awarded by the H.M. Rowe Typewriter Company, for achieving 29 words per minute in typing skills. This added to his journalist credentials, as did his participation on the staff as Feature Editor for the *Willard Echo*, the student newspaper.

He received his Junior High School promotion certificate in June 1947, was recognized by the *Willard*

Echo Staff for his "distinguished service to the school and community," and also received the American Legion Distinguished Achievement Award at Willard for scholarship and leadership. During his junior high school years, Jerry was inducted into the International Order of DeMolay for Boys in the Santa Ana Masonic Temple. He was also baptized and became a member of the First Presbyterian Church, Santa Ana. This was the beginning of a lifetime's affiliation with the church.

Most of his childhood was spent living in the shadow of the Depression and then World War II, but as he lived in a small-town environment (Santa Ana in the 1940s had a population of around 25,000), Jerry remembers a fine childhood in which he was allowed to ride his bicycle anywhere. The small community was a very safe place in which to live.

A world at war

When Great Britain declared war on Germany in September 1939 following the blitzkrieg on Poland, Jerry was only 8, too young to remember any conversations around the dinner table about the developing situation in Europe. However, he does recall the events of December 7, 1941 and the Japanese attack on Pearl Harbor. On that fateful Sunday, Jerry was lying on the living room floor listening to the KFI-Los Angeles (NBC) radio announcer, who was reading the Sunday morning comics to the young audience. After the program ended, Jerry left the radio on. Some time later, the radio broadcasts were interrupted to report the attacks on Pearl Harbor to the stunned nation, an event that brought America into the war. Jerry still recalls the period of rationing. Meat coupons and gasoline stamps were issued at the time, and everyone became involved with "the war effort." As a young boy, Jerry's effort involved raising rabbits and chickens. He sold them to the local meat market, and was able to help the family to save their own meat coupons by supplementing their diet with their own chicken and rabbit supplies. He also toured the neighborhood with his wagon, selling the excess from the family "Victory Garden" and collecting newspapers and baling them for recycling.

Following the Japanese attack on Pearl Harbor, Jerry's parents wanted to support the war effort. His father, Tom, was unfortunately deaf in one ear, and neither he nor his brother Bob, both in their 30s, were drafted, being classified 4F draft status (unfit for immediate call for service, but available if manpower levels sunk low enough). Instead, both brothers took jobs to support the war effort. Jerry's father went to work as an electrician in the nearby boatyards at Newport Beach, where they were building minesweepers. Jerry's mother remained with the electrical company and after some years became an office manager. Her half-brother, William Wright, was also in his late 30s and was also classified as 4F draft status, but Jerry's other uncles were in their late 20s. Victor served as a Navy Seabee (from the initials "CB" – Construction Battalion) at Port Hueneme, California. Jack became a paratrooper with the 101st Airborne Division, saw action in the European Theater on D-Day, and was wounded in the Battle of the Bulge.

Though the war seemed far away from Santa Ana in sunny California, developments were often reported in local papers, on newsreels in the local cinema, or on the radio. Jerry was also able to read of the adventures of his two fighting uncles from letters to his parents and grandmother. One of his Uncle Jack's letters to Jerry's grandmother about his "thrilling experiences" during the D-Day invasion of June 6, 1944 was reprinted in the local newspaper, the Santa Ana Register. In his letter of June 18, Uncle Jack called the whole event "quite a show." As an airborne parachutist, he was one of hundreds dropped behind enemy lines in Normandy prior to the sea-borne landings. The 101st Airborne would be the first Americans to enter occupied Europe, and Jack's regiment would lead the way by parachuting into the area of Carentan in Normandy, on the road from Bayeux to Cherbourg. Like thousands of others that summer day, Jack thought of his home and parents and prayed that everything would work out.

The pilot of the plane was scared to death but, after "a talking to" from Jack, came to understand that he had better do a good job and not make any errors, or else! As the plane crossed the coastline there was no anti-aircraft fire, but over the drop zone, as Jack recalled, "the reception committee greeted us with everything but the band." During the next few hours, Jack gathered and led a force of about 60 men, all from different units, having been separated in the confusion of the drop. In close quarter battles and hand-to-hand combat the casualties were high. The battle continued night and day, with no man getting much sleep for several days. Jack finally got chance to write the letter home to his mother on June 18, describing that he had experienced lots of "close calls" but was not scratched (wounded) yet.

Several days after D-Day, war correspondent Henry T. Gorrell was with US assault troops and Army Chaplain Raymond Hall, known as the Parachuting Parson, watching the fighting on the outskirts of Carentan in France. Gorrell described Jack Borcherdt coming in for a personal report because his portable radio had been disabled. Jack reported, "We were in the leading assault company. We crossed three of the four bridges and were making good progress. But then the Germans rallied, reoccupied prepared positions and began to counter attack with bayonets and grenades. I haven't seen many of my outfit since the last time they hit us." Jack's clothes were torn in many places and his helmet dented by enemy bullets and shrapnel, yet he was still not wounded, although he later wrote, "One bullet went through my helmet and one through my coat collar and burned my neck, in one battle I will never forget."

Uncle Jack *was* eventually wounded, losing part of his calf during the Battle of the Bulge. As he was "wounded in action," he was sent home. After an honorable discharge from the military, he married his pre-war sweetheart and eventually moved to Seattle, Washington, to raise a family. Years later, one of his daughters sent Jerry an email to introduce herself. Jerry's brother, Ron, had taken an interest in tracing the roots of their family and had been in touch with the Borcherdt side, finally ending many years of little contact.

Jerry remembers little about the end of the war (he would have been 13 in August 1945) but the war years had a definite influence on him. He witnessed very little in-fighting because everyone was pulling together for a common cause. There was patriotism in abundance, coupled with a "work hard" ethic to support the war effort.

A yearn for aviation

Growing up with the stories and letters from Uncle Jack from Europe and the newsreels from the Pacific, as well as seeing the latest war planes flying overhead, it was little wonder that aviation and the military were to play a large part in Jerry's life. Indeed, as with most boys of his age, aircraft modeling was a favorite hobby. Jerry started out by wood carving and kit building. Then from the age of 14 or 15 he started flying wire-controlled models. He needed a large field to test-fly the aircraft to avoid getting into trouble by crashing them into people's gardens.

Jerry was determined to be involved in the aviation boom the whole country was experiencing at that time. Several large aircraft companies, including Douglas, had huge plants in the Long Beach area south of Los Angeles. The US Air Force (then called the Army Air Corps) had a training command set up in Southern California, with its Headquarters just down the street from where Jerry was living, and the Marine Corps had also established the El Toro Marine Corps Air Station nearby. This centre of aviation was the beginning of the growth of Orange County, which up to that point had been a fruit-growing centre for Sunkist Oranges. Jerry, his brother and their friends had all played in the rows of orange groves.

JC: "Today, you are lucky to find a few trees in a back yard. The groves are long gone to property development."

When Jerry was 14, his desire to become involved with real aircraft grew. His best friend, Jim Leonard, also shared a passion for aviation, so on Saturday mornings Jim and Jerry would hop on their bikes and cycle the 12 miles to the local Orange County Airport (now called the John Wayne International Airport). They would spend all morning washing old Taylorcraft airplanes for the Eddie Martin Flying Service to earn a free 20-minute ride in one of the light aircraft. Jerry was soon totally infected by the aviation bug. Eddie Martin, incidentally, was reported to be the brother of Glenn L. Martin, the founder of Martin Aviation, which later became Martin-Marietta, then Lockheed Martin, one of the very large aerospace companies involved in the space program.

Jerry wasn't allowed to actually fly any of the planes, but it was enough at that age to enjoy the thrill of flying over their hometown. He also frequently saw the famous movie stunt pilot of the 1930s and 1940s, Frank Tallman, who housed planes near the Orange County Airport. Tallman later joined up with a second company owned by famed stunt pilot Paul Mantz, becoming Tall-Mantz Aviation. They created a museum at Orange County Airport that resided there for several years until it become a very busy commercial airport.

Some of Jerry's other friends from high school included Jim Crowe, Henry Aguilera, Zepheniah Jones, Warren Hupp, Quentin Haug, Gordon Higgins, Jamie Fox, Dick Wilcox, David Haynes, Ron Chandler, Charlie Raeberg, Dan Peterson and Herschel Musick. Herschel happened to be a relation of Pat Musick, later to become Jerry's second wife.

Boy Scout Jerry Carr (1947)

An American Boy Scout

Jerry's other main activity at this time was the Scouting Movement. Both Jerry and his brother Ron joined the Boy Scouts of America program and benefited greatly from the experience. Jerry stayed with the Scouts until he was nearly 18.

Ron, who had followed Jerry through the same schools, had always enjoyed tinkering. One of his favorite hobbies was to restore old automobiles. He joined a group of fellow enthusiasts restoring an old 1929 Nash and for many years worked on an old Chevy. Jerry often joined his brother and soon developed an interest in engineering, mechanics and the 'science' of tinkering. As brothers they got on fairly well, although older brother Jerry usually had to badger Ron to do his chores. On one occasion, Jerry told Ron to do something and Ron said "No." The inevitable fight followed and ended up with Ron throwing a small battery at Jerry, knocking him out for a few minutes. When he came to, Jerry gave chase, but Ron was quicker and jumped on his bike to escape his angry older brother. Ron kept away, waiting until Jerry had calmed down to sort out the disagreement.

Prior to joining the Boy Scouts, Jerry had been a Cub Scout in Pack 120, sponsored by Wilson School of Santa Ana. He was assigned to a 'Den' comprising about 10 to 15 Cub Scouts. The group was managed by a 'Den Mother' who provided the meeting place. The Den Mother, assisted by a Boy Scout called 'Den Chief,' taught some of the curriculum. She provided the snacks as well, which Jerry remembers was the best part of the whole meeting. His Den Mother was Mrs Hupp, the mother of one of Jerry's friends, Warren. She was, as Jerry recalls, "a wonderful bread baker," and each time the meeting was held at the Hupp residence there was always an amazing aroma of fresh bread. Their snacks always included nice large slices of fresh warm bread straight out of the oven.

At the age of eleven Jerry graduated from Cub Scouting and joined Boy Scout Troop 24, sponsored by the Santa Ana Rotary Club, in 1944. Most of the young boys who joined the Boy Scouts were interested in outdoor life and looked forward to learning camping and woodland lore. Because they were called upon to march in parades there were lessons in marching, but there was certainly no military indoctrination. The emphasis in scouting was on self reliance and self confidence, skills that would be valuable later in their lives. Jerry fondly looks back at the valuable lessons he learned in those years, which have continued to play a key role in his later career and everyday life.

After demonstrating knowledge of the Scout Oath and Law, Motto, Sign and Salute, as well as the history and proper care of the US Flag, and proving that he could tie at least 8 rope knots, a scout could became a *Tenderfoot*. Tying knots was an activity Jerry really enjoyed participating in, and today, some sixty years later, he can still reproduce most of the knots he learned in his scouting years. After one month's service and by passing several other tasks, tests and skills as part of the 'scouting way', a Tenderfoot could then progress to *Second Class Scout*. Another few months service and

Troop 24 (about 1945). Jerry is third from right in the front row. Top left is the Sponsor's representative, in the middle is the Scout Council Director, Hugh Wilcox, and on the right is Scoutmaster, Mr. Martini.

passing 13 more requirements (including a swimming aptitude test, operating a semaphore or Morse code, first aid skills, map reading and cooking) would earn a promotion to *First Class Scout*. Jerry achieved this in April 1945.

The top of the scouting tree was even harder to attain. A scout needed to be proposed by his peers, affirming that his skills, conduct, mannerisms, ideals and principles matched the scouting ideal. For *Star Scout* (awarded in March 1946), Jerry needed three months service as a First Class Scout, an excellent record and at least 5 merit badges. As a *Life Scout* (awarded in May 1946), Jerry needed at least three more months with excellent service as a Star Scout and ten more merit badges. Finally, with at least six months excellent service as a Life Scout, and no less than 21 merit badges, Jerry became an *Eagle Scout* in May 1947. After another six months as an Eagle Scout he earned three Eagle Palm Awards (October 1947 Bronze, Silver and Gold) plus several additional merit badges. These merit badge attainments ranged from first aid, to poultry keeping, to personal and public health, to cooking, swimming and lifesaving. Jerry also learned a variety of basic craft skills in wood working, carpentry, leather craft and metal working, as well as public speaking and business studies. These were skills which would prove very useful throughout his life. Studies in nature and the environment gave Jerry an interest in wildlife and animals that continued into adulthood. It was in the scouts that Jerry also undertook merit badge awards in aeronautics, aerodynamics and airplane studies. In all, Jerry earned over 30 merit badges during his scouting years.

One of the more interesting challenges Jerry recalls from his early scout years was learning to build a fire without matches. They were taught three methods. One was to rub two sticks together to develop enough heat, essentially using one piece to 'drill' into the other by using a bow with a loose string on it to create the friction and heat in the hole that would spark off the tinder. The second was to use flint and steel, part of a small kit carried on the scouting belt, creating the sparks into a small ball of tinder by striking one against the other. The third way was by using a magnifying glass – not very useful by night, but during the day a small magnified spot of sunlight could get the tinder smoldering quite quickly.

As an Eagle Scout, Jerry also worked as a Senior Patrol Leader and as a Youth Leader, teaching others about Scouting. Troop 24 met at Fisher Park near Santiago Creek, and Jerry spent many a summer break and weekends at Scout Camp Ro-Ki-Li in the San Bernardino Mountains. The camp was built by three business service clubs, the Rotarians (Ro-), the Kiwanis (Ki-) and the Lions (Li) each giving their names to the club. Situated at about 5-6,000 feet altitude, it provided boys from the flatlands of California an opportunity to have a camping experience in the mountains for a couple of weeks. It was here during three summers that Jerry first learned the skills of wilderness survival, hiking and navigating in the woods, camping and woodcraft. Jerry has always stated that he owed Scouting a debt of gratitude for his ability to succeed in later training. For his fourth and fifth summer at Ro-Ki-Li, Jerry requested to be hired as a member of staff. At this time he was a senior scout and was interested in using his skills and experiences to serve on the staff rather that just attending the camp. Camp Staff remained in attendance for about eight weeks. He worked in the camp store and in the kitchen, washing up the pots and pans after a meal. Jerry received a small salary for being part of the staff. When not working in the kitchen or the store, he provided leadership for activities with some of the younger boy scouts. A camp tradition was a nightly campfire after dinner. The staff would lead the scouts in various activities such as story telling, cheers and songs, before everyone turned in for bed. Jerry remembers one of the 'featured guests' at camp fires was a red-headed fellow named "Red" Knaus, who was in his thirties and a District Scout Leader in southern California. For a couple of weeks each summer, Red was one of the 'entertainers.' Jerry remembers that Knaus had a "rubber face much like comedian Red Skelton, and a really funny patter." He would read from a fictitious camp 'newspaper' he called the 'Wailing Woods Pussy' (the crying skunk). His stories were very outlandish and funny. All a camper or a staff member had to do was make some sort of silly mistake during the day, and he would become a feature in the next reading of the 'Wailing Woods Pussy.' Jerry found himself featured there a time or two.

One of the things Jerry learned at the camp served him well years later at NASA. A scout counselor, an elderly gentleman named Bill Carrithers, was a retired postman. He spent most of his hours sitting at home 'whittling' and wood carving, but he also served as a merit badge counselor who helped boys, including Jerry, gain certain merit badges. He taught them to wood carve and to properly care for their carving instruments. In the summer at Camp Ro-Ki-Li, "Uncle Bill" worked as staff. Each time there was a nice clear night after dinner, Uncle Bill would take five or six scouts for a night time hike to "Slushy Meadows" about a thousand feet higher up the mountain. The name of the meadow derived from its boggy state after a rainfall. Once they arrived, the scouts would unroll their sleeping bags and arrange them in a circle, with the head of the bag towards the center. After an evening around a camp-fire, and with the embers quenched to reduce the light as much as possible, they would retire to their sleeping bags. As they lay there on their backs with their heads together, Uncle Bill would talk about astronomy. As they looked up to the heavens during those very silent, dark and clear nights, he would point out the stars and constellations of that time of the year. Then he related the myths about how the constellations were formed and named. Jerry believes this was his first introduction to astronomy. His interest in the stars began during those summer nights as a young teenager lying on his back looking up towards space. He never dreamed that one day he would be one of the few who would actually journey there.

For some of the older, 14-16 year old scouts at summer camp, there was an organization called the 'Order of the Arrow.' This was an honorary organization in scouting for which certain scouts were singled out for their leadership capabilities. The induction ceremony was very solemn, and occurred once every two weeks during each scout group's tour at the camp. Gathered around the campfire in the still of night, a Native American tom-tom drum would be heard beating from the woods. The whole camp became very quiet. From the woods would appear a procession of six to eight "Indians" dressed head to toe in authentic gear – the members of the Order of the Arrow. They entered the arena of the camp fire, walking down the rows of scouts and staring each in the eye before moving on. Suddenly one individual would be chosen by one of the "Indians" who would slap him on the shoulder three times. This individual had been selected or "called out" for the Order, and he would be required to undergo the initiation ceremony. The initiate was blindfolded and led away from the fire into the wooded area near the camp. At a selected location, he was told to remain there alone all night and was forbidden to move or speak until he was collected the next morning.

This was a coveted honor for the young scouts, and Jerry remembers seeing this occur, envying the older boys who were picked out and then returned the next morning for breakfast, disheveled and tired but very happy. Once initiated, it was the task of the inductee to make his own very authentic Indian costume and become involved in future ceremonies at the camp. Members of the Order were allowed to wear, as part of their uniform, a white sash with a long red arrow on it. It fit over the Merit Badge sash, setting him apart as "a special person."

When it was Jerry's turn to be inducted, he remembers silently praying that he would indeed be selected this time as the Order approached the camp. When they pounded him three times on the shoulder, he was almost knocked down. He was led off and shown where to stay for the night. To make himself as comfortable as he could, he gathered branches and leaves to make his bed for the night's vigil. The next morning at breakfast, he was given his responsibilities as a member of the Order of the Arrow, including the requirement to make himself a good costume. Jerry believes the Order still remains within the scouting movement of America, although the ceremony may have changed. He also believes that this initiation established his leadership potential and inspired him to live up to the recognition he had received. He resolved to pursue his scouting career all the way to his Eagle Scout achievement.

JC: "I was very active in the Boy Scouts, moving 'up the functional leader's ranks.' At one time, I served as a Patrol Leader, and later as a Senior Patrol Leader. As a Patrol Leader I had a number of young scouts in my Patrol. Each such Patrol was named democratically by its members and featured such names as Beaver, Puma or Cougar Patrol. We chose 'Fox' for our name. In terms of leadership, a Patrol Leader looked up to a Senior Patrol Leader, whose purpose was to organize the Patrols and periodically check on how well they were being run. He determined whether or not the Patrol Leader was doing a good job of teaching and encouraging the younger scouts to get on with their advancement. The last job I had before I left the scouting movement was as Junior Assistant Scout Master. The Scout Master of a Scout troop is not a professional Scouter. He is usually a father of one of the boys, or an adult with a strong interest in the scouting movement who is willing to lead a troop of boys. A troop could be made up of five or six patrols, so you could have 30, 40 or 50 in a troop. To help the Scout Master, there was the Assistant Scout Master who was willing to step in to assume the reins of the Scout Master as required, and the Junior Assistant Scout Master who was the senior boy scout of the troop. His job was to make things run smoothly. One might compare him to the Non-Commissioned Officers (NCOs) in a military organization."

Jerry worked in scouting right up to his senior year in high school and continued his association with the scouts as a speaker during his military and astronaut career and beyond.

JC: "Besides the self reliance and self confidence, scouting placed a lot of emphasis on integrity, honesty, truthfulness and patriotism. I think those were characteristics I carried on throughout my career and future years. When I returned from my mission in space, I was designated by the Boy Scouts of America as a Distinguished Eagle Scout and was presented a special ribbon with a silver eagle on it to wear around my neck. There is an organization of Eagle Scouts and Distinguished Eagle Scouts in the United States, and we try to stay in touch and help each other any way we can."

A high percentage of the Astronaut Office participated at some stage in the scouting movement and many reached some of the higher ranks such as Jerry attained. Still, Jerry recalls being told by fellow Group 5 astronaut Stu Roosa, "Hell, I was just too busy camping and fishing to be a scout."

High School

Jerry began attending Santa Ana High School (Grades 10-12) in September 1947. A month later, on October 14, Chuck Yeager broke the sound barrier at Mach 1.06 flying the Bell X-1. Though not avidly following the program (mostly because it was a national secret at the time), Jerry was awe-stuck that Yeager had broken the sound barrier when the news was released. He was not sure what the sound barrier actually was, but he remembers some of the newsreel movies and newspaper articles of the time. Jerry's aeronautical education was not very advanced, but he was very impressed by a person who could fly faster than the speed of sound.

In high school, Jerry further improved his typing skills (reaching 40 words per minute) and tackled trigonometry at Summer School. He also discovered a love of travel, and between 1943 and 1950, he took occasional day trips with his family to nearby Mexico. This began a lifetime love of the country that saw him secure holiday accommodation in Cancun fifty years later. He maintains his fluency in the Spanish language, a study he started during two years of Spanish lessons in high school, and still uses it today when visiting Spanish speaking counties. Jerry found his knowledge and skills in Spanish made it easer for him to learn and develop Italian language skills when he worked in Torino later in life and toured extensively with his wife, Pat Musick, in Tuscany.

Jerry's attainments continued throughout high school. In June 1948, he received his Interim Operators Driving License (Temporary) No 823176E, and in 1949 he received a Certificate of Membership California Scholarship Foundation. In June 1949 he received the California Boys State Certificate of Merit, awarded by the American Legion. He was interested in student government and was selected as one of several students from all over California to participate in a week of introduction and involvement in State government at the State Capital (Sacramento) in June 1949. Jerry considers this week a wonderful experience, meeting top boys from other high schools across the state, and receiving a good introduction to the workings of State government. In January 1950, he was elected as the Student Body Commissioner of Finance in his high school. On a national level earlier in the year, President Truman had signed a bill for a 5,000 mile guided missile test range off the Florida Coast near point Cape Canaveral. A little over a decade later, this became the focal point for sending American missions into space and the site of all American manned launches, including Jerry's Skylab 4 mission, 24 years later.

Jerry enjoyed a brief career in high school football at Santa Ana. He started playing in his Junior year as running guard for the Santa Ana "Saints" B team (for younger and smaller boys). In his Senior year, Jerry became a member of the 1950 Varsity football team, playing second string running guard. He recalls that the reason he was selected for that position was that he, like the first string running guard, was so short that defensive linebackers couldn't see him as he moved behind his own line to be the lead blocker on a play. That year, the Saints started the season well by winning five straight games. They then promptly lost the next three in a row, although Jerry did at least get a mention in one post-game report for blocking a kick late in the game against Newport Harbor. This wasn't quite the game-changing play he might have hoped for though, as Jerry later recalled: "I can't remember much more than receiving the football in the face as I blocked the kick." The Saints lost the game 14-0 as well.

With his growing interest in aviation and his experiences at Orange County Airport, Jerry had decided on an aviation career while at Senior High School. His aim was eventually to enter the regular U.S. Navy and become a naval aviator. In 1949, he enlisted in the USN Reserve (effective October 23 - Serial #384 20 17) before going to college the following year. Jerry was attached to a naval reserve "Weekend Warrior" squadron at Naval Air Station Los Alamitos, near Long Beach, California. He trained as an Aviation Crewman, and, with the rank of Airman Apprentice, was assigned as a Plane Captain for an old Grumman F6F *Hellcat*. His job was to keep it clean, fully fueled and oiled. His first task each weekend morning was to climb into the cockpit, start the engine up and ensure everything was in good running order. He would then go to the Line Shack and sign off the forms certifying that the aircraft was ready for flight so that the assigned pilot could take it out. Jerry refers to these experiences as his "pre-flying years."

For a 17-year-old at that time, just to sit in a real cockpit and start up the powerful R2800 engine was always a big thrill. His biggest fear, however, was that if he inadvertently caused the airplane engine to backfire he would face the wrath of the Line Chief, a fierce, red headed chief petty officer whose language could turn the air blue. Occasionally, though, Jerry got the chance to fly in the back seat or belly of a TBM Avenger torpedo-bomber. Flying out over the Pacific between Long Beach and Catalina Island at about 50 feet above the water was "pretty darn thrilling," as he recalled. As an air crewman in the Navy, he was eligible for 'flight-skins', which meant that if he flew a certain number of hours each month, he would be awarded flight pay. It was a nominal amount of money, and Jerry was more interested in the fun of flying than receiving the cash.

A girl called JoAnn

By this time, he had begun dating JoAnn Petrie of Pomona, California, whose family had moved to Santa Ana. They had met as students at Santa Ana High School in January 1949 during their Senior year, just after the Christmas break. They had adjoining lockers in a school corridor, and their four-year courtship began when Jerry opened her stubborn locker door for her. JoAnn recalled going to the office to get someone to open a former boyfriend's locker (next to Jerry's) so that she could recover her books, but her request was refused: "And so I went back over there, and met Jerry who had the locker next door. He very gallantly gave it a kick in just the right spot to open it for me so I could get my book for class."

JoAnn was born Joan Ruth Petrie on September 20, 1933 in Pomona, the daughter of Arthur James (born 1897) and Ann (née Haber) Petrie (born 1893). She chose at an early age to modify the spelling of her first name to JoAnn. A lot of her weekends in high school were spent behind the wheel of Jerry's first car while he lay underneath it tinkering, or as she called it, "puttering." It wasn't exactly fun for her, but since she wanted to be with him she was willing to put up with it. JoAnn attended Santa Ana College for two years, followed by two years at Long Beach State College, in order to obtain her teaching degree. Her plan was to become an Elementary School teacher, with a longer term aim to become a psychiatrist.

Their graduation year of 1949/1950 recorded the highest number of academic scholarships (12 including Jerry) in the 60-year history of the school. Jerry later recalled that he'd had "a very busy and very happy high school career. I really enjoyed it a lot."

Reserve Officer Training Corps

Jerry had known for some time that he wanted to go on to college, and in his Junior and Senior years he applied for scholarships around the country. He also applied through his Congressman for an appointment to the US Naval Academy at Annapolis, and to the Naval Reserve Officer Training Corps (ROTC) program, which was conducted at university level across the USA.

Several High Schools ran ROTC programs. It allowed students who had an interest in the military to have the ROTC as one of their courses, during which they would wear the uniform and learn to march and drill. Then in the 1950s at the start of the Korean War, a number of American Colleges started ROTC programs for the Army, Navy and Air Force. If a student was accepted as a "regular" Midshipman in the ROTC program, the parent service paid for tuition at the designated university for four years. As well, all text books, a $50 per month stipend towards living expenses, all uniforms and equipment were provided. The student's obligation involved wearing the uniform to school one day a week, on drill day, and taking courses in military science along with his regular curriculum. On drill day, which was Thursday, the afternoon was used for sharpening military drill skills and attending labs on naval equipment and weapons.

Upon graduation, the NROTC Midshipman was commissioned as an Ensign, USN or a Second Lieutenant, USMC with, at that time, an obligation to serve as a regular officer for at least four years. Another way to enter the university ROTC program at that time was to volunteer to take the courses and drills and be designated as a Reserve officer upon graduation. This way the service provided the uniforms and a few textbooks, but the student remained, in effect, a civilian, and was not obligated to continue his military service for more than two years after being commissioned.

By the middle of his high school Senior year, Jerry found himself with several scholarship opportunities, as well as acceptance to the US Naval Academy. In April 1950, he was also accepted as a Regular Midshipman in the Naval ROTC for the 1950 Fall term at the University of Southern California. In May he was offered scholarships at the University of California at Los Angeles (UCLA) and at Berkeley (UC). Further awards also came Jerry's way this year. In June 1950 he earned a Certificate of Life Membership in the California Scholarship Federation for consistent and superior scholarship and service.

Three days later he received the Santa Ana High School Student Government Certificate of Merit for his work as Commissioner of Finance for 1949/1950. Finally, on June 15, 1950, he was presented with his High School diploma. It was time for him to decide which way his career would go.

As a youth, we recall, Jerry had read and enjoyed the fictional stories of Horatio Hornblower of the Royal Navy and of the historical prowess of John Paul Jones in the Continental Navy. He was very taken with naval service and had virtually decided on a career as a naval officer, hopefully leading to assignment as a naval aviator. Consequently, he decided to decline the scholarship offers from the University of California in favor of either the Naval Academy at Annapolis or the Naval ROTC program at the University of Southern California. He was not sure which to accept, so he approached his Reserve Squadron Commander at Naval Air Station Los Alamitos for advice:

JC: "I knew it was probably best for a long-term career to go to the naval academy, but there was this other opportunity at the University of Southern California. I will never forget the (Reserve) Commander, who thought for a moment and said, 'Well, this may seem blasphemous from a Naval Academy graduate, but I do not see any good reason in the world, when you have a choice, why you should lock yourself up in the 'Bastille' (the Naval Academy) for four years when you could have four years of college with girls, study the courses you want and still come out with the same rank. If it were me to do all over again, and I had your options, I would take the Naval ROTC.' So I took his advice and did just that."

1949 Church Youth Retreat photo - Jerry is in the 2nd row from rear, 2nd from left.

Chapter Three

US Marine Corps (1950-1966)

Jerry began studying at the University of Southern California in September 1950. He had entered the USN Reserve as an Airman Apprentice on 23 October 1949 while at Santa Ana High School, well before his graduation in June 1950. He was honorably discharged on 17 September 1950, and the very next day was promoted to the rank of Midshipman, Naval Reserve Officer Training Corps. He had started as a mathematics major, but halfway through his freshman year he discovered that while he enjoyed mathematics, it was not what he wanted as a career. He felt that designing mechanical systems was more to his liking, so he shifted his major to engineering and worked toward a mechanical engineering degree.

Every Naval ROTC group is assigned a Marine Officer and a Marine Drill Sergeant, providing ROTC access to the Marine Corps branch of the armed services. As it is a tradition that the USMC cannot be larger than 20% of the size of the whole Navy, the ROTC class reflected this, allowing only 20% of the ROTC to go Marine if they wished. During Jerry's freshman and sophomore years, the Marine instructors were Major Robert Ervin and Technical Sergeant Anthony Forte. During his junior and senior years, Major Ervin was replaced by Colonel John Finn. It did not take Jerry long to decide that these were the kind of men he would like to emulate. They were fine leaders and role models, who guided Jerry towards a Marine career and developed his interest in the Marine "*esprit de corps*". At the end of his junior year, Jerry decided to take the Marine option and his curriculum changed accordingly.

Summer cruises

The NROTC program also required Jerry to go on naval cruises as a midshipman for about a month during summer vacations. In the summer of 1951 (between his freshman and sophomore years), he was assigned to the destroyer *USS Steinaker* (DD-863), based in Norfolk, Virginia. The ship cruised the eastern seaboard, with ports of call at New York City, Guantanamo Bay in Cuba, and the Panama Canal. During this cruise, Jerry and his NROTC colleagues received intensive instruction on how to handle a ship at sea and how to navigate by the stars. Jerry also recalls that he received extensive instruction on the proper way to chip paintwork and swab decks. He also had the pleasure of experiencing stormy weather off the coast of Cape Hatteras, North Carolina. The crew was restricted to below decks because green water was flowing over the weather decks. The Navy cooks, being a sadistic lot, decided to serve beets for lunch that day, and the galley decks were awash with red beet juice. The midshipmen, of course, were all wearing white uniforms.

Jerry's summer detachment in 1952 (between the sophomore and junior years) was split between aviation and amphibious training, in two-week segments. The first two weeks were spent at Naval Air Station (NAS), Pensacola, Florida and NAS, Corpus Christi, Texas, for a taste of naval aviation pilot training and a glimpse of what a naval flying career might be like. The second two weeks were at the Amphibious Naval Base at Little Creek, Virginia, where they were issued with Marine Corps combat fatigues and taught about amphibious warfare. This culminated in a mock amphibious landing, in which Jerry got "pretty wet." He was introduced to the intricacies of planning and executing an amphibious landing. This included handling an LCVP landing craft like those used in the WWII Pacific theater and in the Normandy invasion. His final lesson was the experience of riding as an infantryman in an LCVP, seeing the front ramp drop and charging through the surf to a position on the beach while desperately trying to keep his rifle dry.

It was at this point that Jerry decided to take the Marine Corps option and look towards aviation. From his experiences on the *Steinaker*, he decided that following in the footsteps of Horatio Hornblower and John Paul Jones was, "not a great way to go." He was not enthusiastic about spending months at sea, especially in a storm, and he'd had enough of a taste of aviation on the second detachment to decide that that was where his future lay, preferring land-based airstrips to a career flying on and off carriers at sea (although this is what he actually ended up doing).

When he returned from his second detachment at the start of the 1952/1953 academic year, he approached Colonel Finn about taking the Marine Corps option. For his third cruise in 1953, however (between the junior and senior years), he would probably be assigned to a ship bound for European waters. He didn't want to miss this opportunity, so the Colonel advised him to make his official decision when he got back.

During the summer of 1953, Jerry completed a "wonderful" cruise on the light cruiser, *USS Worcester,* which visited Naples in Italy, Bergen in Norway and Copenhagen in Denmark. Instruction on this cruise included an in-depth study of the ship's power plant. Many hours were spent below decks mapping out steam lines, boilers, water evaporators and turbines. For a while, Jerry was designated the Midshipman Commanding Officer in charge of 315 midshipmen. This and several other leadership positions were rotated between senior, well thought-of midshipmen about three or four times during the cruise. It gave them the opportunity to experience the responsibilities of command. Jerry returned to USC to spend his final year studying USMC military science and complete his studies for his Bachelor's degree in mechanical engineering.

College activities

In addition to his academic studies and NROTC assignments, Jerry took two part-time jobs at the University of Southern California. Between February 1951 and June 1952, he worked as Assistant X-Ray Technician in the university dispensary, leaving to take a better position as a clerk in the Office of the Registrar in September 1952. He left that post when he graduated from the University in June 1954 with his Bachelor's degree and a USMC regular commission as a Second Lieutenant.

After playing high school football at Santa Ana (left guard), Jerry decided that he was tough enough to play college football, so he signed up with the "frosh pigskinners" at USC. After a brutal season as a halfback and a cornerback, he realized that he was not built for life on a football field:

JC: "I soon learned that I was the smallest player in the entire Pacific Coast Conference. I was a pretty fair halfback because I could quickly get to the holes that our offensive line opened in the opposition's line, and it was difficult for the opposing secondary to see me behind the fannies of our huge 250-pound linemen. The problem was that as soon as I got into their secondary I wasn't fast enough to avoid being mowed down by the defense.

"As a defensive corner back, I remember the awful day when the Freshman team was sent to Bovard Field to scrimmage the Varsity. The Varsity Coach, Jeff Cravath, was furious because the Junior Varsity was just not clicking. The Varsity was running roughshod over them. So he called our coach and directed him to bring us over to give the Varsity a decent workout. Dutifully, we lined up across from the Varsity team. Among their players was All-American running back Frank Gifford. From my cornerback position on the first play, I saw a hole open off-tackle in our defensive line, so I moved in to fill it behind our linebacker. The lead blocker took out our linebacker, and suddenly I faced Gifford at full speed. I hit him, or maybe I should say he hit me. When the stars cleared I was looking out of the ear hole of my helmet, but I had gotten tangled in Frank's legs and had brought him down. That's when I had my epiphany. If I expected to live to a ripe old age, I had better not go beyond Frosh football."

But he did gain membership of the NROTC Rifle Team (1950-1952) that gained top honors at the California Rifle Meet in 1952. The six-man team, manned mostly by members of his fraternity, Beta Sigma chapter of TKE, was placed among the top ten teams in the Hearst newspaper contest after beating UCLA, California and Stanford teams:

JC: "The rifle team participation was certainly rewarding for me. Sgt. Forte was an excellent instructor, and there is no doubt in my mind that his effort was the primary reason that we did so well in competition. Also, it is clear to me that my rifle team experience was instrumental in my becoming an expert rifle and pistol shooter in later years. I never forgot his lessons."

Jerry not only held a key position in the rifle team, but also took lead position in the sophomore class in the NROTC unit in scholarship rankings as well. In 1951 at the end of his freshman year, Jerry was third in the rankings behind fellow student Chauncey "Chan" Miller, but in 1952, Jerry finished ahead of Miller in second place. Jerry and Miller also represented the sophomore unit on the newly formed Naval Executive Board, where they were involved in discipline, finance and athletics as well as social activities. Jerry was also elected Senator at Large for the Associated Men's Students at USC. In fact, he had a very busy social and academic life. He was a member of the NROTC Softball team (spring 1951-spring 1952); IFC Softball (TKE) in the spring of 1952 and the Demolay Softball Team, Santa Ana City 'A' League in the spring of 1951.

Engagement and marriage

Jerry's personal life was also developing nicely. In late 1952 he was one of eight Tau Kappa Epsilon (TKE) fraternity brothers who "lost their fraternity pins" in romance. Jerry "pinned" JoAnn Petrie, his high school sweetheart. This, in the fraternity world, symbolized a strong commitment to one's girlfriend, and was usually a prelude to an engagement announcement in the future. In late 1953, during a dinner and card party for a group of her friends, JoAnn announced her forthcoming marriage to Jerry (with no definite date set). A small white Christmas tree decorated with red hearts on which was written 'JoAnn and Jerry' revealed this fact to all the guests. JoAnn had graduated from Santa Ana High School with Jerry, and while he was working hard at USC, JoAnn attended Santa Ana College and affiliated with Las Gitanas (the Gypsies), a women's social organization at the college somewhat like a sorority. She then matriculated to Long Beach State College to pursue her teaching degree, and graduated in 1954.

During his time at USC, Jerry continued to receive honors and awards. In May 1951 he was elected to Phi Eta Sigma and became an honorary member of the Phi Eta Sigma Gold Legion. June 1951 saw him entering the Tau Kappa Epsilon Fraternity, Beta Sigma Chapter, and in May 1952 he was named 2nd for Best Overall NROTC Scholar (university) and gained a Certificate of Merit for the highest scholastic marks in the NROTC Naval Science course at USC. He also received the 1953-1954 University of Southern California Scroll of Honor for Meritorious Achievements in distinguished contributions to Student Activities, and in April 1954 was appointed NROTC Battalion Commander, USC. He was the 1954 Top Teke of the Year at the Beta Sigma Chapter of Tau Kappa Epsilon Fraternity, and the 1953-1954 Prytanis (president) of his TKE fraternity. Upon graduating on June 4, 1954 with a BS in Mechanical Engineering, he was presented with his USC Graduation beer stein, one of many awards and trophies that adorn his study at home.

Jerry received the Santa Monica Bay Council of the Navy League Award in June 1953. This was a pair of binoculars, presented to the graduating midshipman that had achieved the highest scholastic marks in the navigation course. He was also awarded the Naval Academy Alumni Association Cup, which is retained at the unit, but has the midshipman's name engraved on it. It was awarded for his scholastic attainments, aptitude, activities and drill proficiency. Jerry also received a Benrus watch, awarded to the midshipman of the junior or senior class who had demonstrated the most outstanding all-around abilities, based on scholastics marks, aptitude for the service, extra-curricular activities and drill proficiency.

On Sunday June 20, 1954, sixteen days after his graduation from Southern Cal, Jerry and JoAnn were married at the First Presbyterian Church in Santa Ana. Jerry's brother Ron was best man and members of Jerry's fraternity at USC and cousins of the bride were groomsmen and ushers during the service. Following the service, the guests traveled to Jerry's home on Lowell Street for the reception. The honeymoon was in the mountains at Crestline, CA and then Balboa Island, before they moved to Fredericksburg, Virginia with a stopover in New Orleans. Jerry resumed his USMC career at Quantico, Virginia.

Marine Charm School

Jerry's original orders, dated January 25, 1954, were addressed to Midshipman GP Carr (C-527592, USNR) U of SC NROTC, and, in part, responded to his request for flight training in his letter of November 5, 1953. The orders assigned Jerry to Basic Course 3-54 at the USMC School in Quantico, for what Jerry hoped would evolve into an assignment to flight training. The orders stated, "While under instruction, you will be afforded the opportunity to apply for flight training. If you are found physically qualified and aeronautically adapted for actual control of aircraft, you will, upon completion of your course of instruction, be ordered to flight training as a student naval aviator." In other words, if he could get through the screening process, he could expect to be accepted for further instruction to become a pilot in the Marine Corps. Jerry accepted his commission on July 9, 1954.

Jerry was eager to start his training and reported in good time to the 1st Training Battalion, Basic School, at Camp Barrett in Quantico. The basic school (also documented as the 3rd Basic Course of 1954) was euphemistically known as 'Charm School' and was basically boot camp for officers. As a member of Able Company, Jerry received training on infantry tactics, firefights, team leadership and military exercises, which he referred to as "snooping and pooping" in the woods. He enjoyed shooting and was rated an expert marksman in both the rifle and the pistol, drawing on his experience as a member of the ROTC Rifle Team.

Jerry's experiences of a three-day exercise in the snow and cold, trying to keep warm as well as achieve a military objective, helped strengthen his desire to become a pilot. Although he enjoyed the experience of

Young Naval Aviator candidates at Pensacola. Jerry is front row center

a night-and-day combat situation in adverse weather and drew upon his years of scouting and woodcraft, he still wanted to fly rather than become a Marine Corps "Grunt" infantryman:

JC: "Vivid memories of this period include the snooping and pooping in the woods, the fire exercises and night exercises, where the training marines at the base acted as the 'enemy' during the exercise. One night, we had set up a line of defense and were in our foxholes on watch for an attack. Those enemy rascals infiltrated right through us and we didn't even see them. When they began their night attack we were being fired upon from both directions. They were only firing blanks, but it was a very sobering experience."

Many of Jerry's instructors at Basic School were either World War II or Korean veterans, or both, and were very professional in their training methods:

JC: "I remember one senior instructor, a Colonel Houghton, who taught us tactics and military philosophy. One of the first things he said to us in his lecture was: 'Gentlemen, let's face it. We are professional killers. That's our job. The only reason that we are here is for war. But I think if you were to discuss it with older military leaders, you would find that the people who would least like to go to war are the guys in the military, because they know what it is. So keep that in mind. You are here to do a job and you are not obligated to enjoy it. But it's a job that needs to be done, and if you commit to it, do it well!' I really thought that was a pretty heavy statement. He really rattled my cage when he said 'Gentlemen we are really nothing but professional killers'. That really makes you stop and think about who you are and what you are doing."

Jerry received his new orders on November 24 1954, releasing him from his current station effective January 29, 1955 and ordering him to report to the Marine Aviation Detachment, NAS Basic Training command in Pensacola, Florida no later than February 19, for "Duty under instruction in a flying status, involving operational or training flight as a Student Aviator" for about 18 months. Jerry graduated the 3rd Basic School (in the top 10%) on December 18, 1954 and, following the Christmas holiday, returned to Quantico for a month of postgraduate administration training, for which he received a certificate (graduating in the top 5%). Upon completion of that course in late January, Jerry and JoAnn moved to Pensacola, Florida in early February. They settled in to a small rental home just outside the main gate of the Naval Air Station, and on February 6, 1955 he began ground school and basic pilot training.

During the ensuing months, Jerry learned to play Bridge, became a father and, last but not least, flew his first solo flight. After completion of ground school at the air station, he was assigned to nearby Whiting Field to begin flight training. The trainer plane was the Navy SNJ, a twin of the Air Force's AT-6 Texan built by North American Aviation. Bridge became a time-filler for the flight students because they were required to report for duty at 08:00 every morning, but could expect only one or possibly two training flights during the day. Consequently, there was plenty of slack time available to study procedures for the next flight and to play Bridge.

On July 31, Jennifer Ann Carr was born at the Pensacola Naval Hospital on the air station. It was a grand event for JoAnn and Jerry. During visiting hours one day, as Jerry recalls, "a big hairy-armed navy Corpsman came in the room at feeding time with an infant on each arm and said gruffly, 'Hey, is one of these kids yours? They're hungry.' The staff at the hospital were incredibly sensitive and gentle, so when he said that, we broke up. What a great sense of humor!" JoAnn and Jerry settled easily into parenthood.

John K. Cochran was one of Jerry's fellow students. He was also a Marine Second Lieutenant, and they had met at Basic School in Quantico. John was a bachelor and had become a close friend to JoAnn and Jerry. On occasion he even baby-sat Jennifer so they could break away from parental duties for a couple of hours to go to a movie. He took training well from JoAnn and was even able to change Jennifer's diapers! John and Jerry continued through flight training together and later began their careers in the same fighter squadron.

Flight School

Flying school was difficult for Jerry, who never considered himself a natural pilot, so he worked hard to develop his skills. When it came to Jerry's first solo flight on May 18, 1955 he felt he was ready, but did not think he was ready enough:

JC: "I was polishing off a few things, such as take offs, landings and basic air work, and my instructor did not tell me he was expecting me to solo. So I went out thinking I would be doing a routine training flight, sharpening my skills some more, getting ready to solo. Suddenly, on one of the practice landings, he said, 'Make this one full-stop and let me out.' And that's when I knew I was in trouble. It was an old SNJ aircraft, and as I prepared to take off again my feet began clattering on the rudder pedals. But I finally settled down when I was airborne – that is, until I had to turn downwind and set up for the landing. I got very nervous again, but it turned out OK. I remember also the smell of the Florida paper mills. We flew in the landing pattern with our canopies open, and the unique smell of the mills pervaded the cockpit. It's a smell that I have never forgotten, and when I smell it today it brings back many memories of primary training."

After a few more training hops to build proficiency, Jerry and John were transferred to nearby Saufley Field in July. Here they sharpened their skills further by making their first night flights and formation flights. Then, in September, they moved on to Corry Field for navigation, cross country flight training and carrier qualification. Saufley and Corry Fields were located quite close to the main field at Pensacola and Whiting field, so the commute for the fledgling aviators was easy:

JC: "The night flight was a hoot. We were thoroughly briefed on where to go, and we did so by following the leader. If you were back in the pack you could see the string of lights ahead of you, and all you had to do was follow. Several instructors were also airborne flying around us like sheep dogs to make sure no one got lost. We all made it back in one piece, and I was told that some of the landings were spectacular."

Formation flying was quite stressful. Young pilots who were just beginning to take command of the airplane they were flying were suddenly required to fly them within a few feet of another airplane. It took a bit of convincing to get them to believe that when both planes are going at nearly the same velocity that the relative motion between them was slow and pretty easy to control. Of course there were also the many horror stories of mid-air collisions to keep them on edge:

JC: "I went into the formation flying phase with a good deal of trepidation. Stories of mid-airs and near mid-airs (mostly exaggerated) were rampant, and the threat was ever present of being 'washed out' of the program if you couldn't fly formation. I found that sliding up to another airplane's wing when it was straight and level was not all that difficult, but if the leader began a turn I felt that I would squeeze the juice out of the control stick and throttle as I stayed with him. With practice, however, I became more relaxed and even succeeded in following the leader through mild aerobatics. I always returned from those flights wringing wet with sweat, though. It was a real work out."

Cross country flights in primary training were simple. It was a question of learning how to read a map and navigate from point to point. This was to be the first time that a student ventured out of the air station operating area and navigated his way to two or three other towns or airports. Then he had to find his way home again.

Carrier landing training in early September proved to be another very stressful activity, however. The students were divided up into flights of five or six and were given extensive briefings on the conduct of carrier landing practice. The training was called FCLP (Field Carrier Landing Practice) and was conducted at nearby Barin Field on a runway with an aircraft carrier deck marked out on it in white paint. An experienced Navy carrier pilot was in charge of the activity, and positioned himself at the end of the runway at the corner of the white box. His title was LSO (Landing Signal Officer), and as each airplane approached the end of the runway he used a set of international orange 'paddles' to signal the pilot with directions on how to adjust his approach to the 'deck.' He was also tuned into the plane's radio frequency, so he could talk to (or yell at) the pilot as well. The traffic pattern for FCLP was 600 feet off the ground, and the airplanes were flown at very low speeds (fairly close to stalling speed). As the pilot guided his plane down the final glide slope at landing speed, the goal was to reduce the throttle at the LSO's signal and land the airplane precisely in the area where the arresting wire should be. The landing was to be 'touch and go.' That is, as soon as the wheels touched the pavement, the student was to add full power and take off again to re-enter the traffic pattern.

Once the student pilot was considered to be 'ready' by the LSO, he faced the final test:

JC: "FCLP was really demanding. The classroom was the place where we got our detailed briefings on how to approach a carrier and enter the traffic pattern for landing. We'd had enough formation training so that we could form up after take off with our instructor, who was the leader of the flight. He then led us to Barin Field, where we entered the pattern at 600 feet altitude, and we then became the exclusive property of the dreaded LSO. We were convinced that LSOs were selected for their good eyes and sharp tongues. It took several days of practice and a number of tongue lashings for us to get enough FCLP landings to satisfy our LSO's requirements, but then the day of reckoning came.

"The aircraft carrier *Monterey* was stationed at Pensacola and was to be the ship where we would become 'carrier qualified.' On September 11, 1955, my flight flew out to the carrier (which looked unreasonably small), entered the pattern at 600 feet and began the ordeal. My feet began clattering on the rudder pedals again, and I had to fight to control them. Each of us was required to make six 'OK' passes. Three of them were to be touch and go, and the final three were to be 'traps' or arrestments. I somehow clattered my way through the touch and go landings, each seeming to be a bit easier. But I was not prepared for the sudden stop when my hook caught the carrier's arresting wire. Wow! What a colossal jerk to a stop! The deck crew then repositioned my airplane for take off, I ran the engine up to full throttle and, on signal from the flight deck launch officer, I released the brakes and prayed for enough air speed by the time I reached the bow of the carrier. On the last trap landing I then taxied the SNJ to a parking area and shakily relinquished the plane to another terrified student pilot whose ordeal was ahead of him."

Advanced pilot training

On November 4, Jerry was transferred to the Marine Air Detachment, Naval Air Advanced Training Command at Naval Air Auxiliary Kingsville in Texas, to complete advanced training. So Jerry and JoAnn packed up all of their worldly goods and little Jennifer and moved to a government apartment on the base in Kingsville.

On December 4, 1955, Jerry was promoted to First Lieutenant. Like all promotions, this was a "temporary assignment," until the successful completion of a probationary period, at which point the promotion would be made permanent. The USMC and the USN has shared a system of officer promotion since 1915, in which a selection-board of senior officers determined those suitable for promotion. Those passed over would ultimately retire early if not promoted on the next opportunity. The system was not an automatic path to a higher rank, however. A combination of factors, including efficiency, loyalty and proven capacity to hold higher rank, is constantly reviewed throughout an officer's career.

The Officer Personnel Act of 1947 provides the necessary promotional machinery for all branches of the US Armed Forces. It includes categories for advancement for promotion and also determines how many vacancies are available for each rank. The actual number of commissioned officers on active duty determines the distribution of eligible USMC officers in each grade at any given time. This is computed by Headquarters, Marine Corps as a percentage of total USMC officers in the service and is approved by the Secretary of the Navy. During Jerry's active service with USMC units, this percentage was listed as: Lieutenant (1st and 2nd Grade) 38.5%; Captain 24.75%; Major 18%; Lt. Colonel 12%; Colonel 6% and General Officers 0.75%. The General Officer ranks included Brigadier General, Major General; Lt. General and General.

Jerry's advanced training consisted of instrument training and then transition to jets. The instrument airplane was a two-seater designated the T-28 *Texan*. The instructor pilot flew in the front seat and the student pilot flew in the back cockpit, which was equipped with a cover designed to be drawn over the student to block out any view of the outside. The student pilot spent many hours "under the bag" learning

Marine aviator Jerry Carr VMF(AW) 122

first to fly straight and level and then to do turns and change altitude on instruments with no visual reference to the ground. As he gained experience he was taught how to recover the airplane from unusual attitudes. The instructor would take the controls, fly the plane to a strange attitude and then give control to the student for the recovery:

JC: "I really enjoyed 'unusual attitudes.' It was a challenge to sort out which way was up while under the bag and then to figure out how to get back to a normal attitude without entering into a stall or overstressing the airplane. Most of the time I was successful, but occasionally my instructor had to save my bacon. The ultimate test of the training was to make instrument approaches at nearby airports, and then at about 200 feet above the ground the instructor would take control. I was told to open the hood as he made the final landing. If we were properly lined up and the landing was easy for him, I got a good grade. If not, I did it again until I got it right. Once in awhile he would even let me make the landing from the back seat."

Life in Kingsville for Jerry and JoAnn was better. Advanced training was easier for him, and they began to relax and enjoy their new career. As winter approached, Jerry got a taste of getting into an ice cold cockpit to begin his flights. JoAnn also experienced the rigors of winter on the Texas plains. One day, just as she had finished washing and hanging a load of Jennifer's diapers on the outdoor clothesline, a cold front called a "Texas Blue Norther" blew in. It started with a burst of wind which blew dust and dirt all over the wet diapers. Then came swift, cold rain which drenched them and turned the dirt to mud. Then the temperature dropped to freezing, and the laundry froze. When Jerry came home from the air field that afternoon he found JoAnn furious and tearful. Together they re-washed Jennifer's under things and hung them all over the apartment to dry. It was a while before they could laugh about that event.

In early February 1956, Jerry flew his first jet training hop in the TV-2, a two-seater. For two months, the flights consisted of familiarization, landing practice and instrument training. Then, in March, he transitioned to the F9F-5 *Panther*, a single-seat jet fighter, in which he trained until he completed advanced training in May.

On May 25, 1956, Jerry was designated a Naval Aviator and received his naval aviator wings. JoAnn was given the honor of pinning them on, returning the favor which he bestowed on her at USC in their college years. On May 28, 1956, Jerry graduated from pilot training to be assigned to his first station as a qualified Marine pilot.

Fighter Squadron 114

It was not long before new orders arrived. Effective April 27, 1956, Jerry and his good friend, John Cochran, were ordered to change station upon completion of flight training and were assigned to the 2nd Marine Air Wing at the Marine Corps Air Station, Cherry Point, North Carolina. They reported on June 6, and the air wing transferred him to VMF-114 in Marine Air Group 24 to fly the Grumman F9F-8 *Cougar* aircraft. The Squadron Commander was Lt. Col. H.C. Dees. The *Cougar* was the swept wing version of the *Panther*, so the transition was fairly easy. Jerry remained with the squadron until June 6, 1959:

JC: "I always figured that the reason why I never got sent to El Toro, which is right near my home in Santa Ana, was because somebody put a red mark on my folder and said 'Don't send him to El Toro. It's too close to home, and he would like that too much'."

VMF-114 had been formed at the Marine Corps Air Station in El Toro, California on July 1, 1943.

During the second world war, the unit had completed a number of combat air patrols and escorted bombers on missions in the campaign around the Solomon Islands from Midway. The unit was also responsible for close air support for the Fifth Marine Regiment during landing operations in the battle for the Pacific.

After WWII, the squadron was relocated on the east coast of the US at Edenton, NC, and during the Korean conflict the squadron trained replacement pilots. In 1953, the squadron moved to Cherry Point, which had longer runways and better facilities, and at the same time it received a number of F2H-4 *Banshee* jets. Two years later, these were replaced by F9F-8s, which were in turn replaced by the F-4D *Skyray* in June 1957, one month after the squadron had been designated an all-weather fighter squadron.

As a First Lieutenant, Jerry was assigned by the squadron commander to the duties of Assistant Engineering Officer, Publications Officer and Fuels Officer, and later with additional duties as Engines and Airframes Officer and as Aircraft Accident Investigator. His duty as Publications Officer was to carefully review the flight records and personally inspect each aircraft to ensure its airworthiness and combat readiness. As some of these old aircraft had seen service at sea on carriers, Jerry knew that there could easily be corrosion and metal fatigue on some critical components. To help him investigate these potentially fatal flaws, he supervised a regular program of close visual inspections and regular dye penetration checking of all aerodynamic control surfaces and landing gear components.

In May, squadron command was turned over to Lt. Col. J. W. Ireland.

On September 6, 1956, the Air Group sent Jerry and several young pilots to NAS Cecil Field in Jacksonville, Florida, for an F9F Pilots' Maintenance Familiarization Course. Jerry completed his participation by September 14 and reported back to his duty station at VMF-114. He completed a similar course the following January.

In January 1957, the squadron was joined by a US Air Force exchange pilot named Captain Joseph Guth. Joe was very well liked by all of the squadron pilots and adapted well to Marine Corps flying. By the time he had finished his one-year tour of duty, he was qualified in aerial gunnery and had earned his spurs in carrier qualification. The Marines now considered him a fully qualified pilot:

JC: "I remember one night, shortly after he had finished his F9F training, Joe had been scheduled for his first night flight in the aircraft. About the time he was due to return to the landing pattern, he called us on the squadron ready room frequency and confessed that he couldn't find the landing light switch. Well, the Cougar didn't have landing lights, but he'd assumed that since all Air Force fighters have landing lights, the Cougar must have them too. The duty officer informed him that he would have tough it out and land without lights, which he did quite admirably. But he endured quite a bit of teasing for the next few days."

In March 1957, VMF-114 was ordered to deploy to NAS Guantanamo Bay in Cuba on a 30-day temporary assignment to conduct gunnery practice. Jerry, as Assistant Engineering Officer, left Cherry Point a few days later with his wing man to fly the last two Cougars that had been undergoing maintenance when the squadron deployed. En route, they ran afoul of a thunderstorm in South Carolina and used so much fuel that they had to overnight in Jacksonville, Florida. They finally finished the trip to Cuba on March 24. At Guantanamo, the squadron sharpened its skills in air-to-air gunnery. They did this by firing live ammunition at a rectangular weighted nylon target towed on a very long cable by one of the squadron pilots. That was known as "dragging the sleeve." The ammunition in each airplane had different color paint on the tips of the projectiles. When the sleeve was returned to the field and dropped by the tow pilot, each gunner's performance was determined by the number of holes his colored projectiles had made in the sleeve. The pilot towing the sleeve (which Jerry also did) was always prayerful that his colleagues would keep their guns pointed firmly at the target. On April 14, the squadron completed its assignment and returned to Cherry Point.

During the remainder of April and the first half of May, the squadron conducted Field Carrier Landing Practice at Cherry Point. Then, on May 14, Jerry flew to Jacksonville (Mayport), Florida with the squadron to report to the Commanding Officer of the carrier *USS Franklin Delano Roosevelt*. They were assigned to temporary additional duty for three days of Carrier Landing Qualification aboard ship, the very assignment he'd not wanted to take on. Jerry finished his carrier qualification on May 15. The FDR at that time did not have a mirror system installed so the F9F-8, though a jet fighter, had to be brought aboard by the LSO using 'paddles' just like those used on the *USS Monterey* in flight training. The following day the squadron returned to Cherry Point.

Maintenance training

On May 17, 1957, Jerry received new orders to report to VMF (AW)-115 in El Toro by May 20. Here he was to receive ground school and simulator training and three familiarization flights in the Douglas F4D-1 *Skyray*. He was also instructed to learn as much as possible about maintenance procedures. Each time a pilot changes aircraft, he is given a flight handbook to study, before riding simulators and completing fairly simple flights in the aircraft itself to gradually increase familiarization. Jerry had already trained in rudimentary simulators in flight training at Pensacola, and at Cherry Point had spent a good deal of time in the F9F-8 simulator, so this was not his first experience with them. The F-4D simulator was significantly more advanced than any he had trained in before, however, and he recalls that as technology improved over the years, simulators became more and more advanced and realistic.

Jerry was once again back in the town where he grew up. Although it was a short visit, this was his first tour of duty at El Toro. He took advantage of the opportunity to spend time with his and JoAnn's families and to visit high school and college friends. The course was completed by June 5, and the following day, Jerry reported back to Cherry Point. But he did have another reason to celebrate, as two days before he had been granted permanent promotion to the rank of First Lieutenant at the end of his probationary period (effective date of rank December 4, 1956). Later that month, he and 29 officers and 150 enlisted personnel received new orders for temporary additional duty, reporting to NAS Atlantic City in New Jersey in July for 17 days for more F-4D maintenance training. He completed this assignment by August 6, reporting back to VMF-114 the very same day.

When the squadron began receiving its new F4D-1s in June of 1957 (replacing the older F9F-8s), corrosion was no longer a major concern, but fatigue considerations increased as the aircraft became supersonic. New materials were used, and new methods of bonding were being employed. Jerry also had to assist the Maintenance Officer in determining the method and level of maintenance and repairs. This included both a full review of the damage on the aircraft and of the level of repair resources and facilities available at various levels of repair. It was a constant balancing act to select the lowest possible echelon for repair in order to keep costs down and reduce the time the aircraft was out of service (both important measures in combat readiness), without compromising pilot safety and the operational capabilities of the aircraft.

During maintenance, Jerry closely supervised the repair crews conducting the inspections and repairs on each engine, ensuring that good engineering judgment was made to avoid future premature engine changes or failure in flight. The J57 engine used on the F-4D *Skyray* aircraft required engine and afterburner inspections by dye penetration and involved close visual inspection of numerous components for damage or potential failure of seals due to the high temperatures encountered during engine operation. Engine teardown, inspection and repair were critical to combat readiness because of the time it took to remove and replace each engine. Following the work, the engine was placed in a test area for 'trimming', with the engine run at full military thrust to ensure correct heat balance, that rated thrust was attainable, and that use of the afterburner was both smooth and non-destructive.

As Jerry was assigned to the squadron maintenance department, it was his responsibility to act as a test pilot and flight-certify each aircraft following engine or major component repairs. So he had to formulate a test flight checklist, which was also used by three other test pilots to put an aircraft through its paces and re-qualify it for operational use.

When he became a member of the Aircraft Accident Board, Jerry served as the engineering representative in investigations that reviewed the causes of any crash. His primary role was to evaluate the performance of all the aircraft's systems and their influence, if any, in the accident. For this, Jerry had at his disposal a representative from the manufacturers of the airframe and the engine, as well as a small crew of highly skilled aircraft mechanics and technicians. This allowed him to perform on-site investigations and analyses and to determine whether recovered parts needed to be transferred to depot facilities for in-depth laboratory inspections. Unfortunately for Jerry, he was obliged to serve in his accident board capacity while investigating the fatal crashes of two of his squadron mates. They were pretty demoralizing tasks.

Early in 1957, Jerry's three years of obligated duty following graduation was due to end. It was time for him to declare his intentions. In a February 1957 letter to the Naval Examining Board (Marine Corps), Jerry had informed the Board that it was his intention to remain in the service of the USMC. The Board convened on April 23 and confirmed its decision in a letter dated August 16. It said, 'This board (has) found you mentally, morally and professionally qualified for retention therein, which recommendation was subsequently approved by the Commandant of the Marine Corps on May 2, 1957. The Secretary of the Navy approved the proceedings of the board on June 18, 1957.'

Dawn of the space age

On October 4, 1957, the Space Age that would consume a significant portion of Jerry's life began with the launch of Sputnik 1 by the Soviet Union. At the time, Jerry was too deeply involved in his fighter pilot duties to give the event much thought – beyond the feelings of dismay and disappointment felt by most Americans that their country had been beaten to the punch.

On November 8, 1957, Jerry wrote to the commandant of the Marine Corps requesting consideration for enrollment in the class for Aeronautical Engineering at the United States Naval Postgraduate School in Monterey, California. This was in response to a Marine Corps bulletin dated September 16, 1957 requesting applications from qualified officers for suitable candidates for further education in postgraduate course work. In his application, Jerry listed his qualifications to date, along with the assurance that he would not resign or request inactive duty while attending the postgraduate course and would continue to serve on active duty afterwards for two years, or for the length of the course if it was longer than two years. This was an official letter that had to go through the channels, with recommendations from his Commanding Officers.

Jerry's Commander at VMF-114, J.W. Ireland, stated that the USMC would benefit greatly from Jerry's further education, stating that his level of education to date reflected his excellence in scholastic work, that his degree in engineering had been a positive asset to his squadron, and he had done "an outstanding job as Engines and Airframes Officer." In addition, Ireland recorded that Jerry possessed, "personal traits and mental habits which forecast future growth and development. He is energetic, enthusiastic and studious. His ability to concentrate with a stable attitude in a noisy environment with numerous distractions shows outstanding mental discipline. If given an opportunity for quiet, fraternal, organized study, his effort should be increasingly productive." Though not realizing it at the time, what was being put on record was the character of a man destined to become one of a select group of exceptional individuals chosen by NASA to participate in the pioneering era of American human spaceflight between 1959 and 1975.

The shortlist of officers chosen for participation was issued on February 3, 1958. The list was based upon tentative quotas that were requested by the Department of the Navy, but were not yet allocated to the USMC for Fiscal Year 1959. The list included the names of officers chosen as principals and as alternates, and all officers were encouraged to take refresher courses in preparation for attending the school, based upon the final official allocation. It also stated: "This information is submitted for planning purposes only, for despite the lack of finality, courses requested have usually been approved and it appears advantageous to inform officers of their selection so that they may further prepare themselves accordingly." On the course of General Aeronautical Engineer that Jerry had applied for, there were only four places. Jerry's name was on the list, but only as the fourth alternate in order and not one of the four principal short-listed. It was a tough selection process to gain entry into the postgraduate school and it looked to Jerry that, as fourth alternate, he had missed out for the 1959-60 school year. His listing as an alternate, however, indicated that he might fare better in the next academic year, so he was not disheartened.

Meanwhile, Jerry continued with his assigned duties at VMF-114, which were still coming thick and fast. These included a Group Special Order (117-57), designated Temporary Additional Duty Involving Training and Instruction. Under this order, Jerry reported to Marine Training and Replacement Group 20, and began a two-week course in airborne radar operation. He flew ten flights in the F3D *Skyknight* aircraft in the radar operator's position, learning to control a radar intercept. This training was necessary in order to develop his skills for radar intercepts in the F-4D, which was a single crewman radar interceptor. The pilot had to fly the plane and operate the radar to control the intercept at the same time. This training completed, Jerry reported back to his squadron on December 18, 1957.

On June 3, 1958, Jerry received orders for detachment to NAVPHIBSCOL Coronado (Naval Amphibious Base) in San Diego. He was to take a three-week training course to become a Forward Air Controller (FAC) and then report overseas to the 3rd Marine Division. An FAC is an aviator assigned to an infantry battalion, and his job is to call in air strikes in support of the troops. Jerry and JoAnn were appalled because she was pregnant and expecting Jamee and Jeffrey in July. Jerry's squadron commander, Lt. Col. Ireland, came to his rescue, however, by requesting that his orders be cancelled because he was essential to the squadron. He also asserted that if cancellation was not possible, that the orders should be delayed three months because of his wife's pregnancy. Headquarters Marine Corps relented and cancelled the orders:

JC: "Whew, dodged the bullet!"

Since they were expecting their first set of twins, travel was not a viable option at that time in spite of the fact that moving the household was part of the Marine family life. As JoAnn recalled: "We took off for Virginia for basic training and officer training and we moved around quite a bit until we moved to Houston. We lived in Fredericksburg, Virginia, for a few months, then Pensacola and then Kingsville, Texas. In Fredericksburg we even lived in the upstairs of someone else's house." The landlady liked to retell tales from the Revolutionary War, trying to tell JoAnn that George Washington, whose farm was directly across the Rappahannock River, really did throw a dollar right in their very back yard. "Coming from California, I thought, 'What is all this? Why is everybody so proud that they have cannonball holes (from the Civil War) in their chimneys? Why don't' they fix them? It's been 100 years'!" Finally, at Cherry Point, the family remained in the same home for three years. JoAnn had their first set of twins, Jeffrey Ernest and Jamee Adele on July 3, 1958, born in nearby New Bern, North Carolina.

The young couple thought it would be nice that their children should all have names beginning with the letter 'J', since Gerald Carr was normally called Jerry and his wife was JoAnn. Little did they know when they planned this that they would end up with six children. However, they kept to this idea and watched the family of 'Js' grow over the years. First there was Jennifer, then the first set of twins Jeffrey and Jamee, followed by John Christian on April 4, 1962 and finally a second set of twins, Jessica Louise and Joshua Lee, on March 12, 1964.

JoAnn recalled that while Jerry was learning his new trade, she was learning to cope with the bare essentials of military life. "We didn't have very much. We didn't even have a TV. We couldn't afford it really. We had the bare necessities, the baby furniture and a bed for us; a few pots and pans." When they arrived at Cherry Point, Jennifer was 12 months old, and their first temporary quarters were half of a World War II Quonset hut, known locally as 'Smallville.' The quarters were aptly named because it was literally like living in a small box. "We had to stay there until we got a regular house. I had a skillet, and so the only way I could wash the diapers was to boil water in one of those cans then bring them out, boil more water, wash the diapers by hand, boil more water, rinse the diapers and go outside and hang them on the line." JoAnn was grateful that at that time she had only one child, because any more "would have been the end right there." She did recall the fun of being in a group of Marine pilot wives. When their husbands were away, they all helped each other. It was a small community, with about 30 families and lots of parties because there was nothing much else to do socially.

USS Roosevelt

In June 1958, VMF(AW)-114, as it was now designated, began FCLP training anew in preparation for carrier qualification. In September, the squadron qualified in F-4Ds aboard the *USS Intrepid* and learned that it would be deployed aboard the *USS Franklin D. Roosevelt* in the Spring of 1959 for an eight-month Mediterranean cruise with the Sixth Fleet. The squadron then came under the command of Lt. Col. Mark Jones.

For the FDR, this was the eleventh tour there since her commission in 1945. In August 1958, the Roosevelt was renovated in the Naval dockyards in New York in preparation for its 1959 cruise. When the upgrade was completed, it sailed to its home port of Norfolk, Virginia, for equipment and stores supplies, then sailed to Mayport, Florida, to take aboard Carrier Air Group (CAG) 1. VMF(AW)-114 had been assigned as one of the squadrons in CAG 1, replacing a navy squadron.

In preparation for this cruise, JoAnn, Jerry, Jennifer and the newborn twins, Jamee and Jeffrey, moved home to Santa Ana in August so that JoAnn could be near immediate family during his eight-month absence. Once the family was settled in, Jerry returned to Cherry Point.

On October 1, 1958 (the same day NASA was formed), Jerry received a nomination for promotion to Captain. His temporary promotion was confirmed on February 20, 1959 – three years to the day before another Marine, John Glenn, flew in orbit.

On October 31, new orders saw Jerry flying from Cherry Point to the Naval Air Auxiliary Station at Mayport. The next morning he flew to and landed on the *USS Franklin D Roosevelt*, which had departed Mayport for a 48-hour cruise that involved carrier re-qualification for the VMF-114 pilots. After this deployment, the FDR returned to Mayport on November 5. Jerry flew back to Cherry Point the following day.

In early December, the squadron was returned to the command of the FDR for approximately ten days to participate in carrier air group "shake down" operations. As an example of this type of assignment, Jerry's archives show just what is involved:

29 Nov	1330 hrs	Depart Cherry Point
29 Nov	1445 hrs	Arrive FDR at sea (1 hr 15 min)
8 Dec	0840 hrs	Depart FDR at sea
8 Dec	0940 hrs	Arrive NAS Cecil Field (1 hr)
8 Dec	1940 hrs	Depart NAS Cecil Field
8 Dec	2030 hrs	Arrive FDR at sea (50 min)
9 Dec	0930 hrs	Depart FDR at sea
9 Dec	1050 hrs	Arrive NAS Cecil Field (1 hr 20 min)
9 Dec	1400 hrs	Depart NAS Cecil Field
9 Dec	1500 hrs	Arrive NAS Jacksonville (1 hr)
10 Dec	1030 hrs	Depart NAS Jacksonville
10 Dec	1045 hrs	Arrive NAAS Mayport (15 min)
10 Dec	1530 hrs	Depart NAAS Mayport
10 Dec	1845 hrs	Arrive Cherry Point (3 hrs 15 min)

Squadron Operations Officer

During this "shake down" assignment, Major Bob Minnick, the Squadron Executive Officer, was lost at sea when the tail hook on his aircraft failed and the F-4D plunged into the water during landing operations. Jerry was then assigned as Squadron Operations Officer and relinquished his duties in the Maintenance Department in January.

Jerry witnessed several accidents and experienced two near misses himself during his time with VMF(AW)-114. The first near miss had been in December 1958, flying an F-4D during night carrier qualifications aboard the FDR. Landing on a carrier at night can be even more of a trial than during the day. With no horizon and no visible cues of height above the water, apart from the faint trace of the wake behind the carrier, the only guides are the centerline deck lights, the 'truck light' on top of the carrier, the mirror lights and the glide slope 'meatball' light. This presents a potentially disorientating scenario where the pilot must rely on the illuminated airplane instruments and the carrier reference lights in the void of darkness. To this day, Jerry refers to night carrier landings as the most "terrifying" of all his flight experiences. On the evening in question, it took him about four tries to land. After the third try he became worried about his fuel level, thinking he may have to abort the attempt, return to land to refuel and try again. On the fourth he successfully "trapped," and from then on he had no more problems with night landings. He has referred to the experience as his "trial by fire."

Prior to Christmas, the *Roosevelt* sailed for Guantanamo Bay in Cuba. At that time, the island was described as "a country of unrest" and the crew, unable to take shore leave off base, experienced the true meaning of base liberty. Following home leave for Christmas 1958, the ship visited Port Au Prince in Haiti and the Isle of St. Thomas for three days. VMF(AW)-114 was not embarked at that time. Members had been granted leave to spend Christmas with their families.

In January 1959, new orders were issued for further FDR carrier operations. Over the next six weeks, the crew performed drill operations for a Readiness Inspection at home and flew to and from the ship for various training operations. Jerry left Cherry Point for the last time on February 10, arriving at Mayport and embarking on the FDR the same day. On February 13, the ship left for a two-week cruise towards a rendezvous point with other vessels in the task force. This allowed the pilots to practice further carrier operations. By March 1, the ship had arrived at Pollensa Bay in Majorca, where she relieved the *USS Forrestal,* and on March 3, the tour in the Mediterranean began. Three days later, the *Roosevelt* arrived in Naples, her first port of call.

Cruising the Med

Jerry's second near miss was in an F-4D *Skyray* flying off the *Roosevelt* on February 24, 1959. His mission that day was to tow a DEL MAR target on an aerial missile firing flight. When he catapulted off the deck, his nose-wheel axle broke, and as he climbed away, the nose-wheel went straight ahead and plunged into the sea. Visual observations from the ship and from other aircraft confirmed the situation to

Jerry shortly after he was launched. His nose landing gear consisted of only a strut, with a fork on the end of it where the nose wheel had been attached. Since he would not have any difficulties until he tried to land, the decision was made to let him complete his part in the mission and worry about the problem later. This gave Jerry a lot of time to, "sweat it out," as he thought of the problem facing him. But in flying the mission as planned he was at least able to burn off extra fuel.

He could have returned to the carrier immediately, or to a nearby airfield in Spain or Sardinia. These fields were designated as "bingo fields" ("bingo" being a term for the minimum level of fuel required to safely reach a land-base). The concern here was that the nose-wheel fork might have ploughed into the tarmac causing the airplane to somersault. He was advised of all his options and elected to try to land on the deck of the *Roosevelt* upon completion of the mission. Aircrews prepared the deck of the carrier to ensure that all other planes had landed and were cleared. They removed some of the arresting wires, which could have snagged the nose-wheel fork and flipped the aircraft over, and cleared the deck of all personnel.

By the time he approached, they had also sprayed the center of the deck with a carpet of foam. Jerry flew a normal approach and landing, with the nose-wheel fork skidding in the foam blanket. The plane's nose touched down just forward of number four wire and the arresting hook caught the three wire. Fortunately, the fork did not collapse on landing and he slid to a safe stop. For his piloting skills, Jerry was awarded a Letter of Commendation, dated March 3, 1959, whose citation recorded (abridged):

'In spite of this emergency situation, you carried out your part of the aerial demonstration in an excellent manner. You executed the approach and landing in an outstanding manner (and) as a result of your excellent pilot technique, no additional damage was sustained by the aircraft'. The Citation continued: 'The manner in which you handled this emergency situation, which in the hands of a less capable pilot could easily have resulted in serious injury to the pilot and major damage to the aircraft, is in keeping with the highest traditions of the Marine Corps and of Naval Aviation.' Jerry later went on to complete over 100 carrier landings and earned the title 'Centurion'

From March 11, things started to get busy for the vessel and her crew. Individual ship exercises were completed between March 11 and 13, followed by participation in Task Force 60 in exercise TUNERUP, using NATO procedures. Following shore leave in Cannes and Toulon in France, the *Roosevelt* participated in NATO exercises TOP WEIGHT, GREEN SWING and a significant part of MEDFLEX GUARD.

On April 8, Jerry received good news. He had been selected after all as a primary candidate for the USN Postgraduate School for the Aeronautical Engineering (General) Course for the 1959-60 academic year. In order to arrive back in the US in time and organize his affairs, Jerry was allowed 30 days against his annual leave, in addition to proceed and travel time. While at the Postgraduate School, he would be detached from VMF-114. He was required to supply academic transcripts to the Superintendent of the USNPGS for all previously completed college education. He was also granted clearance for any classified information and material up to and including Top Secret.

At the end of April, between shore leave to tour Venice and a port call in Genoa, Italy, the squadron completed a number of Air Operations. May saw the squadron participate in COMCARDIV TWO Operations 56-59 in the Central and Eastern Mediterranean and Ionian Sea. By now Jerry had taken a real liking to Italy. This was to be the beginning of a lifelong love affair with the country and its people.

During May, while berthed at Athens, Greece, the vessel received the Prince of Greece. An air show, in which Jerry participated, was conducted in his honor. VMF(AW)-114 provided a flight of four aircraft for this event, led by the new Executive Officer, Major "Tex" Montague. Jerry was the second element leader. The flight was privately and appropriately dubbed "The Mangy Angels" in honor of the Navy's premier flight team known as "The Blue Angels." They performed the Blue Angels' trademark maneuver, the Fleur De Lys, which they called "The Florida Leaves." A few weeks later, Jerry flew an F-4D from the *Roosevelt* to the Athens Airport, and was placed under the command of the 7206th Support Group USAFE (AP) 223 US Forces. His task was to participate in Armed Forces Day, with his aircraft on a static display. Acting Sgt James E. Gustafson was sent by the squadron to perform the duties of Plane Captain. He and Jerry were appalled at the curiosity of the Greeks regarding the airplane. They climbed up the tail pipe and down the intakes of the engine and had to be watched carefully when they explored the cockpit. When the event was finished the pilot and the plane captain had to crawl into the same places to make sure that nothing had been left behind. They returned to resume duty on the FDR two days later.

It was at this point that Jerry departed the FDR and returned to the US to begin his studies at the USN

PGS. He left the FDR on June 6 to begin transit back to the US (4 days proceed, 12 days travel) and leave (21 days), prior to transit to USNPGS to report no later than July 26. Jerry left the ship in Livorno. Italy, and was flown to Naples and then to Cherry Point. There he was detached from the 2nd Marine Air Wing and proceeded to California, where he joined his family and prepared for a move to Monterey. He and JoAnn bought their first home in Seaside, California, just north of the Postgraduate School in Monterey, and moved in before the yard had even been landscaped. Shortly after school started Jerry borrowed a dump truck from the contractor who built his home and drove to a place nearby where top soil was available for anyone who wanted it. He had to load the truck the hard way using a shovel, and as he shoveled the dirt he pulled many roots from the soil so it would be ready to spread when he got it home. Unfortunately, he did not realize that the roots were of Poison Oak, so within a few days his hands were covered with oozing blisters. He learned to write without touching the paper, but many of his notes showed signs of yellow and brown stains.

Postgraduate school

The course of study at the Naval Postgraduate School was chosen to provide a background for future selection as a test pilot, the direction Jerry wanted his career to take. Those two years in Monterey involved intensive study in undergraduate and graduate courses focused exclusively on mathematics and engineering. They led to a Bachelor's degree in Aeronautical Engineering.

Jerry and JoAnn converted a clothes closet into an office of sorts where he could do his studies. When he was mentally saturated he pushed in his chair, closed the doors and walked away. He spent many, many hours with his head stuck in the closet office studying his books.

Jerry did well enough during his two years at Monterey to justify going on for a Master's degree. His next assignment was detailed in orders issued in April 1961, in which he was ordered to report by September 11, 1961 to the Commanding Officer of the NROTC Unit at Princeton University, New Jersey. There, as a graduate student at the James Forrestal Research Center, Princeton University, his studies led to a Master of Science degree in Aeronautical Engineering (September 4, 1961 – June 5, 1962). JoAnn and Jerry rented a home in Levittown, Pennsylvania, and began settling in for a year's stay. Jennifer was established in elementary school, and Jamee and Jeff were toddlers. The family experienced its first winter in snow country. On April 4, 1962, the family of five became six when John Christian Carr was born in Bristol, Pennsylvania.

Master's thesis

Jerry completed his Master's thesis project on the design and wind tunnel testing of a winged version of the Curtiss-Wright ground effects machine – a hovercraft. The concept under consideration was the use of an air cushion vehicle as a high-speed landing craft by the USMC land forces.

Two questions were posed in the study: Could the performance of a Ground Effect Machine (GEM) configured as an amphibious landing vehicle be improved by adding wings to allow it to skim above the surf line and obstacles? And could improvement in static stability be achieved with a winged GEM during the ship-to-shore movement in an amphibious landing?

To answer these questions, Jerry and his co-author, Captain John Metzko, modeled a Curtiss-Wright Air Car ACM 6-1, an annular jet-type vehicle. The objective was to use the model in a wind tunnel to determine if the addition of wings to the GEM would indeed improve cruise performance by generating aerodynamic lift. Wings and aerodynamic nose and tail fairings were added to the car. It was hoped that added wing lift would supplement or even replace part of the propulsive power to allow some of that power to be used for horizontal thrust and for attaining some altitude. Since the proposed vehicle was a hybrid of GEM and aircraft, aerodynamic modifications were expected, including horizontal and vertical tail surfaces for pitch control and directional stability. The wind tunnel model tests were completed, and results indicated that the addition of wings was a promising hypothesis, and that the full-sized vehicle should be modified and tested to validate the findings.

Subsequent tests on a modified C-W vehicle after Jerry left Princeton revealed that the addition of wings and a nose and tail fairing had very little effect on hover performance, but they did improve cruise performance and static longitudinal and roll stability. At speeds of about 60 knots the machine was capable of rising a few feet and remaining stable. In subsequent years, the Marine Corps adopted the hovercraft for use in amphibious operations, but the addition of wings was never required nor attempted. In the early part of the school year, Jerry recalls operating a small, ten-foot circular GEM and then the Curtiss Air Car on the runway of the Forrestal Center. Both vehicles were extremely hard to control directionally. It was like gliding on ice or oil. The machines sat on a bubble of air, and in order to translate forward, vents in the back of the vehicle were opened to provide forward thrust. Once up to speed, side vents were used for directional control. In order to stop, forward vents were opened, or the other option was to reduce lift engine power and skid to a stop. There was very little control authority. Everything happened in slow motion, and Jerry felt like he was on the edge of control most of the time.

During his postgraduate studies, Jerry still had to fly at least 4 hours per month to maintain flight proficiency and actually managed to achieve quite a bit more than this. At Monterey, he flew out of the Naval Air Contingent at Monterey airport, flying a T-28, an SNB twin-Beechcraft (C-45) nicknamed the "Bug Smasher," and the T2J jet trainer. While at Princeton, he gained flight time at the Naval Air Station, Willow Run near Philadelphia, flying the T-34 and the SNB.

On February 20, 1962, John Glenn, USMC, became the first American in orbit. The US space effort was beginning to register with Jerry and he recalls being particularly pleased that a fellow Marine should be the first American to orbit the Earth. But it never occurred to him that he might consider a career as an astronaut. The idea was just too far-fetched at this point

As his academic studies were coming to an end, Jerry received new orders in March 1962. These directed him to stand down from Princeton and report to the Commanding General, 2nd Marine Air Wing, Fleet Marine Force Atlantic, pending permanent change of station. The 2nd Wing assigned him to MCAS Beaufort, South Carolina on June 4, where he was officially assigned to Marine Air Group 31 for a few days prior to reassignment to one of the Group's fighter squadrons.

Fighter Squadron 122

He was reassigned on July 12, 1962 to VMF-122, commanded by Lt. Col. Duane Lynch, where he was assigned as a fighter pilot flying F-8 *Crusaders*. He was initially an Assistant Maintenance Officer, but later, when the squadron was deployed to the Far East, he became the regular Maintenance Officer responsible for 24 aircraft, with 180 personnel working under him. VMF-122 at that time was a day fighter squadron, equipped with a high performance aircraft (the *Crusaders*). As with his assignment with VMF-114, Jerry reviewed all flight records and inspected each aircraft on a regular basis to ensure that they were airworthy and combat ready.

On September 6, 1962, Jerry was assigned public quarters on Laurel Bay Blvd, Beaufort, SC. Four days later, a Squadron Special Order gave him yet another additional assignment, as Aircraft Quality Control Officer. By now, his family was expanding and growing up. The eldest, Jennifer, who was 7 at this time, is the one who recalls the most about Jerry's pre-NASA years in the Marines:

"I remember pieces of things, and I remember houses. The first house I can remember was a place called Shell Point. I don't remember the state but I remember the house was in the south and it was my first introduction to roaches. To this day, they just freak me out totally." The next house Jennifer remembered was the permanent quarters at Cherry Point, including one Halloween there in which, while wearing her mask, she ran right into a telephone pole. "The next thing I remember was sitting on my dad's lap and him saying 'What's your name, how old are you.' It knocked me out flat." Jennifer was born at Pensacola and moved to Texas, then back to Pensacola, then to North Carolina, California, Pennsylvania, South Carolina and back to California. When the family moved to NASA in Houston, Jennifer was ten years old. "Since I was the oldest kid, I remember moving every year and a half for most of the first years of my life. We moved here to Houston when I was 10 or 11, and I don't have any inclination to move again."

Crisis in Cuba

Jerry had never experienced actual combat flight, but prepared for it with day and night air-to-air missile training and air-to-air 'dog fight' training. He found this type of work was always exciting and challenging. Like most young pilots, he spent every spare minute "sniveling" for extra flight time, "lurking around like a vulture to get any opportunity to get into the air."

The closest he came to a combat flight during his Marine flying career was during the Cuban Missile Crisis in 1962, when he flew escort to the photo reconnaissance planes taking photos of the missile bases in Cuba:

JC: "The squadron was deployed on six hours notice to NAS Key West, and the situation was very tense. Our families left behind were frightened, with little time to prepare. They were told that, although they were within range of the Cuban missiles, they were unlikely to be in danger. But if an alert was sounded, they were to take mattresses and bedding and retreat to their bathrooms to await an all clear. The Crusaders arrived in Florida to a scene of organized chaos while our maintenance and support people in Beaufort were scrambling to catch up with us. Things settled down after a couple of days, and the Russians began dismantling their missiles. US photo planes continued monitoring the situation, and the fighter squadrons provided cover. We were overhead at 20,000 feet just praying for the sight of a MiG lifting off of a Cuban runway."

For several months ahead of the Crisis, the squadron had been providing planes and crews to man the alert pad at Key West. Jerry and his flight of four had their share of this duty, and when the face-off occurred the entire squadron was deployed to Key West along with elements of the Navy and Air Force. Naturally there was a great deal of competition between the services that had all been thrown together on one air station:

JC: "The Air Force squadron from George AFB in California was flying F-104s. There were so many airplanes on the base that they had to close one runway to provide parking for some of the fighters. We were one of those parked there. The 104s had to taxi down the front of our flight line to get to the take-off position, so they teased us by gunning their engines causing them to make a loud hooting noise. We retaliated by taxiing along their flight line raising and lowering our wings, which was a very unusual sight for them. The Air Force Commanding General's automobile flag quietly disappeared one night and ended up as a bib for our commanding officer at our Thanksgiving dinner. It was just as quietly returned... with gravy stains on it."

For his service during the Cuban crisis in 1962, Jerry was awarded the Marine Corps Expeditionary Medal and the Armed forces Expeditionary Medal, as well as the National Defense Medal.

Aircraft Quality Control Officer

In November 1962, the squadron received a new commander, Lt. Col. Dale Ward, and was re-designated VMF(AW)-122. It began exchanging F-8U-1s for the new F-8U-2NE, which was radar equipped. Under a new designation system, the fighter craft was to be called the "F-8-E." Jerry was back in an airplane in which the pilot had to fly the plane and operate the radar as well. The transition was fairly easy. During the last two months of the year the squadron was again manning the "Hot Pad" in Key West.

In January 1963, Jerry was assigned to Chance-Vought Aircraft Corporation in Dallas for five weeks to review in detail the new F-8E Maintenance Instruction Manual. He joined a select crew of top maintenance personnel consisting of members of both the US Navy and the USMC. They verified the manual with a complete disassembly and rebuild of a new F-8E aircraft taken off the assembly line. Once put back together, the aircraft was flight-tested and, as Jerry recalls, "it flew beautifully, too!" While with VMF(AW)-122, Jerry flight-certified each aircraft after major maintenance with the same test flight checklist he had helped formulate at Chance-Vought. At VMF(AW)-122, his assignments were soon revised again, adding Aircraft Quality Control Officer to his duties as a jet pilot and engineering and airframes officer.

VMF(AW)-122 was directed to begin forming up for a one-year deployment to the Far East in 1964. This was the dreaded one-year separation from the family. All pilots assigned to the deployment were checked in during January 1963, which gave the squadron much longer to learn to work together as a team.

February and March of 1963 were spent flying missile intercept training missions at MCAS Beaufort and on "Hot Pad" alert at NAS Key West, Florida to support air defenses in light of a possible reappearance of Cuban missiles. 'Home' for the pilots at Key West was an air-conditioned trailer, consisting of a ready room and a maintenance office. Each aircraft sat outside, parked in such a way that they could be scrambled in less than three minutes.

Also in February 1963, NASA announced that manned orbital flights of three months duration were being considered to study human reaction to prolonged weightlessness in space. The theory was that if someone could tolerate weightlessness for three months they could probably withstand a one-year mission. (Ten years later, Jerry's mission became the first three-month flight!)

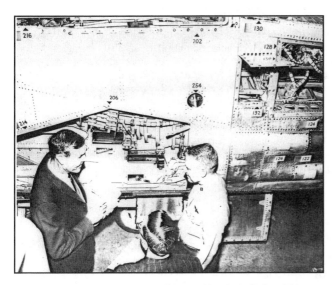

Maintenance evaluation at Chance Vought in Dallas, TX

In April, the squadron deployed to NAS Roosevelt Roads, Puerto Rico and established "Crusader City", for training in aerial refueling, air-to-air gunnery and close air support strafing and bombing training. Crusader City was a tent complex near the runways because base quarters were not available for the squadron. A support detachment from the squadron's parent air group was responsible for setting up the complex, and Jerry had deployed early to Puerto Rico with them to lay out the aircraft maintenance facilities.

Following their return to Beaufort, the squadron continued missile intercept training. In early July 1963, the pilots were fitted and checked out with high-altitude full pressure suits on a two-day course at the NAS in Norfolk, Virginia. The suits looked very much like the full pressure suits worn by the Mercury astronauts. Shortly after his return from the course, Jerry was ordered to proceed to NAS Key West, Florida, for his last watch on the "Hot Pad." Jerry wore the suit just once later in the year. In November, he flew his "zoom" flight – his last flight in the US before deploying to the Far East.

In August 1963, he participated in the 2nd Marine Air Wing's FY63 Competitive Evaluation Exercises (COMPEX) at Cherry Point. Jerry received the prestigious 'Top-Gun' Award on August 5, 1963 for air-to-air sidewinder missile firing. The award was presented by Maj. Gen. Richard C. Mangrum in one of his final official acts as Commanding General of the Wing. This was the culmination of the months of missile intercept training during the spring. In this competition, a pilot from each squadron in the Wing flew only one radar missile intercept armed with a sidewinder missile with no warhead. The objective was to hit a Del Mar target carrying a small heat source towed behind another plane on a half-mile cable. Each contestant was set up into the attack position by ground radar (head-on at a distance of about 11 miles), and then had to use his airborne radar to maneuver into a firing position behind the target. If cleared by a safety observer, he fired his sidewinder and was scored on how close it came to the target. Jerry's missile destroyed it.

Shortly after the COMPEX, the squadron began training again for carrier qualification. This included a program covering day and night field mirror landings (NFML) at Beaufort before completing a final week of "bounce" (mirror touch-and-go landings) near NAS Oceana, on an outlying field called NALF Fentress. Fentress was a well-used site for transient carrier squadrons, as it was equipped with deck-edge lighting resembling a real carrier. Unfortunately, when the squadron put to sea aboard *USS Enterprise* to complete qualification, mechanical difficulties on the carrier forced a hurried return to port. The squadron was reassigned to *USS Forrestal* for qualification instead, with no further delays encountered. The deployment had started on October 3 and the squadron arrived back at Beaufort on October 18, the same date that NASA named its third group of astronauts.

In October 1963, Jerry was assigned the position of Maintenance Officer. The man who was to have had that job in Japan became ill and had to be replaced. The commander, Lt. Col. Ward, declined an outside replacement, and instead chose Jerry for the job. Now he wore two hats – full time pilot and full time chief of maintenance.

Far East Cruise

Flight operations slowed down in November as VMF(AW)-235 took over the equipment and aircraft from 122 in Beaufort. Personal gear was packed and shipped to Japan, and the squadron enjoyed a 30-day leave. After about 17 months in Beaufort it was time to move again. Jerry and JoAnn packed up their household goods, and the young family of six began the cross-country trek to their home town, Santa Ana, California. When JFK was assassinated JoAnn and Jerry and the whole family were driving from the east coast to the west coast. The day before they were due to pass through Dallas they got the news on the car radio and were absolutely dumbfounded by it. The feelings soon turned to grief. That night they stayed at a motel in East Texas on US Interstate Highway 20. The next morning they debated staying on I-20 or circumventing Dallas to avoid traffic, but decided to stay the course so that they could pay their respects to the president, traffic or no traffic. They came through in the late morning, traffic was flowing smoothly and they saw that the banks of the highway were covered with flowers from other mourners. It certainly was a somber moment. In Santa Ana, they bought a home and had a month to get settled. They chose a neighborhood on Louise Street only a few blocks from Jerry's boyhood home and across town from JoAnn's girlhood home. This was to be their home for 2½ years. Jennifer, Jamee and Jeffrey attended their father's old school, Woodrow Wilson Elementary School. JoAnn's parents (Ann and Art Petrie), her sister and brother-in-law (Adele and Norman Hasenyager), Jerry's brother and sister-in-law (Ron and Lois Carr) and Jerry's parents (Freda and Tom Carr) all lived in the area, so the family had a good support system available during Jerry's anticipated one-year absence.

Jerry, however, wearing his maintenance hat, was ordered to head up the advance party, so he left home for Japan on January 5, 1964. The advance party was tasked with the job of inspecting and taking ownership of all the F-8D airplanes it would inherit at NAS Atsugi, Japan. Jerry reported to Treasure Island USNS in San Francisco, California on January 5. Two days later, government transport took him to Travis AFB in Fairfield California, where the USMC Liaison Team made arrangements for transportation to Japan via the Military Air Transportation System (MATS). He left the United States four days later and reported to Marine Air Group 11, 1st. Marine Air Wing for duty. To his surprise, he was immediately ordered to Ping Tung, Taiwan, as part of the liaison team to prepare for a subsequent VMF(AW)-122 deployment there.

On March 12, JoAnn was taken to the hospital in Orange, California by her brother-in-law, Norman Hasenyager, where she delivered Jessica Louise and Joshua Lee. Jerry was notified by dispatch that day. She related the story to Jerry that, when they arrived at the hospital and she was placed in a wheel chair to be taken to obstetrics, she turned to Norman and told him that he really didn't need to hang around, but should go home and get his sleep. Norman apparently responded that he didn't mind and would probably stay around for a little while. The nursing staff probably felt that they were the most casual prospective parents they had ever seen.

In mid-March, the squadron was ordered on an "urgent deployment" to Ping Tung, Taiwan. Jerry and the squadron Operations Officer were ordered to Ping Tung, as an advanced party, to verify that the airfield was ready to receive the squadron. They found that all was well, flew to NAS Cubi Point in the Philippines to rendezvous with the squadron, and then proceeded back to Ping Tung for a multi-week "Crusader City" deployment, living in tents alongside the runway again. This deployment was much like the one the squadron made in April 1963 at Roosevelt Roads, Puerto Rico with Air Group logistical support. The mission was to sharpen up their gunnery and advanced tactics skills, to participate in exercise Backpack supporting the Third Marine Division and Chinese Nationalist troops, and to conduct joint training with other units of MAG-11 as part of the annual group deployment. Located on the far side of the field, VMF(AW)-122 had to endure crossing the runway from the living area to the work area. Working next to the engine trim pad, they 'enjoyed' many a soothing hour of afterburner testing. Home comforts were also rather lacking at Ping Tung. In fact the base was totally lacking in living, messing, recreational and working facilities and a challenge to the ingenuity of the squadron was an added assignment. Some of the squadron were forewarned and brought shaving mirrors, insect repellent and wash basins, Those who were not, quickly caught on. Recreation included volleyball and horseshoe tournaments. Once back at Atsugi every man savored the luxury of a real bed, a hot shower and a host of other things often taken for granted.

During the whole time Jerry was at Ping Tung, he had only one "liberty run" (shore leave) to the nearby city of Kaohsiung. It was an interesting cultural trip for him to have a chance to see how the people of Taiwan lived, but essentially it was a trip into town for a good Chinese dinner and then back to the "salt mines." Jerry was totally focused on his job and had little free time to enjoy the scenery, much to his regret. At the end of March, the MAG-11 participated in a joint air show with the Chinese Nationalist Air Force. The squadron returned to Atsugi, Japan on April 7.

Later in April, the squadron flew back to Atsugi non-stop using aerial refueling. After only three weeks, the squadron moved again, to NAS Cubi Point in the Philippines, where they prepared for carrier re-qualification. This time they re-qualified aboard the *USS Midway*, a sister ship to the *USS Franklin D. Roosevelt*. In contrast to Ping Tung, the facilities here were excellent. There was free waterskiing every day, a frustrating golf course, deep sea fishing and skin diving, and tours of local destinations and viewpoints. The problem at Cubi Point was not a lack of things to do but the lack of time to enjoy what was available:

JC: "Life in Atsugi, when the squadron was there, was quite pleasant. The Japanese people were extremely kind and polite, and we made many friends. The Officers Mess was a well-appointed club, and the meals served there were very good. We also had a "hotsui bath" where one could get a hot bath and massage at the end of a hard day. Now that was luxury! After a long day in the Maintenance Office and a hot bath and massage, then a martini and a good dinner, I was a limp noodle and ready for a good night's sleep. The final touch of the massage was for the woman masseuse to climb up on the table and walk up your back barefooted, using her toes to massage the vertebrae. That was great as long as she didn't weigh a whole lot. As luck would have it one of the ladies weighed about 180 pounds and was nicknamed "Tanko." When she offered to walk your back one would naturally decline. You could always tell when one of the guys had had a martini BEFORE the hot bath. You could hear the grunts, groans and wails from the next cubicle of the one who had failed to decline Tanko's offer."

"The station had an 18-hole golf course as well, and I was surprised when I checked in for the first time. I had a two-wheeled golf cart to pull my bag around and was told that I needed to turn my clubs and cart in at the clubhouse for storage. I was then told that my designated caddy would approach me when I arrived at tee time, and my clubs would be ready to go. Little did I know that the caddies chose the golfers, and that whoever chose you was your caddy for the duration. I also didn't know that the caddies were all old enough to be my mother and were to be referred to as "mamma-san." Mamma-san always selected the club I was to use, and when I chose to use another I got a frown. She also carried, fastened to her belt, a small pouch containing grass seed, and if I made a divot, she would carefully recover it, place seed in the hole and replace the divot – smiling all the time. Mamma-san also refused to use my little cart. She chose to carry my bag on her shoulder – a matter of honor, I suppose. For years after that, when people in the US asked me if I played golf, I would say, "NO. I couldn't find a mamma-san," and then spin my yarn about the golf club at Atsugi. All in all, though we were stationed at Atsugi for a year, we had so many deployments to Cubi, Kadena, Ping Tung, Naha and the like, we probably only spent three or four months of our tour in Japan. I do remember some glorious day and night flights around Mt. Fujiyama. It has got to be one of the most beautiful sights in the world, and it was not far from Atsugi. One of our watch words for flight in heavy weather was, 'Don't buy a piece of Mount Fuji'."

In June, Jerry was given a week's leave and transportation to go to Hawaii for R&R. JoAnn obtained help from family and friends to take care of the children and flew to Honolulu to join him. They spent a restful week getting re-acquainted at Fort DeRussy on Waikiki Beach. The Fort had been a well established and well run R&R location since WWII, and has been visited by thousands of servicemen and their wives in the middle of deployments.

Preparing for flight (Jerry is center), briefing
Bob Marshall and Sam Badiner

June and July also saw the squadron exchange the old F-8D for the newer F-8Es and they also exchanged hangars. The F-8Ds were due for overhaul, so they were delivered to the overhaul facility in Japan and were replaced by the newer airplanes. The only difference between the two aircraft was essentially the radar. The newer E models were, of course, much less of a maintenance problem. The squadron completed the annual Administrative and Material Inspection in the second week of July, which required all aircraft to be scrubbed clean, parts of the hangar painted, new signs made, all gear cleaned and cleaned again. All of the personnel underwent a series of physical readiness tests and personal clothing and equipment inspections. The squadron passed without difficulty, and was rated as one of the top squadrons in the Marine Corps.

Further deployments included Naha, and Kadena, Okinawa, for weapons training and close air support missions in

July and early August. Squadron weapons training continued in Atsugi in mid-August and into September. During that time, VMF(AW)-122 was the number one combat-ready Marine fighter squadron in the Far East. Consequently, they were pretty much on alert from August on. Tension in Southeast Asia was reaching boiling point as the political situation in Vietnam worsened. The squadron was also assigned to assist the Air Force with air defense alert duty at Yokota Air Base. Jerry and his detachment, George Doubleday, Bob Marshall, Bob Morris, Sam Badiner and Dwight Paul, pulled duty manning the "Hot Pad" for several days early in September.

The second and third "urgent deployments" for Jerry without prewritten orders occurred between September 26 and October 12. These kinds of orders were not unusual at that time because of the tensions in Vietnam, and the Fleet Marine Force (FMF) and Wing commands needed the flexibility to deploy their units on short notice. They would follow up later with the necessary paperwork to cover administrative details. These were short deployments for Jerry, his wingman, George Thompson and various other squadron pilots to again man the Yokota Air Base "Hot Pad". There were also 30 days of weapons training at Cubi Point from October 12 and a further 10 days in Okinawa from November 25. This training involved air intercepts, air-to-air gunnery and close air support with rockets, bombs and guns. The squadron also began FMLP practice for carrier re-qualification.

Late in October, they re-qualified aboard the *USS Ticonderoga*, a ship of the same class as the *USS Intrepid*. The deck was smaller than the *Forrestal* so margins for error were tighter, but procedures were the same. November brought Typhoon Louise, with winds of up to 160 knots threatening Atsugi. The squadron then deployed to Kadena to participate in a demonstration of US Armed Forces firepower. During this deployment, the squadron was unjustly accused of insulting the Air Force commanding general at NAHA Air Base Officers Club. It began as simple inter-service friendly ribbing, but escalated when an unknown Air Force officer walked behind the table at which the Marines were sitting and uttered a loud epithet. The general was speaking at the time, and his staff assumed that it was a Marine who swore. The upshot was that VMF(AW)-122 was banished from the NAHA O' Club:

JC: "Thirty years later, in Little Rock, Arkansas, I became acquainted with "Scotty" Scholl, who was an Air Force C-130 plane commander. As we exchanged yarns, we discovered that he had been at that NAHA O' Club that night. He said, 'Lord, I think my co-pilot was the one who got you guys into trouble!' He related that his co-pilot, somewhat inebriated, was on his way to the men's room and had uttered the curse words as he passed behind the Marine table on the other side of a thin screen that divided the dining room from the lobby. We had a good laugh over that, and in 2003 I invited Scotty to attend a squadron reunion in Branson, Missouri. I introduced him as my Air Force friend who had a story to tell them. As he related the incident, it was delightful to look out over the audience of 28 aging fighter pilots and watch the faces light up one-by-one as they caught on to the story. Scotty became 'one of the guys' and underwent a good amount of friendly ribbing for the rest of the weekend."

The tour was not without tragedies however:

JC: "We had three plane crashes while I was with 122. Two of them were fatal – 1st. Lt. Templeton in the US and 1st. Lt. Hudgins during a night training mission on August 24, 1964 in Japan. The third was on a mission flown by 1st. Lt. Bob Marshall on a practice strafing run. The target was a rock sticking up out of the water near Cubi Point. He passed so close to the target that he apparently flew through his own ricocheting bullets and rock debris, causing his engine to fail. He ejected over the ocean, was quickly picked up by a helicopter and was released from the Cubi medical facility with no injuries. He did suffer a good deal of teasing for 'shooting himself down', though"

Requesting test pilot training

On August 11, 1964, Jerry had submitted a request to be considered for the June 1965 USN Test Pilot Class (Class 42). In December, at Lt. Col. Ward's suggestion, he followed this up with an additional supporting letter to Commander of the Marine Air Detachment at Patuxent River, Maryland, detailing his desires and qualifications. With his recent deployment, Jerry had accumulated almost 1,300 hours flying time in jets and a range of experience in his collateral duties. He could back this up with his degrees and his experience in dealing with contractors, as well as his personal experiences of the types of problems Marines encounter in maintenance in the field. Jerry wrote: "I feel that the course offered at the Test Pilot School would be invaluable because of the fact that, in spite of all my schooling, I have never had experience in the practical application of the theories with an instrumented aircraft, (although) I have done some wind tunnel work." In closing the letter for Test Pilot training, Jerry concluded, "I sincerely desire to obtain a test pilot rating, and I honestly feel that I can contribute to the program at the Test Center."

When Jerry returned to the United States, his high hopes of attending the USN Test Pilot School were dashed when a Marine Corps requirement was issued for an all-weather interceptor pilot with an advanced degree to join a radar evaluation group. He telephoned the assignments officer in HQMC and was told verbally that he was the only available pilot who qualified for the job at Marine Air Control Squadron 3 (MACS-3). But he was encouraged when told that his chances for Test Pilot School were very good later on. Jerry ended up back in his hometown, Santa Ana, in California, but was assigned to the unit which was developing a new tactical radar system. This was one of the forerunners of the current approach control radar systems found on most civilian and military craft across the world.

On March 7, 1965, Jerry began what turned out to be his final active assignment in the USMC prior to joining NASA, at the Test Directorate, Marine Air Control Squadron Three, where he remained until April 15, 1966. On September 22, he was promoted to Major.

Testing Officer MACS-3

The new assignment was to assist with the development of a new computer automated radar system for use in the field in a combat zone. The project was in cooperation with Litton Industries Corporation. Jerry was responsible for designing the radar intercept trajectories for various automated aircraft to be used in the project. After plotting the trajectories, he verified that they would be compatible with the aerodynamic characteristics of the aircraft under automatic control and then assisted the computer programmers in developing the software to implement them. Jerry would then fly the intercepts against target aircraft to verify them and make any necessary adjustments. Operationally, the system was used successfully in the Vietnam War and led to the development of subsequent systems by the FAA.

It had all started during a command conference held in November 1965 at 3rd MAW HQ in Santa Ana. The conference stressed the importance of the pilot's evaluation in determining the success of each intercept. It was generally agreed that the Marine Tactical Data System (MTDS) should have qualified pilots on staff who were intimately familiar with Control Center intercept computer design technology. This would aid greatly in the evaluation. Two pilots with such qualifications were nominated, Jerry and Captain G.G. Jacks.

Just prior to Christmas 1965, Jerry took up his new appointment. He traveled to Litton Data Systems Division Inc., in Van Nuys, California. He attended additional meetings at Litton in January 1966 and again for six days in March for acceptance testing. For the engineer in Jerry, this was a very rewarding assignment, adding to his schooling and past experiences. He and Captain Jacks were assigned to a five-day F-4B Pilots Familiarization course (No 1023) and were authorized to checkout in the F-4G which they would utilize to a maximum during the F-4G/MTDS operational evaluation.

In ten years of flying with the USMC, Jerry had logged over 2,100 hrs flying time (including 1,500 hrs in jets), had completed over 100 carrier landings and two overseas deployments, and had flown the SNJ, SNB, T-28, C-47, C-117, F-1, T-1, T-3, F-9, F-6, F-8, F-4 and ground effect machines. The next ten years would see Jerry prepare to fly machines much higher, faster and longer than anything he could have hoped to fly in the Marine Corps. Though he missed out on Test Pilot School, another application was in the system that would result in a complete change of career in the spring of 1966. Jerry was going to fly for NASA.

Chapter Four

The Apollo Years (1966-1970)

Though now classed as one of the nation's new astronauts, Jerry was still very much a serving Marine Corps officer. The largest contingents in the Astronaut Office were members of the USAF and the USN, but each branch of the service in the 1960s used a different management scheme for 'their' astronauts assigned to NASA:

JC: "The Marine Corps assigned its people operationally to NASA, but administratively we were assigned to the local USMC reserve and recruiting district office. They took care of all the paperwork associated between us and the USMC. NASA owned our bodies, but USMC hung on to personnel and pay records."

In later selections, the USAF and USN assigned their new astronauts to NASA for a 'tour of duty' of about five years, after which they could be recalled to their service. This was not the case with Jerry and the USMC in the late 1960s:

JC: "I was not aware of any tour of duty limit, and the Marine Corps didn't mention one to my knowledge. As far as I was concerned, I was supposed to stay at NASA until they threw me out or I'd had enough and wanted to go back to the Corps. As it turned out, when I talked to the assignment person in HQMC on the phone in late 1974 (after the Skylab flight), he told me that I had been out of the normal sequence of duties for so long that they would have a hard time finding a place to put me. There was little chance that I would make General, so he suggested that I finish my Marine career with NASA."

Flight E arrives

Jerry started his NASA career by reporting to the Manned Spacecraft Center (renamed Johnson Space Center in 1973) in Houston, Texas on the 1st of May, 1966. He and the rest of the new recruits began their general astronaut training program on May 9. Today, new astronaut candidates usually complete a 12-month training and evaluation program before they qualify for assignment as a NASA astronaut. In the 1960s, the new arrivals were classed as astronauts from day one, and the term "Astronaut Candidate" (which was coined for the Shuttle groups) did not exist. However, the challenging training program still had to be completed before they could be assigned to a flight crew. Jerry's group underwent survival training in various locations, as with all astronaut selections prior to 1978. Unlike the Shuttle crews, the early capsule crews could not be certain of landing in an accessible place in an emergency. Additionally, as the Apollo spacecraft was designed for water recovery, sea recovery training was a major feature of the training program, along with geology studies for lunar flights.

In the middle of May, the astronaut corps moved into new offices on the third floor of Building 4 at MSC. These offices, codenamed CB in the center's administrative identification system, expanded over the next four decades and are still located in Building 4 South at JSC. In a memo to all astronauts from Alan Shepard on April 26 noting this change, it was stated that the office would be organized into flights, "which represents an effort to streamline and make more stable the administration operation of CB and will in no way affect astronaut technical assignments and responsibilities." The 'flights' (offices) were assigned in alphabetical order, with 'flight chiefs' being drawn mostly from the experienced Group 1 astronauts (Scott Carpenter, Flight A; Gordon Cooper, Flight B; Gus Grissom, Flight C; Jim McDivitt (a Group 2 astronaut), Flight D; and Wally Schirra; Flight E). The five Group 4 members, the first scientist astronauts, were incorporated into Flight E since they would be completing their training with Group 5 after returning from USAF flight school. It seemed both logistically and administratively sound to combine the scientist astronaut group with Jerry's group, allowing them to conduct their academic and survival training as a larger group of 24. Alan Shepard and Deke Slayton had their own offices.

A secretary was assigned to each flight for administrative support, and Flight A's was Toni Zahn. Toni's job was to handle everyone's correspondence and to help arrange accommodation and commercial transportation when required for moves around the country. Requests for appearances were handled by a separate department in the Astronaut Office. These people had to deal, on the astronaut's behalf, with the many political ploys that people used in order to get an astronaut speaker for their function. Astronauts were not allowed to accept any speaking engagements on their own or to endorse any products. As for autograph requests, the Astronaut Office had still another department that managed the mail traffic. For some time the department utilized an autopen to keep up with autograph requests, but autograph collectors soon caught on, and added the phrase, "no autopens please." Some astute collectors learned how to recognize autopen signatures and made a good business out of validating signatures for other collectors. Equally, the Mail Room quickly learned who the commercial autograph collectors were and usually declined to pass signature requests for more than one item to the astronauts. Once an astronaut got a flight assignment and the autograph requests significantly increased, the Mail Room continued to provide that management support. With this support, the astronauts were almost completely shielded from administrative burdens in order to be free to focus on their technical duties.

General training program

The general training program mapped out for the fifth group encompassed 15 months and began with a series of ten science and technology summary courses. Gemini and Apollo operations briefings were also planned:

JC: "Our training was aimed at assigning the astronauts as crewmembers on later Apollo Block II spacecraft. To help enhance our knowledge of these vehicles, we were also to monitor the final Gemini flights and the early Apollo Block I missions."

This first phase was planned to take four months, and would be followed by eight weeks of detailed systems briefings on the Apollo Command and Service Module and the Lunar Module. After that would come environmental familiarization, survival training, control task training and continued geology training, which would take the group to the middle of August 1967. Hopefully, assignment to a mission in a support role would follow, leading to a back-up crew assignment, and finally a coveted seat to space on a prime crew. But that was some years away. For now, it was back to the classroom.

When the group arrived at MSC Houston and began their days in the classroom, first on the timetable was an orientation to the center and current space program activities. Though they were briefed on, and followed, the on-going Gemini program, most of the spacecraft studies they were to complete centered on the Apollo/Saturn system and the lunar landing missions in which they all hoped to participate. This included briefings on the Saturn launch vehicles, the Apollo Command and Service Module and the Lunar Module, launch operations at the Cape, mission control and recovery operations. They also fitted in a program of basic space sciences and technologies, such as geological training, astronomy studies (including a trip to Moorehead Planetarium), upper atmospheric and space physics, meteorology, computers, flight mechanics, guidance and navigation, rocket propulsion and communications, as well as medical aspects of spaceflight.

However, one of the first briefings given to the group by Deke Slayton and Alan Shepard put their chances of flying to the Moon into focus very early. It was stated that these 19 new astronauts were not going to be part of the initial lunar program as enough astronauts were already in training to man all of the seats planned for Apollo at the time. Their best contribution to Apollo would be as support crewmen, with

Group visit to the Moorehead planetarium.
Jerry is 7th from left in white suit

their chances of a flight coming after the initial Apollo landings. The support crews were, at this juncture, the third stringers. That is, first was the prime crew, second was the back-up crew, and third was the support crew. The good news for the support crew was that they were to serve in Mission Control support as Capsule Communicators (CAPCOMS). Jerry and his colleagues were sobered by that announcement since the agency was in the middle of the Gemini flights and no manned Apollo had yet left the ground.

Despite all the talk in the media of creating a lunar exploration program and a moon base by 1980, it seemed that these dreams would remain on paper and most of the new group firmly on the ground. The new group was not made to feel all that welcome. It seemed they were not really

Receiving instruction inside the planetarium,
Jerry is fifth from right

needed for Apollo and were only there because the higher echelons of NASA had pushed to expand the size of the corps, without really needing them. As the new boys in town, they felt the effect of the "pecking order" of astronaut seniority almost immediately. The "Original Seven" Mercury astronauts had almost cult status in Houston, and in parody of that, the new group called themselves "The Original Nineteen". Unlike the later Shuttle era astronaut selection groups, they never designed a "Group Emblem".

To supplement some of the formal presentations, several of the pilot astronauts from the new group (together with the scientist astronauts from the fourth group) organized a series of informal 'bull sessions' to enhance their understanding of certain class lectures. These included sessions by Al Worden on flight mechanics and hypersonics, by Ed Mitchell on guidance and navigation, and by Jack Lousma on propulsion. Joe Kerwin, from the fourth group, handled sessions covering space medicine, while Mitchell and Don Lind presented sessions on the physics of space. It had become clear to the new astronauts that the trainers who were formally instructing them were more engineers than educators. Understanding what was being said was difficult at times, so this extra education from their own peers in the subjects they knew well was an invaluable addition.

The only training equipment available were the Gemini and early Apollo trainers. These, however, were much in demand by the first three groups of astronauts, who were training for missions planned for the next couple of years. The new astronauts were only able to use them during those rare times when the mission-assigned crews were not using them. Most of the initial few weeks at NASA were therefore spent in class rooms, before gradually moving to training sessions on the Gemini docking trainer and Apollo simulators. They also trained in geology, which to Jerry seemed a strange subject to take if there were no plans for the group to fly to the Moon. It seemed that perhaps the plans were not cast in stone.

The group was also informed early on that they had inherited an obligation of exclusivity with Time/Life Corporation to tell their stories relating to spaceflight experiences. The fees for this were to be equally split with all the other astronauts. This did not sit too well with the earlier groups, as there were now 19 more pieces in the pie. After the second lunar landing, however, there was not as much interest in exclusive stories, so this agreement fell into disuse and was eventually terminated.

A home in El Lago

Jerry settled in Houston, initially residing in a motel along with several of the group while they arranged housing for their families. Meanwhile, JoAnn and the children were packing up and selling their home in California to join him. Jerry eventually found property in a community called El Lago that a builder had constructed as a so-called 'spec-house' with no committed buyer at the time of building. The community was situated next to Taylor Lake, hence the name "El Lago" (Spanish for "The Lake"). It turned out to be perfect for the size of the Carr family, so Jerry took photographs and sent them home to Santa Ana. He and JoAnn agreed that it was what they were looking for, so he made the commitment. JoAnn and the children came out to Houston in June, and during July and August of 1966 they settled into their new home before the new school year commenced. The children were quickly enrolled into the local schools, and for a while JoAnn looked after the family, not taking a job at all.

The children all went to the El Lago Elementary School (now named after Apollo 1 astronaut Ed White who had been a resident of El Lago), then moved to the Seabrook Intermediate School. Jennifer, the eldest daughter, spent her first year at Clear Creek High School, before moving to Clear Lake High School when building was completed in time for her second year. All the Carr children would end up at that high school for their later education.

Most of Jerry's training as part of Group 5 was centered at the Manned Spacecraft Center (MSC) at

Clear Lake. The center was opened in 1962. From 1965 it became the site of Mission Control, where ground control for all manned missions was located. The Center resembled a university campus rather than the military base atmosphere of the launch center at Cape Canaveral in Florida. The astronauts and their families resided in the nearby communities that developed over the years as the space program progressed. In 1973, the Center was renamed for former President Lyndon Johnson, who had lobbied for the site in his home state. As Vice President and later President, he strongly supported the space program laid out by President Kennedy in 1961 to reach the Moon by the end of the 1960s. At MSC, the array of simulators and mock-ups of the spacecraft and equipment available included a realistic lunar surface, and various rigs to simulate the one-sixth gravity they would encounter on the Moon.

Focusing on Apollo

Training was given on both the Command and Service Modules and the Lunar Module over eight weeks of detailed systems briefings. These briefings normally lasted four days a week and six hours a day, and where possible (very rarely), instructors would use the system trainers to allow the new astronauts to supplement their course work with hands-on experience. Once the decision on how to get to the Moon was settled, the development of the two spacecraft for the Apollo program was begun and completed over several years. It was a matter of designing the spacecraft to fit that requirement. This sounded much easier than it actually was.

For their training sessions, Jerry's group concentrated on the two main spacecraft in which they hoped one day to fly. It had been decided to launch the manned modules on the Saturn V three-stage booster system. Rendezvous and docking between the parent spacecraft and the smaller vehicle that had landed on the Moon would have to be achieved around the Moon. The Gemini program was used to develop the techniques – and reveal the problems – during ten manned missions in 1965 and 1966. The Apollo two-spacecraft system consisted of the three-man parent craft that would make the trip all the way to lunar orbit and back home again and the two-man lunar landing spacecraft that would support surface operations.

The Apollo parent craft was termed the Command and Service Module (CSM). The conical Command Module (CM) housed the three-man crew, the controls and environmental systems to sustain them and the docking equipment to join it to the lunar landing vehicle. With two hatches and five windows, it was much roomier than the previous Mercury and Gemini spacecraft, especially with the central couch folded away, but was still rather cramped with three men inside it.

Attached to the Apollo CM for most of the mission would be the cylindrical Service Module (SM). This housed most of the propellants, maneuvering engines, fuel cells, oxygen and the main propulsion engine for the flight to and from the Moon. The combined Command and Service Module would also be the vehicle used to take the Skylab crews to and from the orbital workshop in 1973/4.

Apollo CSM system training for the "Original 19" was held at MSC and was carried out by instructors from North American Aviation (later North American Rockwell), the prime contractor for the spacecraft, in Downey, California. The instructors covered a variety of spacecraft systems over a 95-hour course, held either side of the fifth geological field trip (September 12-20 and September 26 to October 11, 1966).

The lunar landing vehicle of Apollo – the Lunar Module (LM) – was an ungainly-looking two-stage vehicle. It was designed only to land on the Moon, sustain the two-man astronaut crew while there, and launch them back to the CSM in lunar orbit. It was the first manned craft designed to operate only in the void of space. The descent stage housed the engine to slow the descent to landing, the landing gear to support the touchdown, and most of the equipment the astronauts would use on the surface. The ascent stage was a compact pressurized cockpit which doubled as the control station for landing and ascent, with the astronauts standing in a restraint harness system. It housed the ascent engine and enough consumables to last two astronauts for about four days away from the CM. The ascent stage was also the astronauts' living quarters for their time on the Moon. With its antennae and spidery landing gear it was more commonly known as "The Bug" (Apollo 9 astronauts christened their LM "Spider"). Lunar Module system training for the new astronauts lasted for 82 hours and utilized the experience of instructors from prime contractor Grumman Engineering Aircraft Corporation of Bethpage, Long Island, New York.

The Saturn family of launch vehicles were also many years in development. The two man-rated vehicles, the Saturn 1B and Saturn V, were highly successful feats of engineering. The two-stage Saturn 1B was used on the first manned Apollo flight and the Skylab and Apollo-Soyuz missions. The three-stage Saturn V propelled all the missions to the Moon and launched the unmanned Skylab space station.

Familiarization with the Saturn family for Jerry's group was covered during a three day (May 25-27) visit to the NASA Marshall Space flight Center in Huntsville, Alabama. Here, the group received briefings on the Saturn 1B and Saturn V launch vehicle systems on which they were likely to ride. After a tour of the center facilities, the group was taken to the Mississippi Test Facility where the huge Saturn V engines were being tested. A static test of one of the Saturn engines gave the group an idea of the enormous power of the launch vehicles on top of which they would be sitting.

These were the vehicles the astronauts would fly, and many hours were spent in the classroom, mock-ups and simulators, becoming familiar with the skills needed to fly these machines. The group also received their first introduction to the pressure garments designed for Apollo. As the group were all jet pilots, they were familiar with the pressure suits used in aircraft, but those intended for spaceflight operations were new to most of them. Representatives of the Crew Systems Division at MSC briefed the group on the suit design and construction, how a crewmember would put it on or take it off, and how much mobility was available (which was not much). They were also shown associated equipment before trying on the demonstration suits themselves.

For their first visit to the NASA facilities at the Kennedy Space Center in Florida, a two-day program (August 4-5, 1966) included briefings and tours of several historic and important launch complexes at the Cape Kennedy Air Force Station for the Mercury, Gemini and the unmanned programs. They specifically visited each of the NASA LC-39 Apollo-Saturn pads, the huge Vehicle Assembly Building (VAB) where the assembled rocket was prepared for the mission, and the launch control centre near the VAB which handled the countdown for the mission. In September, Jerry, accompanied by five of his colleagues, returned to the Cape for the launch of Gemini 11. The six new astronauts joined the prime crew (Pete Conrad and Dick Gordon) for breakfast on the morning of September 10. Unfortunately, that day's launch was cancelled a few hours later due to a malfunction on the Atlas launch vehicle that would have carried the Gemini 11 Agena docking target into orbit. Two days later, the Atlas carrying the Agena, and the Titan II carrying the Gemini 11 astronauts, were successfully launched.

Jerry and five other colleagues have breakfast with GT-11 astronauts Conrad and Gordon on the morning of the September 10, 1966 cancelled launch. Left to right - Gordon, Carr, Brand, Mattingly, Bull, Haise, Evans, Conrad. Jerry is seated behind Gordon, at left of table in white shirt

On a more local visit, across from the Astronaut Office and past the newly planted central pool and garden area at MSC, the group toured Mission Control in Building 30. This visit, conducted between August 18 and 19, 1966, gave the group an overview of each flight controller's console in the Mission Operations Control Room (MOCR – pronounced "Moe-Cur"), as well as the operation of each Staff and Contractor Support Room, the data flow network and real-time operational organization. It was here, at the CAPCOM console, that many of the group (including Jerry) would sit handling communications between the teams of ground controllers and a crew in space during forthcoming Apollo missions.

Pulling the 'G's'

To prepare the astronauts to actually 'fly' the spacecraft, a series of control task training sessions were devised. These would acquaint them with the characteristics of various spaceflight maneuvers, using simulators with 'out of the window displays' of the Earth and representative target vehicles. Over 12 hours was spent in the Gemini part-task trainer, allowing the new astronauts to practice docking a Gemini to an Agena. Though they would not fly such a mission themselves, it was good generic training. The Apollo simulator for docking a CSM to a LM was being utilized fully by the first Apollo crews assigned to the missions planned for 1967 and 1968, so the Gemini simulator was the only one readily available. As the Gemini program wound down, Jerry's group found more and more time available on the Translation and Docking simulator. This simulation gave Jerry and his colleagues experience in attitude control, maneuvering thrust control, retrofire control and terminal rendezvous, as well as planned re-entry control in three modes (pitch, yaw and roll) during two-hour sessions.

To introduce the astronauts to the high 'g' levels possible during a launch into orbit or deceleration during re-entry, each of the group participated in two sessions of centrifuge training during November 7-12, 1966. Each session was separated by at least a 12 hour period. The centrifuge, located in Building 29 at MSC, had no active controls or flight instrumentation except for a 'g' meter and an event timer. A running commentary on the mission events and crew tasks was given to the 'crew' as their run progressed. The crew would make proper motions to simulate tasks on the basis of the commentary from test supervisors. Again simulating a three-man Apollo crew configuration, Jerry completed his runs with Bill Pogue and Ron Evans:

JC: "Those runs were real 'crushers,' but they did give us the experience and the confidence that one could function (although barely) at 'g' levels considerably higher than we had experienced as pilots."

The first session included four runs. The first followed a normal launch 'g' load, which a crew would encounter as they ascended from the pad. This was followed by a run simulating a normal re-entry using the SPS engine on the Service Module, then a pad abort, and finally a high altitude abort using the escape tower.

The following day, they again completed a normal profile run and were then subjected to the entry loads that simulated the use of the Reaction Control System engines only. They completed the session with two aborts using SPS loads, one from a minimum altitude and one at maximum 'g' (15 g). A few days later Jerry and Evans completed a session in the Dynamic Crew Procedures Simulator (DCPS) launch abort trainer.

In the late 1970s, the centrifuge was no longer considered to be necessary due to the lower launch profile pressures expected for Shuttle ascent and landing. It was removed and the room was fitted out to become the Weightless Environment Training Facility (WETF – pronounced "wet-eff") water tank. This facility was used for EVA training for the next twenty years. In the late 1990s the facility was used for post-flight rehabilitation for returning long duration ISS crews and as a personal fitness pool for the astronauts.

Wet and wild

A series of wilderness survival courses were fit into the training program around time spent in academics and in the simulators. Normally, Apollo spacecraft land in the ocean, and the US Navy was to be on hand to provide the recovery of the spacecraft and the astronauts. However, it was recognized that a mission could be terminated at any time (as had been demonstrated by the return of Gemini 8 only ten hours into a planned four day mission), which could result in a landing either on land or water outside of the

SCUBA training course in Key West. Jerry is standing third from right between
Engle and Collins. Neil Armstrong was Jerry's "dive buddy."

planned recovery areas. As Apollo was not then planned for orbital inclinations that would take it over polar regions, winter survival training was not part of the Apollo training program (though it has been for the Russian cosmonauts since 1960).

Dressed not for space but for the jungle, Jerry sets up camp for the night

Members of the fourth (scientist astronaut) and fifth (pilot) astronaut groups undergo jungle survival training. Jerry is kneeling far left behind Weitz and next to Roosa

Jerry (2nd row 2nd from left), behind Worden does not look too enthusiastic about dealing with an evening meal which might bite back!

Jerry and Al Worden are instructed on how to recognize snakes in the jungle.

Water survival training took place over December 8-9 in Pensacola, Florida. This was '*déjà vu*' for Jerry and the rest of the Naval Aviators. They had all undergone this exercise in Basic Training in Pensacola. One half-day of academic training, presented by Recovery Operations Division (November 17), was followed by a day of activities at the Water Safety and Survival School, Pre-Flight School. In use was the infamous 'Dilbert Dunker' cockpit on its track, as featured in the film *An Officer and a Gentleman*. Starting 25 feet up the track above a pool, the cockpit would slide down into the water, where the track guided the unit to a stop in an upside down position. The exercise is designed to teach the disorientated pilot how to

NASA Arab astronaut trainees - Jerry is fourth from left

Desert survival training 1967

L to R Al Worden, Jerry and Joe Engle team
up to conduct an exercise in the desert

Desert survival training. The trio complete their water trap
exercise l to r Engle Carr and Worden

safely exit a submerged craft. The secret is to wait for the motion to stop, calmly unstrap and exit the structure, blow a bubble and watch which way it goes. In this way, you can orientate yourself and find the surface. Jerry had completed this exercise in his Marine Corps years and was easily able to complete the task satisfactorily. This course also featured work in a water tank, both with and without pressure suits, using basic swimming stokes, life raft boarding, and helicopter rescue by sling and seat.

Tropical survival took place during the week of June 12, 1967 in Panama. This was a five-day course supported by the USAF Tropical Survival School at Albrook AFB, Panama. The course was

Geology training 1966 - Jerry is far right in
cowboy hat and sunglasses

split into two days of lectures and demonstrations followed by three days of field training, with the group split into teams of at least three, as they would be in an Apollo landing. They were dropped in the morning for a two day stay in the jungle, with the objective to find or make a large enough clearing which would allow a helicopter to fetch them out. They spent the night in hammocks, with Panamanian natives watching, out of sight, as 'baby sitters', but they had to find their own food. Jerry was actually teamed with three fellow astronauts (Kerwin, Lousma and Pogue) and has fond memories of being taught how to catch, kill and eat a snake, which Jerry decided he would never do. At the end of the training session, however, he and his colleagues were treated to a 'Jungle Buffet' which of course included a tasty slice of snake.

Desert survival training in the week of August 7, 1967 followed a similar five-day pattern to the tropical survival course. The course was held on the high desert at Fairchild AFB, Washington (near Spokane). It was run by the Air Force Survival School, 3635th Flying Training Wing, based at Stead Air Force Base, Nevada. With the course amended by the school to spaceflight mission requirements, the astronauts were instructed on the characteristics of the world's deserts, and appropriate survival techniques, for one and a half days. This was followed by a one-day demonstration program at the field site on the correct use and care of survival equipment, with the astronauts using the spacecraft parachutes as signaling equipment, as tents, for clothing and for protection. Once again teamed into threes, the astronauts spent two days at a remote site putting the theory into practice.

Geology field training

Geology training was spread over a twelve month period from June 1966 through July 1967 and was split into two courses and several field trips. The first course comprised some 56 hours of instruction on basic terrestrial mineralogy, petrology and geological processes, the identification of basic rock structures and geological mapping techniques. The second course included terrestrial analogies of lunar geographic features, geological mapping, geophysical studies and appropriate sampling techniques. The field trips took in study areas around the Americas and in Iceland:

Grand Canyon, Arizona, (June 2-3, 1966). The group studied the history of geology layer by layer as

Geology training 1967 - in Hawaii

they descended the trail, putting the bookwork into practice.

West Texas (Marathon Basin and Santa Elena Canyon) (June 23, 1966)

Bend, Oregon (Newberry Crater and Lava Butte) (July 27-29, 1966)

Katmai, Alaska (Valley of Ten Thousand Smokes) (August 22-26, 1966)

Los Alamos, New Mexico (Valles Caldera). (September 21-23, 1966)

Pinacata Volcanic Area, Mexico (Cerro Colorado and Elegante Craters) (November 30 - December 2, 1966)

Hawaii for the study of volcanology (week of February 13, 1967)

Flagstaff, Arizona (Sunset crater and Meteor Crater). (April 26-28, 1967)

Medicine Lake Area, California (June 22-23, 1967)

Iceland (Askja Caldera and Lake Fissure Areas) (week of July 3, 1967).

The group had the unique opportunity to observe the Surtsey volcano erupting off the coast of Iceland. The volcano had only recently surfaced and was still blowing up smoke, ash and dirty water. Today volcanic action has subsided, and the island covers about one square mile, with a central peak of about 560 feet above sea level:

J.C: "The Iceland trip was really a good one… except for the food in the field. Every meal included mutton for the meat course. No one cared much for the taste of mutton. I felt that it tasted like a wet wool sweater. Anyway, one day the camp cook proudly announced that he was serving hot dogs for dinner. We all cheered 'Hooray!' Dinner was served, and sure enough, the bread was hot dog buns, the mustard, catsup, relish and onions were real… and the wieners were mutton! Argh!"

Geology training 1966 in Katmai, Alaska

Jerry's group found the geology trips to be extremely interesting and a welcome break from classroom activities. They developed some great relationships with the geologists from the US Geological Survey, University of Texas, University of California and Cal Tech. The Earth Observations program that began with Skylab provided an opportunity to renew some of those relationships.

Simulating weightlessness

To simulate zero-g on Earth for weightlessness training is almost impossible. To address this, NASA developed the use of huge water tanks. This would allow the astronauts in pressure suits, suitably weighted and balanced, to be neutrally buoyant. Inside the suit, the astronaut was in 1 g, but the suit relative to the task was nearly weightless. This allowed them to perform mock EVAs. The viscosity of the water was a problem, but not a serious one. The astronauts learned to move slowly in order to minimize its effect. It was proven on Gemini that adequate training underwater was a valuable tool in preparing for EVA in space. It is a technique still employed today, some 40 years later. At the time of Jerry's entry into the astronaut program, MSC did not have a dedicated water tank to simulate EVA, This would not become available until the late 1970s when the centrifuge in Building 29 was removed and replaced with the Weightless Environment Training Facility (WET-F). For the later Gemini 11 and 12 mission training in 1966, a tank was used at NASA Langley Research Center. In 1968, a 22.8 meter diameter, 12 m deep tank was built at Marshall Space Flight Center in the south west corner of Building 4705. Constructed initially for design and engineering development of new space hardware, it became invaluable as a crew training aid in support of Apollo applications (later Skylab). Jerry spent many hours in this tank, both as an astronaut and later as a space station design engineer with the Boeing Company.

The second method of zero-g simulation utilized a KC-135 (the military version of the Boeing 707), stripped out inside and padded. Flying a roller-coaster profile (the Keplerian trajectory), short 25-30 second bursts of zero-g could be experienced by the occupants. The aircraft was pulled up into a steep climb, then pushed over to create zero-g for that short period of time. This ended up putting the aircraft into a steep dive, so it had to be pulled up into a new climb. A program of around 40 sessions of zero-g (called parabolas) per trip earned the vehicle its more common name of "The Vomit Comet." Jerry's group was teamed with the five scientist astronauts from Group 4, and with 24 astronauts flying in groups of three, the two flights per day required eight scheduled flights between January 24-27, 1967. Each astronaut participated for only one of the four days. Jerry never cared much for these flights and was skeptical that they produced enough valid information to justify the cost of the operation. It was fun, however, to feel weightlessness, so as to know what to expect in the future.

Supporting Apollo

For Apollo training, the group had received basic instruction on both spacecraft. However, due to the complexity of the program and the spacecraft, the CB Office and Apollo Program Office decided it would best suit future needs to split the group into CSM, LM, Booster and Apollo Applications specialists. This was done about six months into their general training program, shortly after receiving their first briefings on the hardware. The idea was to have these astronaut specialists in place as soon as the manned missions started flying, which was then planned for early 1967. The booster specialists were blended into the first two groups soon after manned missions began. These specialist assignments ultimately affected the group members' support crew and subsequent mission prime crew assignments later on. The CSM and Booster groups later tended to get Command Module seats, and the LM group, the Lunar Module assignments. Early in the program, the crew designations were Commander (CDR), Command Module Pilot (CMP) and Lunar Module Pilot (LMP). The CDR was a flown astronaut, the CMP and LMP were senior astronauts, some of whom had flown. The reasoning for this was that a senior pilot would be needed to fly the CSM alone for rendezvous and docking. By the time the program had progressed to Apollo 13, rookies from the 5th group were being assigned to the CMP and LMP positions. By then they were fairly senior pilots, though not flown.

The Group 5 'collateral duty specialty' (as the CB called it) group split came out as:

CSM Group: Brand, Evans, Givens, Mattingly, Pogue, (Roosa) Swigert, Weitz, Worden.
LM Group: Bull, Carr, (Duke) Engle, Haise, Irwin, Lind, (Lousma) McCandless, Mitchell
Booster Group (Saturn 1B/Saturn 5): Roosa, Lousma, Duke

At the end of the formal academic training, the group were eligible for assignment to a flight and were "in the available pool" of rookie astronauts awaiting selection to Apollo, alongside the remaining veterans from Mercury and Gemini.

Effective October 3, 1966 Jerry was assigned to the CB technical support branch for the development and testing of the Apollo Lunar Module. Headed by Neil Armstrong, this group was also responsible for the Lunar Landing Research Vehicle and Training Vehicles used in developing the training procedures for LM assignments. The Lunar Modules were being fabricated at the Grumman Aerospace plant in Bethpage, New York. They were built and tested before shipment to the Cape for flight. Jerry, John Bull, Fred Haise, Jim Irwin and Ed Mitchell spent many months there performing tests in the Grumman white room on the vehicles as they were being put together. Though they all worked on all of the Lunar Modules, each was assigned a flight vehicle on which to focus, and Jerry's was LM-6. The LM specialists focused at first on bringing LM-3 along towards its first manned flight, which at the time was designated Apollo 3 and was to be manned by Jim McDivitt, Dave Scott and Rusty Schweickart. Jerry would eventually work on five lunar modules between 1966 and 1969: LM-2 (the ground test vehicle later known as LTA-8), LM-3 (Spider) flown on Apollo 9, LM-4 (Snoopy) on Apollo 10, LM-5 (Eagle) the first landing on Apollo 11, and LM-6 (Intrepid) carried by Apollo 12.

The group spent a lot of time using sleeping quarters located next door to where the LMs were being constructed:

JC: "We were on call day or night to go in there to work with the technicians. As they built the module, we would do the testing of the systems as they were put in, so you can think of it as kind of a layer cake. They put the first layer in, and we tested all those systems. Then they put the second layer in and we tested those systems, and their integration with the one below. As they built up the entire lunar module, we did a lot of the testing work. We were the subjects. We were the people who did the switch throwing and that sort of thing."

In mid-November, the fourteen-day mission of Apollo 2 was cancelled as unnecessary. It was seen as simply a repeat of the Apollo 1 mission planned for February 1967, but with more science experiments on-board . NASA felt that it was more important to move on to the Block II lunar distance spacecraft as soon as possible after qualifying the basic Apollo parent craft design with Apollo 1. As a result, it was decided to fly only one manned Block I CSM (Apollo 1 – Grissom, White, Chaffee) followed by the first manned flight of the Block II CSM and the first LM (now re-designated Apollo 2 with McDivitt, Scott and Schweickart). Apollo 3 would be the first manned launch of a Saturn V and would feature an eleven-day deep space (high apogee) test of the combined CSM/LM assigned to that mission. This crew would be Frank Borman, Mike Collins and Bill Anders.

Towards the end of the Gemini program, the new astronauts were informed that they were to form support crews for the upcoming Apollo missions. By November 1966, crewing for the first flights was gearing up. Flight and back-up crews for the first three Apollo flights had already been in training for several months when Jim McDivitt (training for the complex first manned flight of the CSM and LM) expressed some concern that, with Apollo spread all over the country, some meetings which required a crew representative were being missed as the prime and back-up crew astronauts couldn't cover all meetings when required. In response, Slayton increased the Apollo crew teams from six (three prime, three back-up) to nine by creating a support crew. The support crew would be made up of rookie astronauts who could help the commander as needed and represent the flight crew in important meetings when required. They would follow technical problems and assist on launch day, usually helping as CAPCOM during the flight and supporting the families during flight and post-flight activities. It was a three-tier system that was followed until the final Apollo flight in 1975, and continued Slayton's system of a back-up crew rotating to become a subsequent prime crew. For the support team, it was an indication of being groomed for a back-up position further down the line, which could perhaps lead to a flight assignment.

For the first group of support crewmen detailed to Apollo in November 1966, Slayton assigned nine astronauts (all from Jerry's intake) to work on the first three missions. The first flight (Apollo 1) was supported by an all-CSM team (Evans, Givens and Swigert), as no LM would be flown. For Apollo 2, Worden (CSM), Mitchell and Haise (both LM specialists) got the assignment. Jerry and John Bull now received assignments on Apollo 3 as LM specialists, with Ken Mattingly assigned as a CM specialist. The official announcement came on December 22 and was a great Christmas present. Jerry felt that he might be in the running for a later Apollo flight assignment just eight months after joining the astronaut program.

The same month, Deke Slayton issued the memo that identified new names for the crew positions on Apollo flights. After Apollo 1, clearer descriptions were required. The Commander (CDR – left seat for launch and entry) still retained overall control of the mission and crew safety and handled the actual flying of the LM (from the left station). The Senior Pilot would now be termed a Command Module Pilot (CMP – centre seat in the CM) and, as second in command, would be responsible for CSM operations including the rendezvous and docking with the LM. The Pilot would now be called the LM Pilot (LMP – right seat in the CM and right station in the LM). As Jerry was a potential LMP, the listing of primary responsibilities was of particular interest to him. He would still have to have an operating knowledge of CSM systems (all three crew were cross-trained to a certain degree to support other crew members and, if required, replace a disabled crewmember), but Jerry would now acquire a more detailed knowledge of all LM systems and would be responsible for the guidance and navigation system on the lander. He would back up the commander for descent and landing and for ascent and rendezvous, and he would participate in lunar surface operations and be responsible for the experiments carried out or deployed on the surface.

In December 1966, Alan Shepard issued a memo regarding helicopter training for the new astronauts. This was directly related to Apollo LM training. Not all astronauts received such training, but if one did it was seen as a clear indication that assignment as LMP was likely sometime in the future:

JC: "I had spent eleven years as a Marine fighter pilot, studiously avoiding any possibility that I might be assigned duty as a helicopter pilot. Suddenly, I found myself praying that I would be sent to helicopter training."

January 1967

A month later, Jerry went back his old hometown and school, but this time as a distinguished visitor. On January 16, he visited Santa Ana District TV Department to record two closed-circuit TV broadcasts for the 4th grade science series. These were broadcast on February 6 and 15 to about 2,500 students in over 50 classrooms in schools across his former hometown. The visit was arranged during Jerry's annual vacation which, as a new astronaut, amounted to one holiday a year. This one lasted two days. As well as filming the broadcasts, Jerry took time to visit Santa Ana High School and to return to the classroom of his former elementary school teacher, Mrs. Sophie Scott. The scheduling of the visit had taken five months to organize because of the hectic pace of Jerry's training program, but was part of the on-going series of public relations visits in which all astronauts participated in order to inform and inspire the public about the activities of the space program. Sometimes these activities could be hectic. In just a few days, the guest astronaut hopped from one function to another three or four times a day. These appearances were dubbed in house as "the week in the barrel."

After a day of discussions and final planning, the recording was filmed on January 17. Using models of a Gemini spacecraft and an Apollo CM, and supported by film supplied by North American Aviation, Jerry was interviewed by science teacher Mrs. Bonnie Wiese about the current activities, the astronaut program, travel through space and prospects for the future. As a former local pupil, now an astronaut, Jerry continually emphasized the importance of motivation in education. He hoped to capture some children's imaginations and channel them into exciting and meaningful pursuits. Drawing on his own inspirational upbringing in local schools, Jerry, at the end of the second film, tuned to the camera, saying, "Maybe one of you will be the first to orbit Mars." Jerry indicated that he hoped to fly to the Moon by 1970 or 1971 (which was not a bad guess of what might have happened). In the broadcasts, much was said about the forthcoming first manned flight of Apollo, which was scheduled for launch just over a month away, on February 21, 1967. Jerry expressed his admiration for Gus Grissom and his crew, and was clearly proud to be a member of the astronaut team supporting them.

On January 27, 1967, just ten days after Jerry filmed his two school telecasts, tragedy struck the Apollo program with the loss of Gus Grissom, Ed White and Roger Chaffee in the Apollo 1 fire on Pad 34 at the Cape. During a countdown demonstration test, with the Saturn 1B booster unfueled, an electrical arc ignited the 100% oxygen atmosphere inside the capsule. Fire engulfed the inside of the CM and claimed the lives of all three astronauts. The nation and the astronaut corps were stunned. An accident was always possible of course, but most expected this might occur in space, not on the ground during a test. All crews were stood down pending the investigation, which took several weeks. At Los Angeles airport awaiting a flight back to Houston, Jerry and some other astronauts heard rumors of the fire. In order to get some facts, Jerry called the CBS newsroom, identifying himself and asking if NASA had released a bulletin of an accident at the Cape. When the answer came back "yes" and that three astronauts were killed, Jerry was asked for his reactions but declined to comment. The information they had was sketchy but confirmed the rumors. Jerry recalled this period at NASA as "a pretty terrible time. I think it really demoralized all of us." The aftermath was a sobering situation and Jerry and his fellow astronauts helped whenever they could in the investigations.

Shortly after the accident, Jerry was scheduled to address the awards luncheon of the 51st annual Colorado Society of Engineers convention in the Cosmopolitan Hotel in Denver, Colorado. Jerry presented nine high school seniors with awards in recognition of their accomplishments in science and mathematics. In his speech, Jerry recalled the accomplishments of Mercury and Gemini and stressed that the effort to reach the Moon was continuing, despite the recent deaths of his three colleagues. Jerry said that Project Mercury proved that man could live in space and that a human crew was a necessary and useful thing to have on a space voyage. Gemini proved that man could function in space and that rendezvous and docking could be effectively achieved. The work on Apollo continued as the astronauts awaited the changes that they knew would be part of the recovery from the tragedy. "The tragedy has made us more resolved to go on. There have been no second thoughts; even as pilots we've always lived with this sort of thing."

Astronaut Dad

For Jerry's family, the transition to life in Houston during the latter half of 1966 was both exciting and, at times, worrying. As Jerry's son Jeff Carr, then aged 7, recalled years later: "I grew up in a neighborhood where an astronaut lived on virtually every block. It was really quite a spectacular time. [Gemini astronaut] Ed White became a big hero of mine [the first American to walk in space]. He lived just down the street. It was a very exciting and carefree time. We had an intercom system in our house, with a central am/fm receiver with a two-way communication system throughout the house, installed by our father. It became very intrusive after a while as we became teenagers. But at the time [1966-1967] it was there. Mother used to wake us all up in the mornings. On January 27, 1967, mom was in the kitchen, and I was in my bedroom by myself listening to the radio. The newscaster broke in and said that there had just been a report that astronauts Gus Grissom, Roger Chaffee and Ed White had been killed in a tragic fire on the launch pad. I think that was the first time the reality of the risk sunk in. This program had been polished and shined and held before the American public as some sort of creation, and it struck me right away that this program wasn't about headliners and heroes and stuff. People were the program. I turned my light off and went to sleep. I thought about what I had heard for a long time, and I took a much more serious attitude – a much more practical interest in learning about the program. Why is this worth this kind of risk? What are we doing? What's so hard about this? Why is it worth trying for this kind of thing?"

Jeff also recalled how, by being selected as an astronaut, his Dad had seemed to be granted a "exemption" to the dangers of Vietnam. But the tragedy of Apollo 1 had made him realize that dad's new role was also quite dangerous. "We had left California; he had left the fighter squadron. Many of my father's peers had gone on to combat in Vietnam and some had been killed there. But we had been granted a vacation, an exemption. My Dad didn't have to go and get shot down in Vietnam. He was going into space. And so the Apollo 1 fire for me was a very, very, sobering experience and it made me realize even at that early age that there was more about Dad's new role than I had imagined."

Bereavement

The year of 1967 proved to be very tragic for Jerry and the space program generally. In April, three months after America mourned the loss of the Apollo 1 crew, the Soviets lost Vladimir Komarov in the fatal crash of Soyuz 1. This was the first flight of a new spacecraft that was part of the program to hopefully allow cosmonauts to reach the Moon before the Americans. This tragic event would halt their manned program for about the same amount of time as Apollo 1 grounded the American astronauts. In November, X-15 Pilot Mike Adams died after losing control of the rocket aircraft during the return from a height of 50.7 miles, crashing in the desert near Edwards AFB. He was posthumously awarded the USAF Astronaut Wings for his attainment.

In June, one of Jerry's 1966 class mates, Ed Givens, was killed in an off-duty car crash coming back from a fraternal meeting of USAF airmen. Jerry and the rest of the "Original 19" very sadly attended his funeral in his home town in Texas. They all gathered around Givens' wife and family and gave them all the support they could. A personal family tragedy occurred on July 24, 1967, when Jerry's father, Thomas, died aged just 57. Divorced from Jerry's mother, he had been living in Utah and had suddenly become ill. He refused medical attention for several weeks, until he was admitted to Salina Hospital for two days before his death from internal hemorrhaging. In October, Jerry lost another colleague and friend, 'CC' Williams, the Marine who had inspired Jerry to join NASA himself. CC was killed in the crash of his T-38 near Tallahassee, Florida, returning from Cape Canaveral to Houston. On October 7, Jerry accompanied CC's widow, Beth Williams, to the memorial service held for him at the True Cross Catholic Church, Dickson, near MSC. The Marine Corps officially designated Jerry as her Marine escort, and he and JoAnn stayed close to Beth and her daughter for a long time. Beth was pregnant with their second daughter at that time.

LM-4 and the third manned Apollo mission

Despite all this, Jerry continued working on LM development with the Borman crew in generic training for their mission. His Grumman LM assignments included inside systems checks at Bethpage, then subsystems and mission tests at the Cape. During checkouts at the Cape, lunar landing to take-off tests linked to test computers were carried out, and Jerry took the place of either the CDR or LMP in the tests. Similar activities by other astronauts were being conducted by the CSM group in Downey, California at North American, and at the Cape. The booster group also worked at MSFC in Huntsville, Alabama on the Saturn boosters. Jerry also visited LM subcontractors and other test facilities supporting the development of the lunar lander. This included visits to Arnold Engineering Development Center, Arnold Air Force station, in Tennessee, where tests of the LM engines were conducted.

Most of the earlier unmanned test flights of Apollo-Saturn hardware were not designated with an 'Apollo' flight number, but instead referred to the serial number of the spacecraft or launch vehicle. This changed in November 1967, with the first unmanned Saturn V, which was designated Apollo 4. This was the first flight of the full lunar stack, launched with three stages of the Saturn, the instrument unit, launch adapters and the spacecraft with the escape tower. It was an important qualification of the whole vehicle that

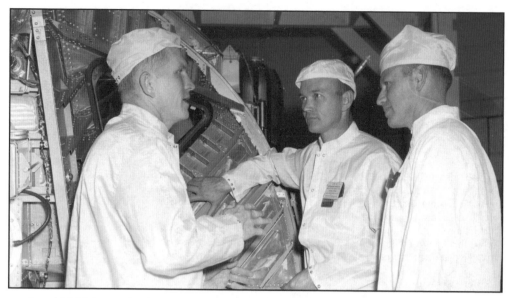

Jerry (right) listens to Apollo mission three commander Frank Borman during the stowage review at Grumman. In centre is the mission's Command Module Pilot Mike Collins

The crew compartment stowage review for LM-4, which was assigned to the third manned Apollo mission. This was held at the Grumman facility in Bethpage, NY in March 1968.

would launch American astronauts (including Jerry, hopefully) on a mission to the Moon. It was a successful mission. The first flight of the LM in space, again unmanned, was accomplished with the Apollo 5 mission launched on January 22, 1968. Despite some small problems during the flight, the LM performed well and the mission was declared a success. The planned flight of the unmanned LM 2 was cancelled. The next time a Lunar Module would fly in space, it would be on an Earth orbit test mission, with astronauts onboard. Apollo 5 was an important milestone for Jerry and the rest of the team preparing to man-rate the Lunar Module.

In March 1968, Jerry accompanied Borman, Collins and Anders to Grumman Bethpage for a crew compartment stowage review with LM-4, which was intended for the then-third manned Apollo mission

Jerry wears the Apollo LMP pressure garment with
air cooler for his participation in the review

Inner gloves are worn prior to fitting
the pressure glove

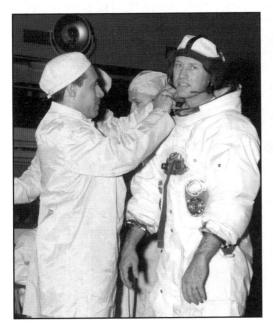

Jerry is fitted with a 'Snoopy' Communications cap

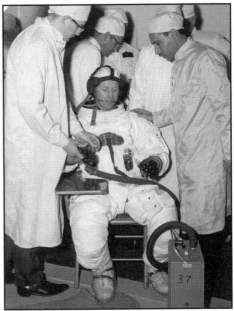

Discussing hand and finger mobility while Jerry's
comfort and safety are being evaluated wearing
the Apollo pressure garment

Inside LM4, Jerry prepares to put on the pressure garment bubble helmet

Jerry peers into the Commander's station forward viewing window of LM 4 at Grumman. Had things developed as originally planned, Jerry could well have been looking out of the LM Pilot's window at a lunar panorama four years after this photo was taken.

commanded by Borman. Jerry participated in some of these tests dressed in the full Apollo LM Pilot pressure garment.

In April 1968, a second unmanned test flight of the Saturn V, designated Apollo 6, was accomplished. Shortly after liftoff, the booster experienced a "pogo" effect, where thrust fluctuations caused the whole vehicle to "bounce" forward and backward during the first two minutes of powered flight. The first stage experienced an early shut down, but extended duration burns from the third stage J2 engine and the Service Propulsion System on the Service Module compensated for the deficiency, resulting in a successful mission. The inquiry and subsequent resolution of the problem resulted in further delays to the inaugural manned flights.

It was during these months that Jerry met one of his favorite film stars – the legendary John Wayne. Wayne was in Houston making the film *Hellfighters (1968)*, loosely based on the experiences of Red Adair, the oil-well fire-fighting expert. Jerry, as president of the Houston area USC alumni group, contacted Wayne, a former USC football star, and invited him to join them at a cocktail party the group was planning at the Shamrock Hilton Hotel where Wayne was staying. The invitation was enthusiastically accepted, but at the appointed date and time he didn't show up.

J.C. "I waited a while and then decided to call his room to remind him. On my way to a telephone I passed another meeting room and there was John Wayne having cocktails with the occupants. So I went in, introduced myself and told him that the USC party was further down the hall. John looked around the room and loudly said, 'Aw, hell. I'm at the wrong damn party!' Later in the evening, Wayne conceded that he had been 'ambushed' and handed a drink as he walked down the hall and thought he was with the USC group. We had a good laugh over it, and he said that ours was 'a hell of a lot better party anyway'."

Years later, Jerry invited Wayne to attend his first launch into space, but unfortunately the actor could not make it. Jerry continued to admire the work of Wayne and his personal video collection includes many of Wayne's films.

Jerry completed the USAF Air Defense Command Survival Life Support Training course at the 4780th Air Defense Wing at Perrin AFB, in Texas on May 27-28, 1968 (Class 68-23), along with thirteen of his Group 5 colleagues and three of the Group 4 scientist astronauts. This was a parachute and water survival training session. They made jumps from the parachute tower to learn how to land properly. In the water, they were towed in a parasail behind a boat and were taught how to deal with a parachute and shroud lines after a water landing.

Christmas around the Moon

In August 1968, McDivitt's Apollo 8 (first LM manned test flight) and Borman's Apollo 9 (CSM deep space flight) missions were swapped, as preparations for the LM's maiden manned flight were falling behind schedule. The Saturn V for Frank Bormán's crew was ready and so were the crew, so Borman actively campaigned for an ambitious flight plan – a Christmas flight to orbit the Moon. After a great deal of discussion, the mission was re-scheduled, and the launch was planned for just before Christmas. Jerry moved with Borman's crew to Apollo 8 and was reassigned as a support crew member for the historic mission.

The following month, Jerry completed his helicopter training with Squadron Eight at Naval Air Station, Ellyson Field, Pensacola, Florida. Between September 9 and 18, 1968, Jerry logged 20.4 hours as trainee pilot, of which 16.7 hours was as first pilot and 3.7 hours as co-pilot, all in a TH-13M type helicopter. (The civilian version was the Bell 47G). He logged 104 landings, soloed on September 13, and flew four more times, logging 6.2 hours. No feet clattering on the rudder pedals this time.

A shining and far distant moon looks down over the NASA Mission Control Center Building 30 at the Manned Spacecraft Center. At the time this photo was taken on December 24, 1968, Jerry was serving as Capcom as the three Apollo 8 astronauts (Borman, Lovell and Anders) were circling the moon.

In October 1968, the Soviet Union launched their first manned Soyuz since losing Vladimir Komarov in April 1967. Test pilot Georgi Beregovoi was the sole pilot of Soyuz 3, and intended to dock with the unmanned Soyuz 2 spacecraft which was launched a few days earlier. Unfortunately, the docking did not take place due to incorrect alignment of the two spacecraft, but it signaled the return of the cosmonauts to space. There were rumors as well of a pending manned launch on a circular flight around the Moon to try to beat Apollo 8 to the prize of the first manned flight beyond Earth orbit. The Moon race was entering its final stages, or so it seemed.

Jerry was a LM specialist, but Apollo 8 had no LM, "So my job became the flight data file (check list library) coordinator for the mission. I worked with the other guys supporting the Apollo 8 mission, getting all the flight data file together. I worked a lot with Bill Anders on getting this done." Jerry also began training as a CAPCOM during the mission simulations with members of the mission control teams. The Flight Directors assigned to this mission were Cliff Charlesworth, Gene Kranz and Milt Windler. One of the characters at this time was Gordon Ferguson. His title was SIMSUP (simulation supervisor), and he always played the villain. During a mission simulation, it was Gordon who threw in all the problems. No one but SIMSUP knew what and when the problems would occur. Sometimes these were formidable, and Mission Control and the crew (including Jerry) had to work out how to get around them. Most of the time they succeeded, sometimes they failed, but the lesson was always learned. Some of the people with whom Jerry worked during his Apollo support days later worked with him in his own engineering firm 20 years later, as they supported Boeing in the design of the International Space Station.

During the mission, Jerry was CAPCOM on Shift 3 (Black Team – the 'night shift') with Brand and Haise. Jerry was on duty as Apollo 8 passed into the Moon's sphere of influence (around 200,000 miles out). "You're in the influence," Jerry informed Borman, who replied that it was better than being "under the influence!" At 68 hours and four minutes into the mission, on December 24, 1968 (at 3 a.m.), Jerry took a breath and calmly announced: "Apollo 8, this is Houston… You are 'go' for LOI (Lunar Orbit Insertion)… You're riding the best bird we can find."

He had spoken several days earlier with Sue Borman – Frank's wife – and asked if there was any special coded message she would like Jerry to send up to Frank during the mission. She quoted an old personal saying they used; "the custard's in the oven at 350" (from their days at Edwards when Borman told his wife; "You worry about the custard, and I'll worry about the flying)." This was Sue's way of saying all was OK at home. Jerry relayed this to Borman as they prepared to round the far side of the Moon for the first time.

The anxiety of waiting for news of a successful LOI burn was one of the most memorable events for Jerry, as he sat listening to the silent down link while Apollo 8 passed behind the Moon. Everyone was acutely aware of the narrow margin for error. If they under burned – shut down too early – they would not be in a proper orbit and might skip out. If they over burned – continued too long – there was the possibility of impact on the lunar surface. At the appointed time for the spacecraft to reappear from the far side of the Moon a radar contact appeared, and Jerry called repeatedly over the radio: "Apollo 8… Apollo 8… Apollo 8, this is Houston." But the burn was OK, and Jim Lovell's voice came back: "Go ahead Houston… burn on time." Apollo 8 was in lunar orbit. Jerry was also on CAPCOM duty for another memorable event in the flight – the historic reading of Genesis from Apollo 8 on Christmas Eve:

JC: "I really was thrilled to be part of Apollo 8. It was really a monumental mission; very gratifying to be part of it."

Jeff Carr also recalled the flight from a different perspective. "The [Bill] Anders kids lived behind us, and these are the kids I played with and went to school with. I will never forget the night that the Apollo 8 crew was around the Moon. Dad was on duty as CAPCOM, and he was at Mission Control at the time. It was evening, and they were making their first turn around the Moon. I was riding my bike down Lakeshore Drive, coming from Glen Anders house over to my house, and I remember looking up and seeing this incredible, bright, beautiful Moon in a clear sky in Houston that night. My dad was in mission control, and Glen's dad was up there circling the Moon. I thought, 'How privileged can we be? This has to be the most exciting, coolest place to be on Earth right now.' It was spectacular. I will never forget that bright image of the Moon. I just kept riding my bike from my buddy's house to my house while his dad was circling the Moon and my dad was talking with him. There are things you take for granted as a kid, and I did take it for granted to an extent… but I remember filing away a memory at the time that I knew I would appreciate later… and I do."

Apollo's golden year

There was no response to Apollo 8 from the Soviets, at least not a manned mission. An unmanned, adapted Soyuz spacecraft flew a looping mission around the Moon as part of the Zond program, but lunar cosmonauts remained firmly grounded. In January 1969, the Soviets finally achieved what they had tried to accomplish in April 1967, a manned linkup of two Soyuz craft (this time Soyuz 5 and 4). Then they did an EVA transfer of two cosmonauts from Soyuz 5, in which they launched, across to Soyuz 4, in which they landed. It was headlined as a demonstration of a mini space station and space rescue capabilities. It was also a technique that cosmonauts were planning to use to transfer from the main vehicle across to their lunar lander.

In March 1969, Apollo 9 tested the LM in Earth orbit with a crew aboard for the first time. Apollo 10 flew a dress rehearsal of the lunar mission (all except the actual landing) around the Moon in May 1969, paving the way for the first landing attempt on the surface with the next mission – Apollo 11

In July 1969, during the historic Apollo 11 mission, the Carr family "supported" the Aldrin family. Jerry remembers Joan Aldrin (Buzz's wife) almost "squeezing the juice" out of his hand, listening to the NASA 'squawk box' and watching on TV as Neil and Buzz landed. At one point during the moonwalk, the astronauts remained motionless. Joan Aldrin asked Jerry what they were doing. Jerry quipped, "they have to go to the bathroom… they're looking for a bush!" At the time, Jerry was already supporting the crew for the second landing, but his son Jeff again recalled the heady days of the first moon landing.

Jeff Carr: "Apollo 11 was really interesting. We went to church at the Webster Presbyterian Church, which is the most important epicenter of our family. Buzz Aldrin was a member there, and so I became friends with Andy Aldrin, who was my own age. During the mission, I spent a few days over at the Aldrin house. My Mom and the wives would all get together for support groups to help each other and take care of things. My mom was on the Aldrin support group, and so I got to tag along. Since Andy was a friend of mine at church, I had the opportunity to be there. Andy and I were more interested in entertaining the photographers and reporters who were camped out all over the neighborhood than we were in what was

Jerry in Marine whites in 1969

going on in the mission. At one point, we had decided to set up a high jump pit in the back yard. Andy's dad [Buzz Aldrin] had an old foam runner, some sticks and sacking and things in the garage, and we were able to set up a high jump in his back yard. We were out there entertaining ourselves, setting the bar higher and jumping over. We thought, 'This is not nearly as much fun as if we had some audience', so we opened the gate and invited some of the media into the back yard. Lo and behold the next day, the covering media printed a picture of Andy Aldrin and a friend jumping for the Moon. I was banished as a result. It was a few days after my banishment that my mom woke us all up and called us downstairs to sit down and watch the moonwalk. I remember sitting in our family room watching this ghostly figure. Squinting your eyes trying to make sense of what you were seeing, it seemed strange to me. I can remember distinctly thinking that, after the Apollo fire, it was OK. Now it was OK. I also wondered if Andy was up and watching this. It was quite amazing, but it was a difficult event to connect with. I could, at that young age, intellectually rationalize this had happened and this was a good thing, but it was still hard – probably for all Americans, and people around the world – to actually connect with the experience and, beyond your sense of awe, get practical realization that it had happened. And my friend's dad was up there making it happen."

Supporting Apollo 12

On April 10, 1969, Jerry had been named to the support crew of Apollo 12 for November of that year. He also trained as CAPCOM for the mission (as Shift 1 with Scott and non-astronaut Warren). His primary training task during the summer of 1969 was following LM-6, the Lunar Module he had baby-sat through its gestation at Grumman, through its shipment to the Cape and the testing program undertaken prior to installation on the Saturn V launch vehicle:

JC: "As a support crewman, I had a piece of machinery that I was responsible for working with (the LM), and that made it even more gratifying to work Apollo 12. Working with Pete Conrad, Dick Gordon and Al Bean was just a joy. There was always light-heartedness. There was always a lot of fun alongside the hard work that was going on while we were working there."

Jerry was also the crew's Flight Data File (checklist) contact. After a simulation, he collected the checklists from the prime crew. They were all marked up with their comments and recommendations for improvement. He then worked with engineers and training staff to see if the list could be changed to accommodate the way the crew wanted to do things. This was not uncommon, as each crew needed to modify things into routines that they could most easily work with. Every change, however, had to be reviewed by the system experts to make sure that the change wouldn't get them in trouble.

As the launch CAPCOM in Houston, Jerry was the crew's voice contact with Mission Control. Seconds after launch, at about the time the booster cleared the tower, lightning from a nearby thunderstorm struck the very tip of the vehicle at the Launch Escape Tower (LET). Static electricity had attracted the lightning from the clouds, and the resultant electrical charge passed through the LET, through the CSM and through the Saturn V booster system and rocket exhaust gasses to ground itself on the launch pad. On its way through the 'stack', the charge shut down the electrical system and the CSM computers, and batteries had to pick up the load. Jerry describes the Apollo 12 launch as a "wild and woolly lift-off." In Mission Control, flight controllers were reporting systems failures to the Mission Director at the same time as the flight crew were reporting problems on the downlink to Jerry. All of the caution and warning lights and alarm buzzers came on at once. Thinking they were about to lose communication with the crew, Jerry was told to instruct the crew, "SCE to Aux." When Pete Conrad replied, "Say again?" not remembering where the switch was or that it existed. Jerry said, "Configure the Signal Conditioning Electronics (SCE) switch on Panel 8 to AUX." Eventually they found it, communications improved and the launch proceeded nominally into orbit.

From the on-board tapes (not broadcast on air-to-ground), Conrad was heard to say, "What's going on? What the hell's going on?" as the lightning hit, while Al Bean was shouting "Still got power; there's power on the BUS. We're OK with power," because the batteries had done their job. Meanwhile, back in Mission Control, the only console that was calm was BOOSTER. They kept reporting, "The booster's OK, trajectory is good. Stay calm." As it turned out the Saturn V electrical and computer systems were not fazed by the lightning, so the spacecraft was delivered into a perfect Earth orbit. The crew and Mission Control had time to restore the electrical system, re-boot the computers, upload new data and check out the systems. Everything looked good, so Apollo 12 was given a "Go" for Translunar Injection (TLI). Looking back on the incident, the exchange between Conrad, Bean and Gordon sounded quite comical.

It is interesting to note that the Apollo 12 landing site was a short distance from the unmanned Surveyor 3 spacecraft that had landed there in April 1967. Conrad and Bean were able to use bolt cutters to remove some pieces of the vehicle to bring back to Earth.

Jerry briefs Mary Haise, wife of Apollo 13 astronaut Fred Haise, on the revisions to the flight plan of her husband's mission following the onboard explosion several hours earlier.

Less comical was the Apollo 13 aborted lunar mission in April 1970. Jerry undertook LM simulator work, utilizing scenarios he had helped develop during his LM support crew training and assignments, which helped to bring the crew safely home. He was also family escort for Mary Haise during her visits to MCC. Jeff Carr recalled Jerry leaving home after receiving a phone call from NASA. Jerry told JoAnn that they had a big problem. Jeff turned on the TV and, like almost everyone else, sat glued to the news for the next few days. "I sat there petrified. I was thinking about Ed White, and about Fred Haise who lived down the street from me. After Apollo 11, you did not have this sense of 'Oh we can do anything,' more like a sense of 'Wow, that's amazing.' But after Apollo 13, we *did* have the sense that we could do anything. We could bring these guys back under such trying circumstances. And I think I carried a sense of invulnerability in my thinking about the space program after that. I was absolutely confident that when my father got his flight, it would be the safest space program you could imagine. I felt much more relaxed about my father's participation after Apollo 13."

Lunar roving

In January 1970, following Apollo 12, Jerry and Jack Lousma received a new assignment as CB representatives in the development of a roving vehicle for later Apollo missions. The Boeing company had worked on an LRV design for years, and after Apollo 11, a new impetus was made in 1970 in securing the design and development contract to provide extra mobility for landing crews on Apollo 15 and subsequent landings. General Motors AC Delco Electronics division partnered with Boeing in the undertaking and Marshall Space flight Center was assigned management of the program. Jerry and Jack provided crew involvement for early development of crew station layout and subsystems design. Later on, both worked on the development of training and mock-up vehicles and assisted the assigned flight crews in training for their missions with the LRV.

Jerry rides on the back of the Lunar Rover development vehicle driven by Jack Lousma, during one of his Apollo technical assignments in 1970

Apollo 15 astronauts pose fully suited by a training model of the Lunar Roving Vehicle.
Jerry's work on the development of this vehicle was one of the unsung assignments
astronauts often perform in between flight assignments.

Jerry and Jack spent a lot of time working with engineers at Marshall in the design phase, then during fabrication at the Delco plant in Goleta near Santa Barbara, California. Their purpose during its development and construction was to supply input from the Astronaut Office point of view.

Jerry became friends with Al Haraway at Boeing, who remembers their participation fondly. "Several astronauts had different ideas or ego trips. Jerry was like a breath of fresh air, along with Jack Lousma. There was a problem with the steering gear, and Jerry and Jack, with their engineering and mechanical minds, listened to the Boeing people, and broke down walls before they were built, helping to solve the problem. Jerry's work on the LRV helped get his company (CAMUS) contracts with Boeing years later, due to his hard work and no nonsense approach. He was approachable and methodical."

Jerry was always the fighter pilot though, as Haraway recalled from one meeting at Santa Barbara. "As with most young marines and astronauts, office meetings were never a favorite. Jerry left early one day to return to Houston, and he announced his departure to the members of the meeting by flying his T-38 six inches, it seemed, off the top of the air conditioning plant on the roof of building six."

Some of the challenges facing the rover designers included the wheels, which were made out of stainless steel mesh, allowing the wheel to absorb the impact of hitting small lunar rocks; the T-shaped hand controller, allowing a pressure-gloved hand to manipulate the controller; and the inertial navigation system, affording a more direct route back to the LM instead of following the tracks made earlier. As Astronaut Office (CB) LRV points of contact, Jack and Jerry helped work out these and many other problems, representing the crew and voicing their concerns or limitations with each system or test.

To evaluate the LRV, they used a 1g mock-up Mobility Test Unit (MTU), powered through umbilical connections to a pick-up truck, to develop steering and control procedures and to test the response of the mobility subsystems, wheels and steering. The tests were completed between June 9 and 24, 1970 at the AC Electronics Defense Research Lab in Santa Barbara, and on the sand dunes at Pismo Beach on June 25. The latter was strictly a contractor affair with no formal NASA participation or requirement. Jerry and Jack were invited on an informal basis to provide CB input early in the development cycle.

The dunes at Pismo Beach were thought to be the closest to the consistency of the lunar surface. In evaluations of the wheel design, Jerry and Jack completed a lot of ingress and egress testing in full EVA suits, evaluating the level of effort needed for a crewman to sit on the seat swivel, strap in, and get his feet

NASA formal photo of Jerry in Marine dress uniform circa 1971

on the pedals and his hand on the T-controller. Working in a full pressure suit in zero-g is difficult, but in 1g in desert conditions, the situation was very tough. Liquid cooling worked well and provided a useful bonus test for that system, but it was still very tiring.

Pismo beach in the early 1970s was a dune-buggy drivers' paradise. During one ingress/egress test, Jack was on the LRV in full suit with Jerry standing by watching:

J.C: "I suddenly heard a great roar and knew that a buggy was on the other side of a nearby dune. The buggy then flew over the crest of the dune, and its driver saw the scene for the first time. He must have applied the brakes in mid air, for as he landed on the LRV side of the dune, he stuck into the sand in a dead stop. His face took on a classic cartoon expression of wide eyes and dropped jaw, watching a space-suited man-looking at him in the middle of a California beach. He slammed the buggy into reverse and disappeared back over the dune never to return."

The Soviets had remained remarkably quiet about their intended lunar program since Apollo 11, emphasizing that they were never in a Moon race and instead intended to develop the first long-term research base in space, more commonly called a space station. Soyuz 6, 7 and 8, launched a few days apart during October 1969, became the first group flight. Soyuz 6 was a research mission that tested space welding techniques in space, while Soyuz 7 and 8 were intended to dock, but could not, due to a faulty rendezvous system, another bitter blow for the Soviets. In June 1970, two cosmonauts broke the Gemini 7 record of 14 days in space by staying aloft for 18 days, setting a new spaceflight endurance record (without docking to a space station) that remained unbroken until the era of space stations.

After his LRV development-work, Jerry became involved in a preliminary assignment to a back-up crew for a later lunar mission. Slayton had planned for Jerry to be assigned as LMP on the Apollo 16 back-up crew with fellow Group 5 colleagues Fred Haise as CDR and Bill Pogue as CMP. With the normal rotation that saw crews miss two missions and fly the third, Jerry would wind up on the Apollo 19 flight crew as LM pilot, which suited him just fine. The Moon suddenly became much closer, at least for a while.

Chapter Five

Almost The Moon

From entering the NASA astronaut program in 1966 until the summer of 1970, Jerry had worked hard supporting the Apollo lunar program. He had also come closer than most in the office to making the coveted flight to the lunar surface himself, as Lunar Module Pilot of Apollo 19.

Though never officially named to the crew, being identified even as a potential member of a flight crew was recognition of Jerry's hard work and commitment to the huge team effort that was Project Apollo. All of the astronauts were equally committed. They all wanted to fly, but there were limited seats available. While some would become known as "Apollo astronauts", others would deal in their own way with the disappointment of not flying by putting on a brave face and committing themselves to support their friends and colleagues, helping to make sure that each mission was more successful and safer than the previous ones. There are 24 astronauts who flew on Apollo missions to the moon, of whom 12 walked across its surface and into history. There is also an additional group of astronauts who had tentatively been named to the last few Apollo flights before these were changed or cancelled. That select group, including Jerry, could rightly claim to have *almost* reached the moon.

The loss of the Apollo 18, 19 and 20 lunar missions in the 1970 budget cuts came as a bitter blow to the astronaut office. They also signaled another setback for NASA. Even before the first landing on the moon, the agency was struggling to find political and public support for the grand future plans they had hoped to implement.

Since Apollo 19 was cancelled long before any real mission preparations had commenced, it has always been difficult to piece together what the mission would have achieved or featured, had it flown. The thin paper trail and a time span of over 35 years has made piecing together the potential mission even more challenging. However, using the limited resources available, it is possible to summarize what Apollo 19 could have achieved had the mission progressed to flight.

Until Apollo had achieved its primary goal of landing Americans on the Moon and returning them safely to Earth, there were no firm decisions made for the follow up missions. There were, however, planning documents for these later flights, mainly for administrative purposes and for scheduling the hardware. NASA planned the flights so that if Apollo 11 had failed in its primary objective and recovered the crew, then either Apollo 12 or 13 could still be launched to achieve the primary Apollo goal before the end of 1969. With the huge success of Apollo 11 on the first attempt at landing on the lunar surface, however, planning for follow on missions could be better defined.

The paperwork for generating these future missions originated from the Advanced Mission Program Office at MSC, under the title, "Future Apollo Lunar Exploration Planning." A team of engineers, headed by Joe Loftus of the Program Engineering Office, evaluated what could reasonably be achieved using the hardware currently available. This team was designated the Group for Lunar Exploration Program (GLEP) and identified nine additional missions following the initial landing, making Apollo a ten lunar-mission program. In conjunction with the various safety, technical, engineering and scientific elements within NASA, the group utilized advisory committees of scientists, contractors and support teams across the United States to plan what Apollo 12 to Apollo 20 might be able to achieve.

This plan built upon several years of study about what would follow the first landing. Though heavily dependent on adequate funding and the availability of hardware, a multi-phased program was eventually proposed. Details of these phases changed constantly over the 1966-1970 period, but by 1969 what was planned for the remaining missions in the "basic" Apollo lunar program (and what was hoped would follow under Apollo Applications) was becoming clear.

Phase I (1969-1971) was to begin with the first landing (Apollo 11), with limited scientific objectives and a short surface stay time. This was to be followed by additional "basic" Apollo mission profiles (Apollo 12-15), each with expanded scientific objectives and hardware but still limited to two days on the surface. Phase II (1972-1973) was to feature five extended-duration Apollo missions (Apollo 16-20). Here, surface time would be increased to three days, with the capability of supporting multiple (3-4) surface EVAs on each mission. The crew would also have the Boeing Lunar Roving Vehicles (LRV) that Jerry had helped develop available to extend their exploration range.

The early missions were targeted to landing sites within what was known as the Apollo Landing Zone. This zone was identified as early as 1965 as a band of lunar terrain 10° wide and 90° long, located along the central equatorial region of the nearside (between 5°N and 5°S latitude and 45°E and 45°W longitude). This equated to a narrow strip that measured about 190 miles wide and 1700 miles long.

This narrow rectangular area was considered to contain the safest areas to attempt the first landings on the moon. This decision was derived from a number of considerations, including the level of available consumables, lighting conditions and the engineering capability of the hardware. Early planning focused on these areas, although photographs from the unmanned Lunar Orbiter probes and by the crews of Apollo 8 and 10 highlighted other areas "of geological interest" outside of the Apollo landing zone. Such areas were analyzed for later missions. It was assumed that once Apollo had proven its capability of landing on the moon and getting the astronauts safely home again, mission experience and confidence would allow future missions to extend the range of landing sites. These spacecraft would also receive some modification, allowing for more time around or on the moon and for carrying a larger scientific payload, thus maximizing the return from each mission. The last five "Apollo" missions were manifested and designated the "J" series of "super-science" Apollo missions. Improvements to both the CSM and LM, the provision of the LRV, and upgraded pressure suits with added mobility, would allow the astronauts to work with a larger, more sophisticated scientific payload than on the earlier landings. Not only were the surface experiments expanded, but one of the equipment bays in the Service Module (to be called the "SIM bay") would be fitted out with a set of science experiments, so that the CM pilot could study the moon and space from orbit while his colleagues were on the surface.

Less than one week before Apollo 11 flew to the moon in July 1969, NASA's Apollo Lunar Exploration Office, in conjunction with BellComm Inc., presented a sequence of ten landings based on accessibility. For the fourth J mission (Apollo 19) and ninth Apollo landing mission, the target was Rima Prinz I in the Harbinger mountains, northeast of the Marius Hills and deep in Schroeter's Valley. The plan envisaged the crew spending about 70 hours on the surface. There were also suggestions that the landing crew might attempt a descent into a deep crater to determine cause of mysterious "red flashes" seen there by astronomers.

The success of Apollo 11 rapidly changed the plans and desires of the scientists for later missions. Based on the scientific knowledge gained from Apollo 11, and taking into account the plans for Apollo 12 and subsequent missions, the Manned Space Flight Weekly Report for July 28, 1969 indicated that Apollo 19 would now be targeted for a July 1972 mission, near to the Hyginus Rille in a linear canyon and volcanic cratered area. This time, there were suggestions that the EVA astronauts could possibly descend into the rille to retrieve samples from inside the feature.

There would be many more meetings between the scientists, the Apollo Site Selection-board and members of the GLEP team to define the landing sites of the nine landings following Apollo 11. Sometimes heated debates ensued, but to allow time for a detailed flight plan and time line to be developed, to factor in the crew training program and for instruments and a geological surface plan to be devised, it was clear that landing sites had to be quickly narrowed down. However, until the previous mission had flown, each nominated landing site was for planning purposes only and subject to change based on the success (or failure) of the previous mission, as well as its findings. On August 27, 1969, a memo was issued citing "the official desire of the Apollo Site Selection-board ." This "official and revised" schedule now revealed that Mission J-4 (Apollo 19) was targeted for the Marius Hills, in a valley floor between four hills. Over the next four months, sites for all nine remaining missions were debated and identified, in order of scientific merit and accessibility. When the final report was published in December 1969 (after Apollo 12 had flown), the Apollo 19 landing site had changed once again.

Mission J-4 was now planned to visit one of the most impressive landing sites of the program. It was to be near to Hadley Rille in the Apennine Mountains on the eastern boundary of the Mare Imbrium. The mountains form a triangular-shaped elevated highland region, surrounded by the Mares Imbrium, Serenitatis and Vaporum and cut by a v-shaped sinuous rille called Rima Hadley. This impressive feature winds in a north-easterly direction parallel to the Apennine front for over 30 miles until it merges with the Rima Fresnel II in the north. It would be an impressive site, and until the cancellation of the mission, remained a primary target for the Apollo 19 flight. Indeed, the importance of the Hadley-Apennine region was so strong that, after the rescheduling of the four remaining Apollo missions in September 1970, the first J mission (now designated Apollo 15) was targeted to that very area.

In January 1970, Jerry had completed his work on supporting the flight of Apollo 12 and had been assigned to development work on the LRV. He was in the middle of this assignment when news came that Apollo 20 was formally cancelled, with some of its hardware being reassigned to AAP. With no sign of any follow-on lunar program, Apollo 19 became, for a while, the final mission to the moon. Debate regarding its landing site now included consideration for sending the mission to the huge Copernicus crater that had been the intended landing site for Apollo 20. This could have been a landing on the floor of the crater. However, Apollo 19s primary site remained the Hadley Rille area.

In February 1970, the Orbital Workshop of the Apollo Applications Program became officially known as Skylab. For many of the veteran Apollo pilot astronauts, flying a long space station mission did not have the appeal of a lunar flight, and their thoughts turned to new goals. Since the primary goal of Apollo had been achieved, many of those not assigned to the remaining lunar missions began to leave the program after fulfilling their current assignments. Deke Slayton decided to bring in some of the unflown astronauts to fill out crew requirements through to the end of Apollo, in order to make use of their extensive support experience. Several of these assignments were dead end roles, with no prospect of a mission at the end.

The crew rotation system devised by Slayton and used for several years would normally see the back up crew for Apollo 13 fly on Apollo 16. Therefore the back up crew for 16 would then be in prime position to take the last Apollo (19) to the moon. The back up crews for Apollo 14 and 15 were already in training and therefore in line for Apollo 17 and Apollo 18 respectively. With the cancellation of Apollo 20, and the aborted flight of Apollo 13, there remained only three flight seats in Apollo, those of Apollo 19. If the pattern was continued, those astronauts would first serve as back up to Apollo 16.

As this was to be the final landing mission, Slayton looked for a combination of experience and potential for its crew. Apollo 13 Commander, Jim Lovell, had already decided to retire before '13' left the ground, so Slayton assigned his LMP Fred Haise as Commander of the Apollo 16 back up crew instead of CMP Jack Swigert, who would normally have rotated from a prime CMP position to a back up Commander role. This was probably due to Haise's length of LM and surface training for Apollo 13, and Slayton's decision not "to send poor Jack around the moon again." However, *if* Apollo 13 had landed on the moon and Apollo 19 had remained on the manifest, it is unclear whether Haise would have received the back up Commander's position on Apollo 16 and the possibility of a second landing on the moon on Apollo 19.

Fred Haise, LMP of Apollo 13, holds a solar wind composition experiment during training in February 1970. He could easily have been working with Jerry on similar tasks in preparation for Apollo 19 surface activities had the mission proceeded as planned.

Assigned with Haise to Apollo 16's back up crew were Bill Pogue as CMP and Jerry as LMP. Both had extensive experience in support work for Apollo. The assignments reflected the modification of the requirement that the CMP should be a flight experienced astronaut, a practice changed with the Apollo 13 crew selection. Bill had been a CM specialist and supported the flights of Apollo 7, 11 and 13 (and was working on the support crew of Apollo 14), so he had extensive CSM experience. Jerry's selection to the flight was a logical LMP assignment. He was an LM specialist in the office, working on the processing for LM 4 and LM 6 and having worked on the support crews for Apollo 8 and 12. Additionally, having developed the crew procedures for the LRV, his experience and skill would be especially useful on the challenging flight plans scheduled for Apollo 16 and proposed for Apollo 19. The back up crew for Apollo 19 was never identified but would probably have been recycled from the Apollo 17 prime or back up crews, filling out dead-end assignments.

Unfortunately, as a "crew," the trio of Haise, Pogue and Carr conducted very little formal training during the summer of 1970. They snatched what simulator time they could in between the Apollo 14 and 15 crew training sessions. Almost twenty years later, Fred Haise alluded to the limited activities the Haise-Pogue-Carr crew actually did: "Apollo 19 was never official and a crew would not have been named [formally] until after Apollo 16 flew. Also a [final] landing site was not chosen. The development of spacecraft names and such things as a crew patch was something to work on after the Apollo 19 crew was officially named. Tradition would have had myself, Bill, and Jerry as that crew. The training [late spring - late summer] in the early phase consisted of many things that were both 'training' and the development of the mission. Some of the activities included flight plan development, training in the mission simulators, training in the lunar roving training vehicle, my checking out in the Lunar Landing Training Vehicle (LLTV), geological field trips and classroom lectures covering the newer [J series] modified Lunar module and Command and Service Module."

Concerning call signs, Jerry thought that the LM call sign might reflect the US Marine Corps, since both he and Fred Haise had served in that service. Bill Pogue was an Air Force pilot, however, and Jerry thought he probably would not think much of that idea unless he could reflect his own parent service in the choice of a CSM call sign. The availability of training photos for this period is very scarce. Further commenting on the mission in 1999, Haise recalled seeing photos of Jerry and himself sitting on the LRV trainer while on a field geology exercise in the summer of 1970. Bill Pogue was also in the picture. That image hasn't been found for this book, but records indicate that three such field geology exercises were completed by the original back up crew of Apollo 16.

These were:

July 8-10, 1970, San Juan Mountains, New Mexico (Young, Duke, Haise, Pogue, Carr)

July 23, 1970 Medicine Hat, Alberta, Canada (Young, Duke, Haise, Pogue, Carr)

September 1-2, 1970 Colorado Plateau (Young, Duke, Haise, Pogue, Carr)

By reviewing the training cycles for the last three Apollo missions, it is possible to approximate the overall training load for the Apollo 19 crew. Had Jerry remained in the crew after fulfilling his Apollo 16 requirements in the summer of 1971, he would have spent another year undergoing mission-specific preparation. This would have included nearly 4000 hours in simulations, 3000 hours on special purpose training, 400 hours on procedures, 700 hours on briefings and 400 hours in various test procedures. These totals would have encompassed about eighty simulations of lunar surface activities, including pre- and post-EVA operations, deploying the experiment packages and collecting samples (training with and without the EVA suits). About a dozen geological field trips would have been completed, with around twenty-eight days spent inside the CM and LM simulators and the combination CM/LM simulator.

Based on the December 1, 1969 mission definitions document (the last known update for the lunar mission), Apollo 19 was planned for a launch on July 14, 1972 to its primary landing site at Hadley-Apennine. The mission's objectives were to include geological sampling of the area around the landing site and the deployment of a scientific experiment package. The lunar orbital science survey would also have been conducted, using a package of scientific instruments in the Service Module. As it would have been the final mission to the moon, future landing site photography probably would not have been included, though in reality the crew would, no doubt, have taken photos of opportunity from orbit. The hardware planned for the mission was manifested as the three-stage Saturn V (serial number SA-514). The "Apollo 19" spacecraft would have consisted of CSM 115 and LM 13, with an LRV for use on the surface.

This July 1971 launch of Apollo 15 to Hadley Apennine is the reality, but it could have
been what the launch of Apollo 19 might have looked like the following year, as
the mission was originally targeted for the same landing site.

With ignition of the first stage of the Saturn V, the sequence of the 'Apollo 19' mission would probably have followed this typical 'J' series profile. After 2.5 minutes of flight, the vehicle would have reached an altitude of about 200,000 feet (38 miles) before the second stage took over, boosting Apollo 19 to an altitude of 606,000 feet (114.5 miles) over a six-minute burn time. Shortly after ignition of the second stage engines, the inter-stage would have been separated followed by the launch escape tower and its boost protective cover over the CM being jettisoned at GET 3 min 17 seconds. Once the second stage burned out and separated, the third stage would take over for a 2.75 minute burn taking the spacecraft into Earth orbit. Traveling from zero to 17,500 mph in about 12 minutes as one of a selected few to experience a Saturn V flight into orbit would have given Jerry the ride of a lifetime.

Once in orbit, the crew would not have had much time to look out the window as they would have to complete a star sighting (probably by Bill Pogue) and checkout of their CM, before being given a 'Go' for Trans Lunar Injection burn 1.5 orbits later. This burn, of approximately six minutes, would place them on their required trajectory towards a rendezvous with the Moon three days later. Pogue would have taken control of the CSM and, by firing the reaction control engines on the side of the SM, moved the spacecraft out about 100 feet in front of the S-IVB stage. He would have then completed a 180° turn and gently brought the docking probe on the nose of the CM into the docking cone on the roof of the LM. This was known as the transposition and docking maneuver. Then the bolts securing the LM to the booster stage would be severed, releasing the combined spacecraft to journey on to the moon.

The flight from Earth to the moon was expected to take about 110 hours, during which time the crew would have taken regular navigational star sightings to ensure their correct trajectory. Mid-course corrections would be made as necessary, firing the Service Module reaction control system or main engine to refine the flight path if required. Crew activities to and from the moon on an Apollo flight were fairly relaxed and often featured live TV transmissions of the crew, demonstrating life in space and conducting tours of their spacecraft for Earth bound viewers. Jerry would have accompanied Haise into the LM to check out some of its systems and ensure that it had not suffered damage during the boosted ascent from Earth.

After entering lunar orbit, the crew would spend the next 34 hours ensuring the system integrity of both spacecraft and making observations and telecasts from orbit. On landing day, Jerry would have probably preceded Haise though the docking tunnel into the LM. After sealing the hatches between the spacecraft, they would have put on their pressure garments and attached their harnesses to the LM structure, allowing them to stand in an upright position for the landing. Descent would have been from the perigee of a 60 x 8 nautical mile orbit, firing the descent stage engine to head towards the surface. Despite being designated the LM Pilot, Jerry would not have actually flown the vehicle. That was the role of the Commander from the

If Jerry had made it to the moon. then this image of an Apollo 15 astronaut saluting the American flag at Hadley Base could well have been a famous image from Apollo 19 instead.

left station. Instead, occupying the right hand position, Jerry would have supported Haise with computer readouts on rate and angle of descent, altitude, velocity and fuel levels (in seconds) as the commander looked out of the window. Using two hand controllers to control the attitude of the vehicle as it approached the prescribed landing area, Haise would be ready to take over from the computer if he saw the vehicle heading to an unsafe area. Once any of the contact probes on the landing legs touched the surface (there were three of these; the fourth leg held the ladder for the crew to descend to the surface), an electronic signal would have illuminated a lunar contact light and the crew would have shut off the descent engine. Their LM would have dropped the last few feet to the lunar surface about 144 hours after leaving Earth.

Apollo 19's LM was scheduled to remain on the surface for at least 54 hours, sufficient time for Haise and Jerry to complete three surface excursions. Traditionally, Haise would have exited first followed by Jerry. Normally as LMP, Jerry would be the first back into the LM at the end of each excursion. This was necessary because of the design of the forward hatch, which opened inwards towards the LMP position. This meant that the Commander had to get out first, and in last, to allow the LMP to move inside the cramped area of the LM cabin with the pressure down and hatch unlocked.

Instead of jointly holding the US manned spaceflight endurance record for over 20 years, Jerry may well have become the 16th man to walk on the moon, had Apollo 19 remained on the manifest. Using equipment similar to Jim Irwin in this frame, Jerry would have collected samples from the Hadley Apennine area for analysis back on Earth

Landing site coordinates for the mission were planned for as close as possible to 2° 27'E and 24° 47'N. This area contained a small, 3.5 mile diameter crater called Hadley C, which was of great interest to the scientists. This crater covered the rille, indicating that it was a younger feature. It included a raised rim and ejecta

An LRV would certainly have been used as part of the surface activities during Apollo 19.
In that case it would have been Jerry, rather than Jim Irwin from Apollo 15,
standing alongside the LRV in front of Mount Hadley in this image.

blanket across the landing area. Samples collected from this area would yield important clues for age dating the material from the crater, rille and surrounding areas. This information would help identify the origins of the sinuous rille and the controversial crater. As previously mentioned, the site was deemed so important to scientists that it was chosen as the landing site for Apollo 15 in 1971 following the cancellation of Apollo 19.

The first EVA would probably have seen the collection of preliminary geological samples close to the landing site (in the event of early termination of surface operations) and deployment of some of the surface experiments. With six full science packages deployed on earlier missions, there was already a network of stations across the moon, so the remainder of the surface experiments could have been deployed on EVA 2 or 3. There would have been a short mobile sampling traverse using the LRV. Planning documents suggest there would have been three, five-hour EVAs, with a contingency plan for a limited foot traverse should the LRV become non-operational. At several points during any Apollo lunar mission, contingency alternatives were built into the procedures to maximize the yield in the event of an abbreviated mission. Mission safety rules were, of course, paramount and could not be infringed upon.

The three rover traverses would have taken Haise and Jerry to the base of the Apennine ridge, the lip of the sinuous rille, and close to Hadley C crater. Though not fully approved, the proposed Apollo Lunar Surface Experiment Package (ALSEP) for Apollo 19 would have included experiments flown on previous missions, as well as new ones, to increase the network of science stations emplaced on each landing. These stations were planned to operate long after the astronauts returned home. For Apollo 19, the proposed surface experiment package would have included a lunar dust detector, a passive seismic detector, an electro field gradiometer, a gravimeter, a mass spectrometer and a radiometer, a low energy nuclear particle detection experiment and a water detector. The crew would have had a range of cameras, sample return containers and geological hand tools to assist in their collection of samples and recording their finds, as well as a "Lunar Survey Staff" for detailed topographical measurements across the landing area.

After about 54 hours on the surface, the ascent stage of the LM carrying Haise and Jerry would be launched back into orbit to re-dock with the CSM flown by Pogue. After docking and transfer of the sample return containers and film cassettes, the two men would return to the CM. Upon closing the hatches, they would separate the LM. The empty LM would likely have then been deliberately crashed on the surface to provide a man-made seismic event that could be recorded by the network of ALSEP seismic instruments.

Apollo 15's Falcon 'on the plain at Hadley'. Apollo 19's unnamed and unflown LM was originally targeted to sit serenely on the lunar surface here.

While Haise and Jerry were on the surface, Bill Pogue would not have been at a loss for work. He would be orbiting the moon every two hours, operating the Scientific Instrumental Module (SIM-bay) experiments and taking terrain photos. He would probably also have been operating several small technological experiments inside the CM. The array of survey instruments in the SIM-bay for Apollo 19 were proposed as an S-band transponder, a 3-inch mapping camera, a sounding radar used for mineral location and topographical mapping, an electromagnetic sounder to identify subsurface structures, an infrared scanner to measure surface temperatures and thermal anomalies, and a laser altimeter. There would also be a sub-satellite to be deployed from the SM, carrying a magnetometer and a Far Ultraviolet Spectrometer to identify the composition and density of the lunar 'atmosphere'.

After returning to the CM, Haise and Jerry would have assisted Pogue in these studies and observations for about another 72 hours, changing the orbital plane of the CSM to achieve the desired photographic and scientific coverage. In all, about 61 hours of data collection from the SIM-bay was planned, which would have been completed by approximately 8 hours before leaving lunar orbit.

The firing of the service propulsion engine on the SM signaled the start of the long flight home, which normally lasted about 100 hours. During the return trip, Jerry would have performed a stand-up EVA, filming and assisting Pogue during his deep space EVA to retrieve SIM-bay film cassettes and any exposed samples from the Service Module. The only other activities on the way home would be mid-course corrections, packing loose gear away and TV transmissions, before splashing down in the Pacific about 384 hours after leaving Earth. The splashdown and recovery would have completed the thirteenth manned Apollo lunar mission, the eleventh to the vicinity of the moon and the eighth landing. It was, however, a mission that only ever existed on paper.

The official announcement of Haise-Pogue-Carr to the back up crew for Apollo 16 had been expected shortly after the successful completion of Apollo 13. However, the in-flight explosion on Apollo 13 and the tense recovery of the crew, together with the official investigation into the accident, had a profound influence on future Apollo plans. With NASA's budget increasingly restricted (not least by the great cost of the war in Vietnam), the repercussions of the Apollo 13 accident would see the Apollo program constricted still further. Subsequent missions were delayed and further crew announcements were suspended until after the Apollo 13 Accident Review Board released its findings. Overshadowed by rumors of further budget cuts throughout the summer of 1970, nothing concerning the crew assignment for Apollo 19 was ever formally released prior to its cancellation (along with the original Apollo 15) in September 1970. In the re-scheduled manifest that followed, Apollo 14 was now targeted to Apollo 13's intended landing site at Fra Mauro, while the J-series science missions Apollo 16-18 were re-designated Apollo 15, 16 and 17.

Jerry had hoped to become the 16th man to walk on the moon, but it was not to be. "It was a bad day at black rock for the three of us," Jerry recalled years later. Slayton decided to recycle flown Apollo astronauts to the dead end assignments left in the program, so, after Apollo 14, Stu Roosa stood in for Bill Pogue and Ed Mitchell for Jerry as the Apollo 16 back up crew. Fred Haise would continue to fulfill his assignment because of his previous training. This released Jerry and Bill to move to new assignments.

Thus ended Jerry's connections with the Apollo lunar program. Through most of his first four years in the astronaut office, he had focused solely on the Apollo lunar effort, but following the loss of his seat on Apollo 19 he received a new challenge. Two weeks after the cancellation, Tom Stafford told Jerry that he had been selected to command the third and final Skylab mission of 56 to 90 days. Asked if he could work with Bill Pogue and scientist astronaut Ed Gibson, he replied firmly, "Absolutely." He was flabbergasted that he was being entrusted with the command of a crew for his first mission. This would be the first time since March 1966 (when Neil Armstrong and Dave Scott had taken Gemini 8 into orbit) that an all-rookie American crew had been selected to fly a mission. It was not the moon, but it was a mission into space.

It is clear that, even if Apollo 19 had remained on the manifest, the scientific payload and possibly the landing site may well have changed. Certainly, the timeline would have been refined as the missions approached and clearer scientific objectives were defined. Taking all the flown Apollo landing missions into consideration, the three J missions of Apollo 15 through 17 contributed to 76% of the total time spent on the surface by Apollo astronauts. They accounted for 74% of all the samples returned and, with the use of the LRV, 94% of the traverse range. *If* Apollo 18 and 19 had flown and matched the return of Apollo 17 (22 hrs EVA time; 35 km traverse distance; 514 kg experiments; 110 kg returned lunar samples) these ratios would have increased the experiments, sample return and surface time to about 87%, and traverse distance to 96% of the overall program. What is very sad is that the hardware to fly these missions had been built, and most of the astronauts who would have flown them, trained. Therefore, the additional cost of mounting these expeditions would have been minimal compared to the extra scientific gain. Instead the flights were cancelled, the hardware dispersed and the astronauts reassigned.

The Saturn V assigned to Apollo 19 was placed in storage and assigned to public display at NASA centers. The S-IC first stage was subsequently used as the back up first stage of the two-stage Saturn V that could have launched Skylab B into orbit in 1976. Instead, the stage was taken by barge to Houston in 1977, arriving at the L.B. Johnson (formerly Manned) Space Center on September 21, where it was put on display with other stages as a composite full-sized Saturn V for public viewing. By 2004, this stage (along with the others) was showing the signs of years of exposure to the Texas weather, and plans were announced to renovate the stages and enclose them in a structure similar to the one at Cape Canaveral, to protect them for future generations to admire. The intended Apollo 19 second stage (S-II) and third stage (S-IVB) were taken out of storage and placed on display outside the VAB in 1976. In 1996 they were transferred along with the other parts of the display to a new indoor Saturn V center. The CSM for the Apollo 19 was used in the JSC Saturn V display, while the LM was only partially constructed and was placed in the Cradle of Aviation Museum at Long Island, New York.

The debate over what Apollo 18-20 (and beyond) could have achieved continue to draw interest and speculation. In 2002, the mission designation re-surfaced, this time in a fiction novel titled *ICE* by Shane Johnson (Watercress Press, July 2002). In this plot, Apollo 19 lands near the Aitken Basin, near the lunar South Pole, in February 1975, following the discovery of a vast quantity of water ice near by. Drawing on actual Apollo plans, the fictional crew used a MOLAB to explore the area, but in true science fiction tradition, equipment fails and a rescue crew is dispatched in a reactivated 'Apollo 20'. From here the story journeys from the realms of science fact into that of science fiction. Interestingly, the cover design for this book was painted by Astronaut Al Bean.

At the time of the debate for keeping or cancelling the three Apollo missions, alternatives included a proposal to fly the Apollo 19 mission in 1974 or even 1975, after the first Skylab had supported its three manned missions. Consideration was also given to flying an extended lunar orbital or Earth orbital mission using the already constructed hardware. As it became clear that the Apollo missions would end at Apollo 17, and the last Saturn V would launch Skylab, hopes lingered for using the leftover hardware for a second Skylab in 1976. This, too, came to nothing. By the end of 1970, Jerry had left the Apollo Astronaut Office and moved to the Skylab Office headed by Pete Conrad. He was working to make his first mission into space as successful as he could. Despite losing the opportunity to fly to the moon, his mission to Skylab would become one of the most important milestones in the development of long term space endurance and exploration. Jerry might not have made his own "small step" onto the lunar surface, but his Skylab mission would prove to be a "giant leap" into the future of space station development.

Chapter 6

A Laboratory in the Sky

On July 22, 1969, just two days after Neil Armstrong and Buzz Aldrin had landed on the Moon during Apollo 11, NASA announced a major redirection in the Apollo Applications Program that would follow the initial lunar landing missions. The S-IVB Orbital Workshop would be launched, fully fitted out on the ground, on top of a two-stage Saturn V. This configuration was termed a 'dry workshop' and utilized the first two stages of a Saturn V originally intended for the lunar program. In the original definition, Skylab was to have been launched as a fully fueled operational rocket stage. Once in orbit the residual fuel would have been drained and gasses vented into space. This would enable a space suited crew to enter the cavernous and now empty hydrogen tank for fitting out on orbit, prior to full habitation by the first crew. This was termed the 'wet-workshop' configuration. It was determined to be too complicated and expensive to use this system, so in 1969 the dry workshop concept was adapted. The Apollo Telescope Mount included in the configuration would be based on an unmanned structure, attached to the workshop at launch and operated from within the workshop.

The primary focus for the American manned space program in the closing years of the 1960s had been in getting a man on the Moon. With this achieved, and repeated in November 1969 with Apollo 12, new horizons were being explored beyond the Apollo program. NASA had grand plans for extending the lunar exploration program and using surplus Apollo hardware to fly the AAP OWS missions. They also wanted to create a lunar base, develop a reusable space launch system and construct a huge 50-man space station, as well as dispatching the first human expedition crews to Mars – all by the end of the 1980s. But the grand plan lacked political and public support and would remain only on paper. The budgetary cuts that trimmed the lunar landing program, and eventually cost Jerry his trip to the Moon, also began to eat into the Apollo

The Skylab space station and the trio of three-man crews

Artist's impression of the Apollo CSM approaching the Orbital Workshop. The illustration depicts a fully-deployed Apollo Telescope Mount and twin solar arrays, as well as the EREP package.

Applications Program. On February 24, 1970, the AAP program itself became known as Skylab, but by then it had been trimmed from two workshops to one. Instead of up to seven manned missions to the two stations, there would be only three visits to one facility. Out of NASA's grand plan, only the possibility of a short joint flight with the Russians remained on the manned launch manifest between Skylab and the maiden flight of the Space Shuttle. Everything else would have to be left for future generations.

The loss of three Apollo missions to the Moon was a very personal blow to those who were scheduled to fly them. JoAnn Carr recalled the disappointment of losing the Moon in the family household: "Sometime after Apollo 11, we were beginning to see a pattern in who was assigned to what flight. So were pretty sure Jerry was going to fly Apollo 18 or 19 and walk on the Moon. Because he had worked on the development of the lunar rover, it was pretty much a given that he was one of the leading contenders to reach the lunar surface. Then, one night in the summer of 1970, Jerry came home. Jerry is not a very emotive person, but he was that night, and he actually had tears in his eyes. He said that his flight to the Moon had been cancelled. That was really a heartbreaking time for both of us because I was fully invested emotionally and every other way in the space program for the long haul. I was ready to support Jerry on a flight to Mars. I just thought it would be so wonderful, to get off the planet. It was really heartbreaking in view of the fact that he had worked so hard for four or five years now there was to be no payoff."

The CB Skylab Office

The availability of flight seats for the astronauts dropped dramatically between 1969 and 1970. For some, the prospect of waiting years for a possible flight on Shuttle was a mission too far and they began to leave the program. Others decided to stick it out. Fellow classmates of Jerry, like Joe Engle, Bruce McCandless and Don Lind would eventually reach orbit, but not until the early 1980s, some 15-19 years after joining NASA. In the short term, however, with sufficient crews in training to fill out the prime crew roles on Apollo, the nine flight seats assigned to Skylab suddenly gained new importance in the Astronaut Office.

Artist's cutaways of the Orbital Workshop area of Skylab - home for
Jerry and his crew for three months.

SKYLAB ORBITAL WORKSHOP

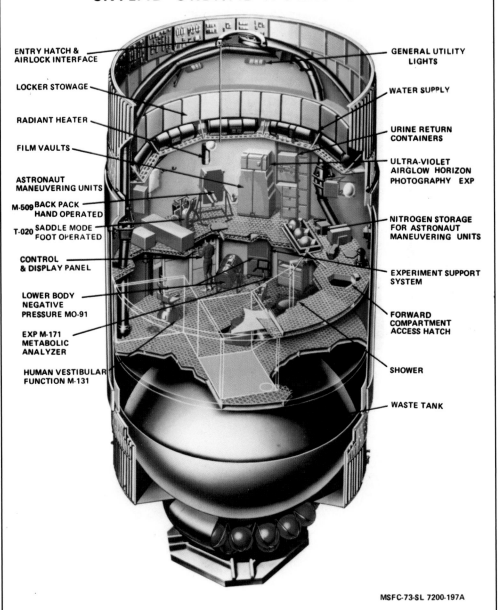

ENTRY HATCH &
AIRLOCK INTERFACE

LOCKER STOWAGE

RADIANT HEATER

FILM VAULTS

ASTRONAUT
MANEUVERING UNITS

M-509 BACK PACK
HAND OPERATED

T-020 SADDLE MODE
FOOT OPERATED

CONTROL
& DISPLAY PANEL

LOWER BODY
NEGATIVE
PRESSURE MO-91

EXP M-171
METABOLIC
ANALYZER

HUMAN VESTIBULAR
FUNCTION M-131

GENERAL UTILITY
LIGHTS

WATER SUPPLY

URINE RETURN
CONTAINERS

ULTRA-VIOLET
AIRGLOW HORIZON
PHOTOGRAPHY EXP

NITROGEN STORAGE
FOR ASTRONAUT
MANEUVERING UNITS

EXPERIMENT SUPPORT
SYSTEM

FORWARD
COMPARTMENT
ACCESS HATCH

SHOWER

WASTE TANK

MSFC-73-SL 7200-197A

Artist's cutaways of the Orbital Workshop area of Skylab - home for
Jerry and his crew for three months.

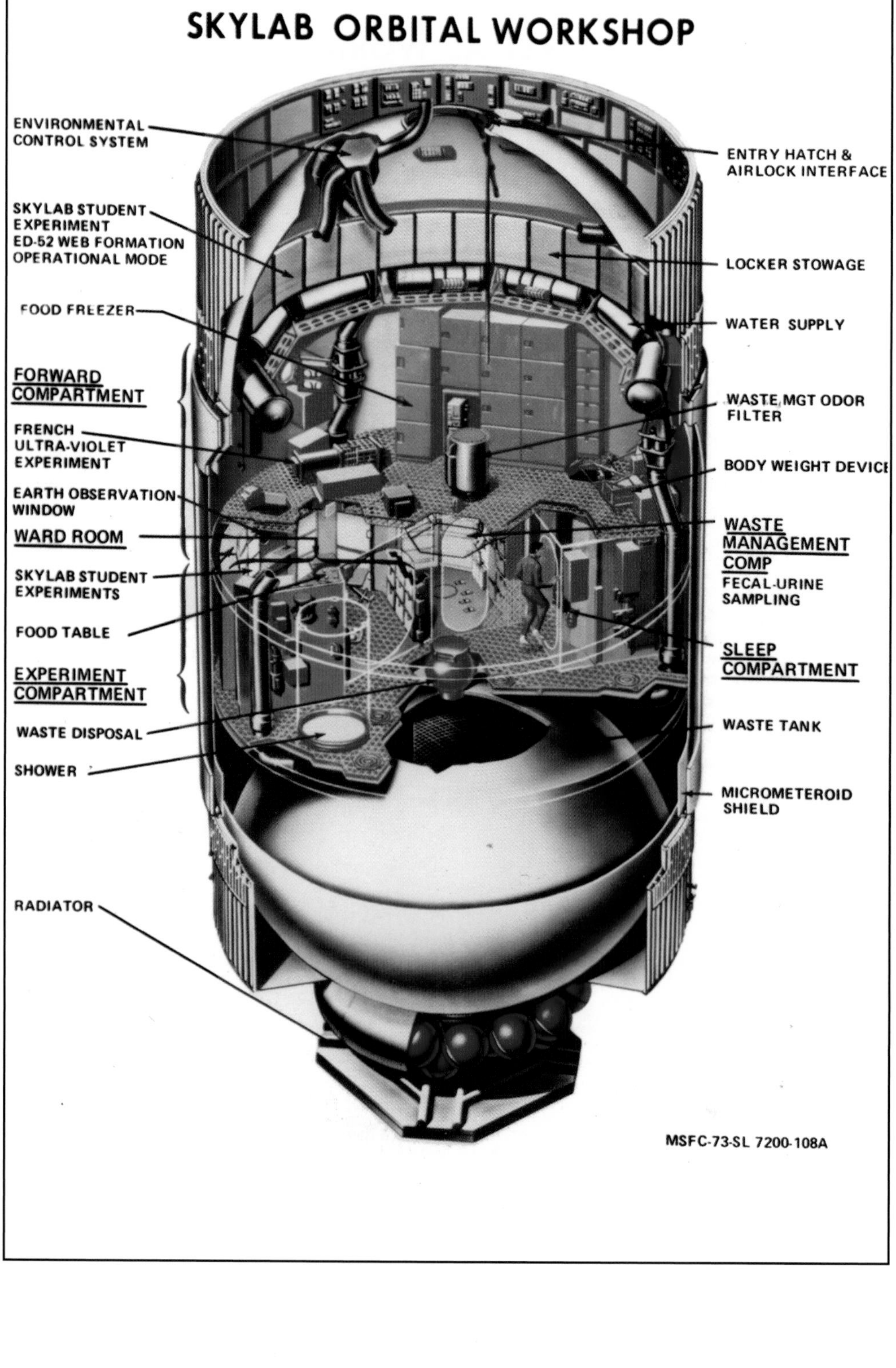

SKYLAB ORBITAL WORKSHOP

ENVIRONMENTAL
CONTROL SYSTEM

ENTRY HATCH &
AIRLOCK INTERFACE

SKYLAB STUDENT
EXPERIMENT
ED-52 WEB FORMATION
OPERATIONAL MODE

LOCKER STOWAGE

FOOD FREEZER

WATER SUPPLY

FORWARD
COMPARTMENT

WASTE MGT ODOR
FILTER

FRENCH
ULTRA-VIOLET
EXPERIMENT

BODY WEIGHT DEVICE

EARTH OBSERVATION
WINDOW

WARD ROOM

WASTE
MANAGEMENT
COMP
FECAL-URINE
SAMPLING

SKYLAB STUDENT
EXPERIMENTS

FOOD TABLE

SLEEP
COMPARTMENT

EXPERIMENT
COMPARTMENT

WASTE DISPOSAL

WASTE TANK

SHOWER

MICROMETEROID
SHIELD

RADIATOR

MSFC-73-SL 7200-108A

NASA highlighted and promoted Skylab as a unique place from which to observe the Earth, given the new global awareness of the planet's fragile environment. Never before had such a unique facility been available to American astronauts. After the short missions and somewhat cramped confines of the Mercury, Gemini and Apollo spacecraft, the chance to live and work for several weeks in a vehicle comparable in volume to a three-bedroom house would be a unique opportunity to really explore the phenomena of living and working in space. Though not as attractive to the general public as a mission to the Moon had been, the work conducted on Skylab would have a far more important and long term role in future exploration of space, though many did not recognize this at the time. Over 30 years later, the legacy of Skylab is now being realized on the International Space Station (ISS).

When Jerry arrived at NASA, the Apollo Applications Program Branch Office within the Astronaut Office (CB) structure had been in existence for some months under the leadership of astronaut Al Bean. When Bean was reassigned to replace CC Williams on Conrad's Apollo crew, he in turn was replaced at AAP by Mercury astronaut Gordon Cooper. The CB involvement in the program had in fact been in existence for some years, initially involving pilot astronauts who had a more scientific background than one in operational flying. Rusty Schweickart (with a background in upper atmosphere physics, star tracking and a thesis in stratospheric radiance), Walt Cunningham (a former scientist at Rand), Bill Anders (experience in nuclear engineering) and Donn Eisele had all worked on future NASA programs, experiments and the early phases of AAP before the first group of scientist astronauts were assigned to the program in 1965. Several of the scientist astronauts also worked on AAP issues after completing their astronaut training from 1969, while others supported the remaining Apollo missions.

Most of the work in the AAP Branch Office during 1966-70 involved attending meetings and reviewing hardware development, procedures and program development. Some of the 1966 intake of pilot astronauts were assigned support roles in the AAP program, but Jerry was assigned to Apollo issues and never worked on AAP until reassigned from Apollo in the summer of 1970:

JC: "I paid very little attention to AAP. Bill Pogue had had a short assignment, like I did, with the lunar rover. He worked Apollo Applications for a little while, so he was a little more familiar than I. But most of us had been totally focused on the Apollo lunar program."

Jerry recalled that fellow astronauts assigned to support the LM/ATM development for Apollo Applications were at Grumman at the same time as he was, but he had no knowledge of that side of the program at the time:

JC: "In those days I was strictly an Apollo lunar landing man. I had no knowledge of, or interest in, AAP because I was completely focused on Apollo things. So I really was out of touch with any of the AAP goings-on. I guess, as far as Skylab astronauts are concerned, I was a sort of Johnny-come-lately."

In June 1970, the Soviets returned to manned space flight operations by flying the 18-day Soyuz 9 mission. This was a precursor to their own space station program, which was under development to replace their failed and struggling manned lunar program (finally cancelled in 1974). The two Soyuz 9 cosmonauts, Andrian Nikolayev and Vitali Sevastyanov, remained in space long enough to surpass the American Gemini 7 record of 14 days set in 1965. Though a limited program of scientific research was completed, the cosmonauts were somewhat lax in following their daily medical exercise routine. As a result, their recovery after spaceflight took longer than planned and was more painful. Despite this, Soyuz 9 gave the Soviets baseline data with which to plan longer flights on their space station program. A space platform with military applications had been planned by the Soviets for years under the Almaz (Diamond) program, supported by military versions of the Soyuz ferry spacecraft. However, when the hardware being developed for the program began to slip in its launch schedule, it was decided to modify an Almaz station with parts from Soyuz, in an attempt to put a space station in orbit prior to Skylab.

By the summer of 1970, former Apollo 12 commander Pete Conrad had recognized that his only option of making a fourth flight into space rested with Skylab. For over a year, Apollo 7 astronaut Walter Cunningham had been Chief of the CB AAP/Skylab Branch Office and was hoping for the command of the first mission. In August 1970, when Conrad became the new branch chief for Skylab he was also penciled in as Commander of the first 28-day mission. Having seniority in the office by means of being selected to NASA a year earlier than Cunningham, Conrad was always a strong candidate for commanding the first crew to the space station. Cunningham's expectations turned to command of the second or third mission, until Conrad's navy colleague and Apollo 12 crewmate Al Bean returned to the Skylab office and bumped him down the queue again. This upset Cunningham to the point of him leaving the program in 1971.

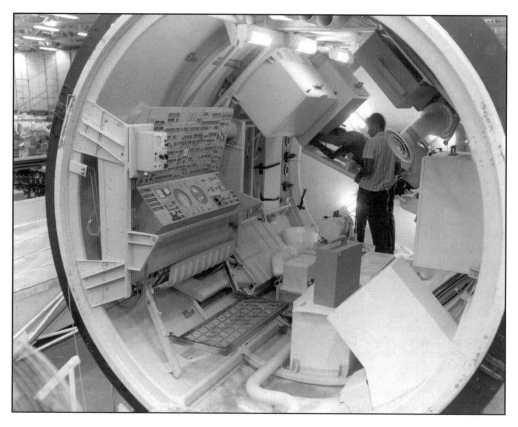

This early 1970 mock-up of the Skylab Multiple Docking Adapter displays the location of the ATM console and the position of the EREP package, both of which the crews spent a significant portion of their time working with. Paul Weitz in the background.

With Conrad heading the Skylab astronaut team, and with the cancellation of two additional flights in the Apollo program the following month, there were sufficient experienced astronauts to fill the remaining nine flight seats. There were also more than enough scientists from the 1967 selection and the recent 1969 transfer of former Manned Orbiting Laboratory Program pilots to fill out back up, support and CAPCOM roles.

With over 20 astronauts assigned to the program by early 1971, this was a significant increase over the team in 1968 when there was but a handful of scientist astronauts and rookie pilot astronauts assigned to the office. Flight experienced command candidates for the three missions included Conrad (Gemini 5, 11, Apollo 12), Bean (Apollo 12), Cunningham (Apollo 7) and Rusty Schweickart (Apollo 9). Rookie pilot candidates were Jack Lousma and Paul Weitz (the leading LMP and CMP candidates from the cancelled Apollo 20), Jerry and Bill Pogue (formerly the LMP and CMP for Apollo 19), and Don Lind and Bruce McCandless from the 1966 selection.

The Science Pilot candidates had pretty much been chosen for some time. They were Group 4 scientist astronauts Joe Kerwin (medical doctor for the first long duration mission), Owen Garriott (Chief of the CB AAP Branch between Cooper and Cunningham) and Ed Gibson (who had, like Garriott, worked on AAP issues since returning from flight school). Members of the 1967 scientist astronaut selection who would provide support roles in the program included Don Holmquest, Bill Lenoir, Story Musgrave and Bill Thornton. From the 1969 MOL transfer, Bob Crippen, Bob Overmyer and Dick Truly were assigned to engineering support roles. Despite a call for two scientist astronauts on each crew, due to the increased scientific payload and objectives, Slayton and Shepard favored assigning one scientist and two pilots to each flight.

Joining Skylab

Following the loss of his assignments on Apollo 16 and Apollo 19, Jerry was obviously disappointed: "We had lost our opportunity to go to the Moon. I remember we moped around for quite a few weeks." But after Tom Stafford (who was acting Chief Astronaut while Alan Shepard was in training as Commander of Apollo 14) had informed Jerry that he would be the commander of the third Skylab mission, there was a new goal to focus upon. The disappointment of not getting to the Moon was still there. In fact, coming so close to the flight and then not going was probably more difficult than not being part of a mission to the Moon at all. But Jerry now had a new mission – to America's first space station.

Jerry was delighted to finally get a flight seat "What delighted me the most, I think, was the fact I was going to be working with Al Bean and Pete Conrad after supporting them on Apollo 12. That was a really wonderful thing. I was also awed by the fact that, as a rookie, I had been selected to be a Commander. That hadn't happened since Gemini."

By now, JoAnn Carr had lost interest in the space program for some time, but when Jerry told her of his assignment to Skylab, and as Commander of the third mission, she was pleased for him. "Jerry, who had not flown before, was named commander of the third and last flight, which was pretty good. Not prestigious exactly, but it was recognition of what he could do. So that was better than nothing, a whole lot better than nothing."

At a pilot meeting at JSC in either late 1970 or early 1971, Deke Slayton released the initial flight crew assignments for three prime crews and two back up crews. The latter might, depending on real-time evaluations, have been able to fly a fourth and fifth mission. Jerry was identified to command the third mission, with Ed Gibson and Bill Pogue, without having been officially assigned to a previous back up crew. This was a huge honor and compliment to his character and performance in the CB over the previous four years. The initial announcement saw Conrad, Kerwin and Weitz assigned to the first mission, with Cunningham, Story Musgrave and McCandless as their back up crew. The second flight would be crewed by Bean, Garriott and Lousma, with Jerry's crew third. The second and third missions would be backed up by Rusty Schweickart, Bill Lenoir and Don Lind. Cunningham's crew was assigned to the possible fourth mission and Schweickart's crew the fifth.

Soon after this was announced to the astronauts, Cunningham resigned and left the program. Slayton moved Schweickart up to replace him on the first back up crew and brought in Vance Brand, recently off the Apollo 15 back up crew, to replace Schweickart for the second and third missions. The official announcement was not made until January 19, 1972, when the nine prime and six back up astronauts were identified to the world's press. By then, they had been in training for some months.

There was also a plan to have the next Saturn 1B and CM on hand as soon as possible after each launch to support a possible rescue mission, should the flight require an early termination and the crew or CSM be unable to affect a nominal recovery. A two-person CSM would be launched to the station to return the three-man station crew in an unprecedented five-person recovery mode. Astronauts Vance Brand and Don Lind trained for this contingency. Though a rescue mission was nearly called for during the second residency, this option was not pursued. Neither was the prospect of a further short visiting mission by the Brand-Lenoir-Lind crew after the third mission.

At the same press conference, the roles for each crewmember were explained. The commander would have overall responsibility for the success and safety of the mission and would be the expert in Command and Service Module systems. The Science Pilot would be an expert in all medical equipment, as well as the Apollo Telescope Mount and associated equipment. The Pilot would be the expert in the systems of the Orbital Workshop and the electrical sub-system on the station. All other experiments would be divided between the crew members according to their availability and choice.

By the time the Skylab crews had been announced, the Soviets were recovering from the tragedy of losing their first space station crew. In April 1971, the world's first space station, Salyut, had been launched. It was supposed to restore pride in the Soviet program after losing the race to the Moon. However, the first mission, Soyuz 10, could only soft-dock to the station, and the crew was unable to enter the workshop. Having trained for months for a long stay on the station, they were back on the ground just two days after launch. Crew illness resulted in a change to the next crew, with the Soyuz 11 back up crew flying the mission. They had a difficult three weeks, due in part to their rushed assignment to a mission they had not expected to fly. Tragically, at the end of the mission, an air equalization valve on their Soyuz failed,

The official Skylab 4 crew photo - l to r Jerry (Commander),
Ed Gibson (Science Pilot) and Bill Pogue (Pilot)

resulting in the loss of internal air pressure and the deaths of the three cosmonauts during the automated re-entry and landing sequence. Instead of finding three heroes, rescuers found three dead cosmonauts. For the second time in four years, a nation mourned its lost cosmonauts. The Soviet manned space program was grounded pending investigations and improvements to the failed system.

Training for Skylab

Training for the three missions was to be more intense than for the lunar missions. Each lunar trip had taken less than two weeks, but the Skylab crews were to spend one or two months in Earth orbit. The Apollo missions had of course involved several rendezvous and docking maneuvers and the challenge of landing and working on the Moon. Skylab, on the other hand, would involve weeks of coordinated scientific activity with over 50 experiments, several EVAs and extended long duration residency in space. This would bring new challenges and hurdles to be overcome.

In 1975, during an interview for the official Skylab History published by NASA in 1983, Jerry reflected that although Skylab training included simulations in the Apollo CSM, as had the Apollo lunar crews, the main focus was on training in the workshop mock-up. Without the full-sized 1-g simulator (duplicating the wardroom area, experiment floor, airlock and docking adapter areas), Jerry strongly believes that the work in space would have been much harder than it actually was. But he also compared the 1-g trainer on Skylab to the lunar surface training that Apollo crews completed at geological field sites or on simulated lunar surface terrains. It was useful, but was a far departure from actually working in the environment they trained for.

JoAnn Carr was very disappointed that Jerry's work was not recognized with a flight to the Moon: "This is a heads up guy who really worked hard, and if anybody deserved a flight, he did. There were a lot of politics played out, and he does not play politics; it's not his style." When he was assigned to Skylab in 1970, the launch was still three years away, and JoAnn assumed that the training would be less intense than for Apollo. She was wrong. During this time preparing for Skylab, JoAnn's father died, and with all her children at school, for the first time in 15 years she was at home alone, with time on her hands: "I have been totally consumed with trying to be the best housewife, the best mother, the best wife that I could be." Now, with time for herself each day, she began to look for ways to occupy it. She began taking more tennis lessons, and undertook volunteer work in the depressed areas of Houston, a continuation of her church volunteer work with the senior high school group. At the time, the war in Vietnam was not going well for America, and the question of why the country was involved at all was being raised more frequently. JoAnn was asking it in the Carr household: "Having been a military wife, I was used to not asking any questions. Now I was beginning to." This began to add to the stress she encountered in her charity work and the increasingly demanding Skylab training program Jerry was undertaking. It all began to create problems between them. In seeking help to sort out these problems, JoAnn tried to seek counseling, but due to the pressure of Skylab training, Jerry always found it difficult to attend these meetings. Instead, JoAnn made a determined effort to support Jerry in his quest for spaceflight. "So that was how we dealt with the next three years. It was basically me absorbing any impact the family might have on what he was doing. In my thinking, it was truly a matter of life or death for him that I kept things running smoothly at home with six children. Then he would be able to pay attention to what he was doing. I felt it was a really important job." It was one that occupied her time while Jerry focused on training for his spaceflight.

Jerry's official portrait in suit, circa November 1971

At the beginning of the Skylab training program, the most noticeable thing to Jerry was that they didn't *have* a training program. Along with Bill Pogue, Jerry worked to develop a formal training syllabus for all of the crews to work through, while the training hardware was being developed as they went. Ed Gibson, meanwhile, was involved in the development of the scientific protocols for the missions. As the group needed to wait for the Apollo program to draw to a close, time in the CM simulators was limited for any of the three Skylab crews. With Conrad and Bean's crews ahead of him, Jerry's was left looking for scraps of simulator time: "We were working with cardboard mock-ups trying to figure out what to do." On reflection, however, Jerry thought that the time spent with the staff of the training department to create the Skylab training program was an "excellent use of our time." It was invaluable in allowing them to prepare for their own mission in a shorter simulation timeframe, once the first two crews had flown. The three years training time from announcement to flight was just about sufficient to allow Jerry and his crew to learn about the CSM, the workshop and all the experiments.

Formal training begins

The formal training program commenced in November 1970 with background training, providing general information and orientation on spacecraft systems and experiments. They also participated in spacecraft testing, reviews of flight plans and procedures, and began their training in solar physics. By the time the astronauts had been formally identified to the world's media in January 1972, almost all of their simulators and mock ups were in place. They had already begun their specialist training for individual activities, such as experiment operations, simulations of certain aspects of the flight, and Intra-Vehicular Activity (IVA) and Extra-Vehicular Activity (EVA) training. In February 1972, an Integrated Crew Training program began, with simulators and training specialists assigned to the astronaut group. Integrated Mission Team Training commenced in November 1972 with the astronaut/flight controller teams. This gave each astronaut experience in working with the rest of the team – those with whom he would fly and those on the ground with whom he would communicate. It was during this phase of the training that the mission simulators were linked with Mission Control Center in Building 30 at JSC to simulate actual elements of the mission they were to perform in space. This training included the introduction of specialization in each crew, which reduced their overall individual involvement in every aspect of the flight.

As commander, Jerry had responsibility for the CSM, the launch phase, rendezvous and docking, the EVAs, undocking at the end of the mission, the de-orbit, and entry and landing. Aboard the station, he was responsible for navigational guidance and associated systems and hardware. He was also responsible for the overall success of the mission, in terms of attaining personal and mission objectivies, and for the health and well-being of his crew in terms of motivation, safety and management. Jerry's science pilot, Ed Gibson, took responsibility for the ATM, EVA and medical and solar physics experiments. Having authored a text book called *The Quiet Sun*, he naturally became the expert on solar issues for the crew. Bill Pogue, Jerry's pilot, took responsibility for the Airlock Module, the Multiple Docking Adapter, the Orbital Workshop, the fluid systems onboard, and the Earth resources experiments. Each of them received sufficient cross training to ensure completion of all the primary aspects of the mission. Though each man focused on their specific areas, Jerry and his crew structured their program so that all three of them could operate anything, but if anything went wrong there was always one expert on hand.

The majority of the simulator training was accomplished at JSC, with the exception of EVA training at the neutral buoyancy facility at the Marshall Space Flight Center, in Huntsville, Alabama. They participated in spacecraft testing and did launch procedure training at the Kennedy Center at Cape Canaveral. The training for Jerry's Skylab mission was, like the two missions preceding it, broken into three primary training elements. Firstly, the theory of the experiment had to be reviewed. The astronauts had to have a basic understanding of the objectives, what was being conducted, and how the data was to be gathered, in order to be a competent operator. As the pilots were not astrophysicists or astronomers, there was a limit to their academic understanding because their training program focused on other aspects of the mission. Jerry reckoned he understood more about solar physics by observing than by trying to become a solar physicist in the classroom. Secondly, operational briefings were conducted on specific systems and experiments. Finally, operational procedures training, which involved simulators and mock-up trainers was conducted.

A training plan was published early in 1971. Highly flexible, it was constantly updated and revised as experience began to reveal what worked and what didn't. A major review of Skylab training was conducted on December 8, 1971 at JSC, and highlighted the aspects of training and the present status of both hardware and software to support the training program. For the planned 2,000 hour mission, each of Jerry's crew spent about 2,150 hours in training. This was not counting the numerous meetings, private studies at home and flights across America to visit contractors and experimenters or to participate in public events.

Initially, in some cases, the experiments had no hardware with which to train. The crews worked with cardboard mock ups until the Principal Investigator (PI) delivered a training item. When it was delivered, the crews usually found that the investigators had little or no concept of working in weightlessness, and Jerry and his fellow astronauts worked hard with the PIs to configure their experiments to allow productive operation in space:

JC: "I think they appreciated our willingness to do that. For us it was important, because we wanted success."

The three scientist astronauts assigned to the flight crews (Joe Kerwin, Owen Garriott and Ed Gibson) helped develop a handbook called the Joint Operating Procedure (JOP), which evolved into the astronauts' experiment bible. The PIs had to agree that the JOP would be the only document that everyone would follow. In the event of an experiment failure or inability to perform an experiment, there would be no in-fighting. Program planners would try to insert the experiment or procedure later in the flight plan. Gibson and his two colleagues bore the brunt of much heated discussion between the scientists and the astronauts, as Jerry recalled some years later:

JC: "Experimenters wanted to make sure that we understood that they had invested a lot of money – probably most of their budget – on this thing, and they didn't want it screwed up. There were a few experimenters (or Principal Investigators) out there that had absolutely no confidence that we could do anything. They probably considered us to be not much more competent than a bunch of chimpanzees up there trying to do their very important experiments."

At the end of Jerry's missions these same experimenters came forward and thanked the crew on a fine job, admitting they had misjudged the ability of an astronaut in space. They were, in the end, exceptionally happy with the work completed and the results obtained.

About half way through the Skylab 4 training program, the crew decided that they should take extra film supplies up to Skylab to support the Earth resources program. They requested additional briefings from experts across the globe on aspects of Earth phenomena, so that when asked a question the crew would be able to reply with more than: "Yeah, we saw it. Sure was pretty." Earth observations had occurred on every mission, but it was during Gemini that the first serious program was established. Further work was conducted during Apollo 7 and Apollo 9, but Skylab was the first program since then for which a sustained and dedicated program could be developed. This became the genesis of the larger program continued on Shuttle and today's ISS. Jerry wanted his crew to be "intelligent observers of Earth," when they were not doing other things. Program directors William Schneider and Kenneth Kleinknecht agreed to grant their request and provide extra film. A total of twenty experts, in fields such as seismology and earthquake fault zones, ice formation, desert formation, meteorology, land use, and oceanography, were each given two hours to brief the crew on what important features could be observed and photographed in their fields of interest. They lectured on how to recognize these features and how to record the most relevant data:

JC: "We enjoyed those forty hours of training. That turned out to be probably the most exciting and the most rewarding of all of the experiments that we did. We had the opportunity to ad lib, and to ad lib intelligently."

The extra film had been stowed aboard the workshop before it launched, so all three crews had an opportunity to participate in the ad lib photography activities. Unfortunately, a lot of the film was ruined when the OWS insulation was torn away during the launch phase. Luckily more than enough film had been stowed, and each crew brought up as much as they could, so they were able to overcome this setback and still obtain some spectacular and important results. Jerry believes that the Skylab astronauts were probably the "fathers" of the Earth Observation program carried out on the Shuttle and ISS.

Practicing recovery techniques

Most of the design of the Skylab configuration had been worked out prior to Jerry joining the program, when Skylab was termed the Apollo Applications Program (AAP). The astronauts assigned to AAP were instrumental in evaluating the crew's role in the planned flights, as well as working with the science community to define the research objectives and hardware. When Jerry arrived in the Skylab Astronaut Office, the training was already delineated into operational training on Skylab systems and science training with the experiments. The Principle Investigators (PI) would develop the experiment and work with the training staff, who would then define the basic training requirements for that experiment. This meant that the training protocol for the experiments was not available to the astronauts until after the operational systems training. Therefore, early in Skylab training, the astronauts concentrated on operating the systems of the orbital workshop, such as environmental, electrical, waste management, communications, and guidance and navigation. Initially, the crew worked with checklists defined by the training team from the sequence of events devised by the designers of the systems. As the crews trained, they modified the procedures to fit their style of operation. Then the trainers verified that they had not overlooked anything and certified the new procedure. Crew training on just one system or phase at a time was termed Part-Task Training. The integration of this disconnected training came much later, once the crew and Mission Control became more proficient.

Training continued on the ascent, rendezvous, docking, descent and recovery phases of the mission, using Apollo CSM simulators. In addition, a mock-up CM was placed in a water tank (in Building 30) for the crew to practice landing in a Stable 2 position (upside down, with the spacecraft's apex underwater) instead of the nominal Stable 1 (apex up).

With Jerry, Bill and Ed dressed in their pressure garments inside the sealed CM, a crane flipped the CM over to begin the training session. NASA had told Jerry that their families were invited to watch this particular exercise if they wished, which they did, filling up the viewing galleries. Inside the CM which had stabilized in the inverted position, a switch was thrown that was supposed to initiate the inflation of three large air bags on the nose which, at that time, was pointed at the bottom of the tank. These would change the center of buoyancy of the spacecraft and right it, allowing a nominal exit. However when the crew threw the switches, nothing happened. This was the failure that they were training for.

When the failure occurred, the crew knew they had to evacuate the spacecraft by swimming out through the tunnel. They first sealed their suits with rubber neck and wrist dams to prevent water rushing into the suit. They then had to open vent valves to allow water to enter the CM to a prescribed level. This could be compared to floating an empty jar neck down in the water. It will reach equilibrium when the trapped air inside provides enough buoyancy to keep it afloat. Once the trapped air in the CM brought it to equilibrium the crew was to open the tunnel hatch and then swim out and up the side of the spacecraft. It

The Skylab 4 crew appear before a gathering of news media during a press conference in December 1972

was back to the 'Dilbert Dunker' training again for Jerry!

With all their pre-evacuation procedures completed, the valve was opened and water began to enter the CM. Patiently waiting inside the CM for the water to reach the point where it was supposed to stop due to trapped air in the spacecraft, Jerry noted that the water kept on rising. After another 15 to 20 cm (3 to 4 inches), he quickly turned to Bill and Ed: "Guys, I think we'd better step up the pace a little bit, because something is wrong." All three made it safely out to the surface and into the rubber raft. Apparently, a small drain valve had been left open or had failed to close, allowing the trapped air to escape and resulting in excessive water intake. Thus the CM could sink to the bottom of the tank. The spacecraft was quite low in the water, and the crew noted that divers and support people were pretty excited when the CM did not stop sinking.

JC "I am not sure if our wives and kids knew what was going on, but we got out. That was probably the most exciting period of training that we had the whole time."

When the procedures checklists and mock-ups became available, the crew started working out the sequencing and timeline of each system or experiment, with trainers and support staff monitoring their procedures and activities. Though very rudimentary, this created a baseline from which to work in order to refine or detail each step or sequence required to activate, deactivate, or operate the hardware. Crew input was very important at these stages, to determine what could be done better, or in some cases what could not be achieved, so that amendments to flight procedures or hardware could be implemented in time. This would make for smoother operation once in orbit. Jerry recalls that experiment hardware configured as flight hardware did not become available until late 1972.

Even though there were crews named, it was not always the case that they trained together. With Apollo 14 through 17 still on the manifest, there was limited time available on the CM simulators, hardware was still being developed for Skylab training, and a lot of the procedures remained to be defined, especially the experiments.

For the first crew, a lot of time was spent in the CM simulator defining procedures for launch, rendezvous and docking, as well as entry and landing. The second crew, in the beginning, also helped Jerry's crew work on workshop systems and procedures development and experiment work. In the early days (1970-1971), crews were more fragmented. Often, Jerry recalls, Ed Gibson and Owen Garriott flew off to a university or a science laboratory to review the progress of a given experiment or to talk with a PI regarding the development of procedures for operating the experiment. The Skylab astronaut team worked as a team of nine or six, rather than as defined crews of three.

With everything seemingly being developed at the same time, it appeared to the astronauts that the whole program was unfolding as they progressed through their training. As a result, there was still a lot of traveling to do, although the crews tried to minimize this as much as possible.

With two Skylab crews to fly before him, Jerry found that the only way his crew could fit in simulator time was to fly to the Cape and use the CM simulator there whenever possible. With the Conrad crew taking priority on the simulator at JSC, the Bean crew would use the Cape simulator. Jerry's crew had to use any free simulator they could until Conrad's crew flew. This freed up one simulator (usually the Cape) until Bean's crew began their mission. Then Jerry's crew had full access to the one at JSC.

A lot of time was spent at the McDonnell facility in St. Louis where the Multiple Docking Adapter was being fabricated, along with the airlock facility. Work on the S-IVB workshop was handled at Huntington Beach, California. In Huntsville, Alabama, an engineering mock-up was available for astronaut review, which incorporated the latest updates and developments. One of the real concerns for the astronauts was that the mock-ups could get out of step with the flight hardware. It was worrisome that hardware and systems they would find in space might not be what they trained on. But both the second and third crews spent a lot of time ensuring this did not happen.

Daily flight plans

After 18 months of piecemeal training, "trying to soak up as much information as we could," as Jerry recalled, the first crew started training to daily routines in better mock ups. Initially, this was simulated system failures in part-task trainers. About 18 months prior to launch, the mission controllers become involved in training the crews and began planning a basic daily activity schedule based around rest, hygiene, work and housekeeping. These were called "mini sims," lasting from half to a full day.

Trainers had defined what they thought could be achieved during half a day, and the crew would arrive at 6 am (the planned Skylab wake up time). A set routine in the Skylab simulator was then followed up to lunch time. This allowed the bugs to be worked out of the procedures to determine if the planned timeline was achievable, or whether a conflict in activities could potentially cause a problem in orbit. Planned failures were not normally inserted into these mini sims, the idea being to allow the crews, controllers and trainers to determine how a basic working day would proceed nominally. Jerry's crew completed the majority of these sims, as they were the most available. It was especially useful for the rookie crew, as it gave them confidence in hardware and procedures that helped make up for their lack of actual flight experience.

Over 18 months, the third crew determined whether or not there was time to complete a particular activity before moving on to another task in a daily routine, turning the paper timelines into actual schedules. They ate breakfast in the 1-g trainer using Skylab food and habitation equipment, blending into a 'typical' flight day where the operational and experimental activities were time-lined. Activities such as learning to work the cameras, selecting the menus, and small procedures were slipped into later training sessions as the program developed. Towards the end of the third crew's training cycle, these mini sims lasted a whole day from breakfast to bedtime. The one thing that Jerry's crew could not factor in based on actual experience was the extra time it would take to do the same task while weightless. That lack of experience would "come back to bite them" later on orbit.

Integrated simulations were conducted in which the computers on the CM simulator were electronically locked into those at mission control, allowing the crews to practice countdown, launch, ascent, rendezvous, docking and entry into the OWS. They were able to fast track the long gaps between activities in order to focus on the key sequences of events. As these simulations progressed, failures and contingencies were included which forced the astronauts to learn to overcome difficulties and setbacks with equipment or procedures.

Jerry does not recall ever having to take work books home or having to pore over reams of paper to cram for the next day's activities. Skylab was very much a hands-on training program. It was a case of grabbing a checklist, following it to do the work and evaluating the outcome. In many cases prior to getting to the checklist stage, many hours had been spent in bench tests handling and looking at the equipment to see how it worked or fitted in with other items or procedures. The PI or designer, and an assigned training officer, were there to support questions and provide guidance for the astronauts so they could become familiar with a given item or experiment. The astronauts scribbled notes like mad as the session progressed.

There was a main training coordinator for the crew, who was the single point of contact. There was also a training officer for each item of equipment or experiment. These people, in conjunction with the designer or PI, were responsible for the development of each checklist which was to flow into the system.

Interestingly, unlike some of the Soviet preparations for long space missions, the Skylab crew never spent longer than a day in the simulator. In fact, Jerry's crew never spent the night in the 1-g mock up. The longest stay in the 1-g trainer was from 6 am to 10 pm following 'daytime' activities without evaluating the sleeping facilities or arrangements. In 1972, a 56-day ground simulation, called the Skylab Medical Experiment Altitude Test (SMEAT) was performed in a vacuum chamber at JSC, but this was primarily to gather baseline medical data. However, the three astronauts involved (Bob Crippen, Karol Bobko and Bill Thornton) were able to evaluate the habitability systems, exercise equipment and protocols during the simulation, which proved very beneficial to the later manned missions. In fact, Bill Thornton has stated that without SMEAT, the Skylab missions would probably have not been as successful as they were. The three crews who were to fly to the workshop received a daily report of the activities in the SMEAT simulation, and as often as possible, Jerry and the other Skylab flight astronauts would check in at the simulation in the 20-foot chamber to see how the SMEAT crew was doing.

There were no vacuum chamber tests on the OWS or CSM, except for some EVA suit verification work. CM training was accomplished in the CM simulator or 1-g mock-ups, not in the altitude chambers. As the training developed, so did the identification of problems, or things that might pose problems once in orbit. Generally, if the crew wanted something changed, it got changed, but the astronauts also realized they had to be reasonable about any changes. There was no point in creating a whole load of work over relatively minor issues. Any change suggested by one crew would have an effect on the others, so it was usually a group decision, as they tried to focus on efficiency rather than convenience.

A meeting was held once a week (normally Thursday or Friday) to discuss developments in the procedures, experiments, hardware and systems as required. With Apollo still underway and with planning in process for the joint mission with the Russians, Deke Slayton and Alan Shepard never really had much hands-on involvement in Skylab training activities. As Jerry recalled: "They were our bosses, and they depended upon on us to do our job, do it right and keep them informed." Every Monday morning at the weekly Astronaut Office meeting, the Skylab team would brief the rest of the astronaut group. These meetings usually lasted all morning. If a crew was close to (or had just completed) a mission they would report first, then each commander of the follow-up missions would update the status of their pending mission. Once the first and second Skylab crew had flown, there was little need for such meetings, as the training turned into operational experience. After the first and second crews came home, a crew debriefing sheet in an agreed format was followed. The crews had seen this prior to the mission, as it had been created by the training team, and it allowed them to focus their evaluations and comments effectively to the questions asked on the sheet. Shut away in a briefing room, the recently returned crew, the subsequent crews, plus Deke Slayton, Alan Shepard and JSC Skylab Director Ken Kleinknecht, went through the sequence of questions. The process covered the whole mission, in stages, from launch to rendezvous and docking, significant events during the mission, and for undocking and recovery.

Jerry found the briefing sheets from the first two missions useful as they were broken up into topics, covering subsystems, procedures and the experiments. For each system or procedure, a member of the crew was the spokesman, with additional comments coming from the other two crewmembers as required. This spokesman was also the prime person for the repair issues in flight on that item or experiment. Though each astronaut was certified in emergency or contingency procedures and operations sequences, if something went wrong, one of the three was usually more qualified to investigate a problem or repair than the other two. In the days before video conferencing, the meetings and briefings were usually taped by the audio branch of Media Services at Public Affairs and then transcribed for the record. The astronauts themselves usually chatted, making notes as necessary. But, as Jerry recalled, "just filing things away in the head was the normal method of retention."

There was no plan to completely simulate the whole mission. Most of the training program had focused around part-task training and then integrated mission simulations that lasted no more than one working day. Planning teams in the Mission Operations Directorate at Houston assigned Activity Officers to sketch out a basic 24-hour schedule (sleep/meals/hygiene/ experiment time/EVA) for each flight day. But real-time planning was essential as the mission progressed, and this, as Jerry and his crew soon discovered, caused its own problems; something which no training could prepare them for. The training coordinator for Skylab 4 was Bob Williams, a former Martin-Marietta employee. Other key figures in the training preparations for Jerry's mission were: Dick Creasy at Martin-Marietta, who handled reference documentation (including the Skylab Operations Handbook and experiment support documents); Hugh McNeese at Rockwell for CM training; Bob Kohler, the training coordinator for the Orbital Workshop; Stan Isom of McDonnell Douglas, who worked with Jerry's crew on the systems of the workshop; and Earl Thompson, also of Martin-Marietta, who worked on airlock issues with them.

Skylab flies

The Soviets had lost a second space station in a launch failure in July 1972, early enough in the mission that it never received a formal Salyut designation. Just weeks prior to the launch of Skylab, the next station, which *was* designated Salyut 2, had problems with its main engine system shortly after entering orbit. It re-entered a short time later. Salyut 2 was in fact the first military-orientated space station (Almaz), and further flights of that vehicle were put on hold until the cause of the failure had been determined. However, another non-military station was ready for launch in early May 1973, with the first crew planned for launch on May

The unmanned Skylab 1 OWS is launched from Pad A, LC 39 at KSC on
May 14, 1973 by the final Saturn V launch vehicle.

14 – the same day that NASA planned to launch Skylab and its first crew. This time the station's attitude control system failed before it could be called a 'Salyut' and it, too, re-entered a short time later, having never been occupied. With both the civilian and military versions of the Soviet space station in trouble, the way was clear for America to grab the headlines with its own space station. In May 1973, Jerry's home in space was ready for launch. He would not be visiting it for another six months, and two other crews would complete tours there first. It was assumed that the workshop would fare better than the Soviet ones. It did, but not without difficulties of its own.

On May 14, 1973, the final Saturn V left Earth carrying the Skylab Orbital Workshop aloft. Before the two stages had placed the workshop in orbit, there were signs that something had gone wrong. Shortly after the vehicle had reached the period of maximum dynamic pressure (known as Max Q), there were indications that the micrometeoroid shield had deployed prematurely. Then, once in orbit, it appeared that the twin solar arrays that provided electrical power to the workshop had not deployed fully. Skylab was in trouble even before a crew had left the ground. Jerry, Ed and Bill were at the Cape to witness what looked like a beautiful launch, but by the time they had returned to their quarters they were being told something was wrong. "We had another bad day at Black Rock, because that looked like it might be the end of the program," Jerry reflected in his 2000 oral history. Former astronaut and fellow Marine pilot John Glenn was at the Cape at the time and told Jerry to keep cool, as NASA would think of something to solve the problem. He was right. Over the next ten days, an ingenious recovery plan was devised in order to try to deploy the solar arrays (at that time it was not understood that one of the arrays had been completely ripped off the workshop during ascent) and to deploy a solar shield to thermally stabilize the vehicle and return it to an operational condition.

The prime, back up and support crews of the first mission pitched in to help solve the problem and develop procedures to deploy the solar wings and sun shields, and members of the second and third prime crews helped out wherever they could:

JC: "We lent support, and we went down to watch every once in a while to see what was going on. But if my memory serves me correctly, we didn't get involved too much in finding the solution, because we would have just been in the way."

After Conrad's crew completed repairs and finally settled in aboard Skylab, Jerry's crew's training picked up speed. In addition, they were very attentive to everything Conrad and his crew did on Skylab, following the daily activities. This continued with the Bean crew when they launched. In his 2000 NASA oral history, Jerry was asked if anything from the first two flights was learned that applied to the training program and planning for his own mission:

JC: "We drew conclusions from what we saw there. I think that the most important conclusion that we drew was that when the first crew came back after twenty-eight days, they were pretty wobbly, pretty weak. So the second and third crews decided to bump the exercise periods up, and we doubled the exercise from half an hour to an hour. Al Bean's crew went up, and exercised for an hour a day. It turns out that an hour didn't appear to be enough either, so we increased it again to an hour and a half for my crew.

"We watched the way the experiments were being done, and some procedures were modified based on what the first two crews had learned. One of the things that we noticed on the second crew was that they were really hauling it all the time. The rate of work, the rate of activity from them was extremely high. We began telling some of the managers that we didn't think that rate of work was going to be wise, over a ninety or eighty-four day period. We weren't sure we were going to be able to sustain that. We thought that the workload should be leveled off some and there should be more rest.

"Everybody agreed with that, and the experiments that were on the schedule were slowed down and spread out quite a bit. But unfortunately they added several new experiments at the last minute. We allowed ourselves to get trapped into a situation with those added experiments and all the new problems that would be associated with them that weren't taken into consideration. So when we got up there, the first thing we found was that we were again overcommitted, just like the first crew, but that we were going to have to sustain it for eighty-four days."

Countdown for the family

With Skylab repaired and two crews completing two highly successful missions of 28 and 59 days, the focus turned to the third crew and the final mission to the workshop. After over seven years in the wings and having lost a flight to the moon, Jerry was heading up the next crew to fly in space on a mission that was to challenge all three men. It would be immensely rewarding and personally gratifying, and would push the boundaries of both men and machine. It had been a long and hard road towards launch, but by the fall of 1973, Jerry, Ed and Bill were the next crew on the manifest. Meanwhile, in the background, Jerry's family had come to terms with his loss of a mission to the moon and with the realization of him finally being more than just an astronaut. Now, it was their husband or father that they would witness sitting on a rocket and being blasted off the Earth. And this mission was different. Jerry wouldn't be up there for a few days or a couple of weeks, but for up to three months. That would include Thanksgiving, Christmas and the New Year holidays as well.

The Carr formal family photo taken in October 1973. l to r are Jeff (15) Jennifer (18), Jerry, Jessica (9), JoAnn, Josh (9), Jamee (15) and John (11) The family pets Rags the dog and Priscilla the cat are also included

Jerry relaxes on the running board of the Astronaut Transfer Van during a visit to LC39B to view the Saturn 1B launch vehicle that would carry him and his crew into space. Beside him is a representation of the crew emblem. The patch features three main elements - a hydrogen atom symbolizing the sun or technology, a human form symbolizing mankind, and a tree symbolizing nature and the environment. The rainbow, a biblical symbol of reconciliation, surrounds the human and represents human responsibility to reconcile nature and technology. The patch was designed, approved and fabricated when Skylab 4 was designated Skylab III. Sometime later the NASA Administration decided to "clarify things" and re-designated the flights as SL-1(OWS), SL-2 (Skylab I), SL-3(Skylab II) and SL4 (Skylab III). This has been a source of confusion for years.

The three members of the Skylab 4 crew in the Part Task Trainer at the Mission Training and Simulation Facility at JSC. Jerry (on right) is seated at the simulator which represents the controls and displays console of the Apollo Telescope Mount. On the station, this was located in the Multiple Docking Adapter. Bill is working at the STS Control Station, while Ed is perusing the JOP (Joint Operations Procedures).

A couple of weeks before the launch of Skylab 4, JoAnn asked Hank Hartsfield, one of the Skylab CAPCOMS, to sit down with the children and tell them about their father's flight. She wanted them to know what to expect in every phase and what the dangers were, so that they could ask questions if they wished. When friends heard of this, JoAnn was asked why Jerry could not talk with them. JoAnn explained that the training schedule was so busy and intense that he simply did not have the time. Jeff Carr was in high school at the time and on the track team. Every week, a call came into the Carr house from the coach to see if Jerry could help out with the timing, but he never could. After several weeks of this, JoAnn 'volunteered' instead. She was instructed by the father of one of Jeff's best friends about what Jeff was doing, so that she could stand in for Jerry timing the events and practice sessions. "That just illustrates the fact that there was no Dad here," JoAnn recalled.

In those days, the sheer intensity of training was almost all-consuming. In Jerry's case, coming into NASA in 1966 meant two years intensive training, over two years supporting Apollo and then three years training for Skylab. In these seven years, although he was at home more than during his years in the Marines, he was essentially also away "in the space program." He was still, to a degree, separated from his family growing up, without being "away" like he was on military detachment in the Marines. However, though he was gone during the week most of the time, he was home nearly every weekend, and the family took vacations together as well. Jerry seldom brought work home with him and tried to give his family his full attention while he was there.

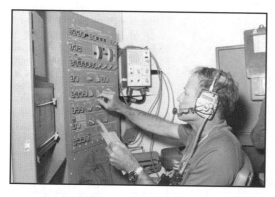

Jerry changes a dial on the control and display panel for the Earth Resources Experiment Package I in the 1G trainer mock-up of the MDA.

Science Pilot Ed Gibson enters a notation in the JOP while seated at the ATM control and display panel during a 1G simulation at JSC

Pilot Bill Pogue works in the Skylab 1G simulator in the Mission Training and Simulation Facility (Building 5) at JSC during Skylab 4 training.

When he was at work, NASA got his full attention.

In 2002, JoAnn reflected on what the astronaut's family went through during preparations for a flight. For the shorter duration Apollo flights, with milestones every couple of days, there were lots of diversions for the wives because friends gave them lunch, and there were interviews and trips out to dinner. For ten to fourteen days, it all went pretty fast. Watching from inside the astronaut wives' group, JoAnn was aware that the astronaut's children were pushed to the background (particularly by the media) simply because of the time constraints in those few days Dad was in space. She was determined that when Jerry's time came to fly (in Apollo at the time), their children would not be left out. She wanted them to experience the event and get as close as they could, and when the lunar flight was lost she continued that desire into Skylab.

In recent years, during some of the Apollo era wives' reunions, recollections of the family at the time of the mission were discussed. According to JoAnn, NASA at the time considered that "the wives were window dressing, and the kids didn't matter... just keep them quiet and out of the way." This meant several of the children of astronauts had a harder time growing up as teenagers and became attention seekers, not really handling the temptations of youth well. JoAnn agreed that the wives were making the situation worse by walling them off from their fathers as they prepared to fly in space.

"You have a husband who is getting ready to do a flight. Every flight has its own risks and its own dangers, and you do not want to be distracting your husband from learning everything he needs to learn. So when you have problems with the children, you don't tell him. What you are effectively doing is saying to the children 'You are not important to your father. You are not important enough so we are not going to tell him about this.' So we made choices for the guys without them ever knowing it – we isolated them from their children. Our intention was to protect them (the husbands) and insulate them, and in effect that was not exactly the same thing. So that's the way we lived with our kids. We insulated their fathers from them and some of the kids rebelled. My way of dealing with that was to bring our kids in as close as I could. If I had an interview to do, they were there with me."

Jennifer Carr recalled that at the time of losing Apollo 19 she owned a horse named 'Sugarfoot'. "I basically checked out of the

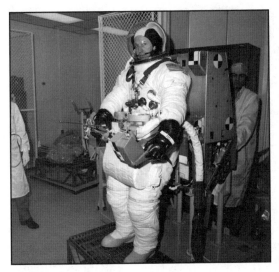

Jerry participates in a suited operation of the M509 manned maneuvering unit experiment during a hardware fit check at Kennedy Space Center in August 1972.

Ed Gibson adjusts a setting on the control box of the S190B camera, one of the components of the EREP. The single lens Earth Terrain Camera took five inch (13 cm) photographs

Jerry (center right) and his crew participate in egress training in the tank in Building 30 at JSC in September 1973. This is the one that almost sank.

family. I was always with that horse. I got him when I was in seventh grade, around 13 years old. After school I was at the barn, and at weekends – all weekend – Saturday morning to night and all day Sunday. So I was pretty oblivious to what was going on." When Jerry was assigned to Skylab, Jennifer admitted she was not looking forward to him sitting on top of a rocket and being blasted into space. "It was very scary that he was going to sit on all that fire power to thrust him up into the sky, but it was kind of exciting too because he had been waiting so long for this. It meant a lot to him. I don't remember feeling 'I don't want you to do that' or being afraid for his life. It was just a scary thing that he was going to do."

John Carr recalled that he was never "one of the kid's who bragged, 'my dad's an astronaut.' I didn't tell anybody because I didn't think it was the kind of stuff to brag about. I didn't mention it to anybody outside the space community, and I never really have." John never realized how much losing the lunar mission affected his Dad until years later: "I didn't recognize it at all until the last few years how upset he was. I just don't think I was old enough to realize what it meant at that point because when I was old enough Apollo was almost over. I'm almost glad it happened because the only thing on Apollo 19 that would have made him any more famous than he is now is if he had died on it. To me, what he did on Skylab is 100 times more than anything that happened on Apollo and I think it was a godsend that it happened. I was the outdoor type and I didn't start getting concerned until we started getting all the media attention. I wasn't really concerned at all. Dad was Dad, and I thought he was indestructible. I never thought about what would happen to him… To me, Dad was Superman, so there was no fear of an accident."

Jamee Carr remembered that having Dad as an astronaut was not worrying at first; he was just 'Dad.' Before and after his spaceflight people used to say to Jamee "Oh, your Dad's a hero," but for Jamee it was just his job. When the family moved to Houston in 1966, they moved into a community in which nearly everyone worked in the space program, "so we all just worked together and I didn't see any difference." She found the idea of 'sharing' him with the public difficult, as she "never saw him as anything other than *my* Dad. He was there pretty much for birthdays, Christmas and stuff like that… Yes he was gone a lot, but it was just part of his job." The prospect of Jerry flying into space on Skylab was "interesting" to Jamee, but this "was Dad and really Dad has always been a hero to me. I look up to my Dad because of the times I have

Jerry, fully suited enters CM 118 (Skylab 4 spacecraft) at the start of a high altitude chamber test at KSC in August 1973.

Bill Pogue (left) and Ed Gibson, both fully suited, prepare to take part in a high altitude chamber test at KSC in August 1973. Jerry is already inside the Skylab 4 CM (CM118)

seen him stop and assist someone in trouble. I saw him help some guy who had hit his head on a metal post and gashed himself, and Dad was there giving assistance immediately." When Jerry lost Apollo 19 and a flight to the moon, the risk of mishap did not hit Jamee until Jerry was to be launched to Skylab: "Jennifer and I had gone out to lunch, and we had some journalist shove a microphone in our faces and ask us how we felt about the risk of Dad's flight into space. I remember just bursting into tears. I was 15 at the time, and I got sick in my stomach that night with stress. I was so worried."

Jeff Carr remembered a family announcement that his Dad's lunar mission had been cancelled and that NASA was moving on to the next program, a very exciting program called space station. "I remember thinking 'That's only half way to the moon, what's the deal?' It struck me that it was an odd thing. Why would you just stop flying to the moon? It was just getting started, and it struck me as an odd thing to stop doing it so soon. But the fact that he was going to lead the first all-rookie crew to the first American space station and live in space... I thought that was pretty damn cool. I was not as disappointed about the cancellation of the moon missions, as I was very, very excited that my Dad was going to get his chance. My Dad was going to get to do something new. He was going to command a mission... *Command* a mission. I was very proud of him and I was very excited about the program."

Jeff recalled that his Dad had entered a pretty intensive training program. "It was frankly difficult for him to get anyone more than his crew (and family) intensively enthusiastic about what he was doing. But he brought me books, information, pictures and a model. He knew I was interested and I appreciated what he was doing. I was a little bit older at this point. Most of my friends whose Dads flew on Apollo were in Grade 6 or Grade 7, but I was glad I was older when Dad flew. I took an interest right off, and learned much about the program. I was grateful for the opportunities NASA gave us, like going to see the mock up or to watch him train."

There was also a realization of events outside of the home, Skylab or NASA, which Jeff recalled with frustration. "There was a lack of national interest in what Dad was about to do. I was struck by the contrast. The contrast of the program that was focused on exploration, discovery, achievement and excellence when our country was at the same time beginning to lose its trust, its faith in its government, as a result of failure of the Vietnam War. Based on the slow pace at which our country was coming to grips with the civil injustices that were brought to public view during the 1960s. Based on emerging corruption in our federal government leadership and the Watergate scandal. With all this going on I believed that there would be inspiration from Skylab, hopeful that this was a good thing. I struggled with why our country couldn't focus on what is good. My sense of frustration grew with the fading popularity of the space program, the loss of vision of our national leadership, of congress, and the presidency. It truly

Skylab 4 crewmen (l to r) Carr, Gibson, Pogue, fully suited inside the SL4 during high altitude pressure chamber test runs at KSC. In part this was for qualification for use as a possible Skylab 3 rescue vehicle as well as the SL4 flight vehicle.

JoAnn and the children, with their escort, tour some of the training facilities with other crewmember families. Here they look at the egress training tank located in Building 30 at JSC. Jerry was in the tank at the time.

began to crystallize and take shape as frustration just about the time when my Dad was aimed at his mission. I thought it was a wonderful opportunity for people to focus on something exciting, when there were so many negative things about ourselves, our country, our role in the world. That there was not more interest in Dad's mission was very frustrating to me. I was young. I was 13, and yet here I was being attentive to national issues."

The intensity of three years of preparation for Skylab, after four years preparing for Apollo lunar landings, certainly filled the diaries of Jerry and his colleagues. However, there were allowances for annual vacations and time off for illness. No astronaut works seven days a week or 52 weeks a year, though it may seem like it sometimes. Training intensity certainly increased during the final run up to the mission, but if Jerry wanted a vacation his application was submitted to Alan Shepard and considered accordingly.

With six children, it was always going to be a challenging event to organize a family vacation, but Jerry, with his military background, overcame the hurdle – though not always as planned, as his family recalled.

Jennifer: "We had a big old *Flexible* bus. It was not that big, but we would sleep eight people in there. We all had to go to bed in order to allow everyone to fit in. There were two bunks in the back and by putting a board in between we could sleep a couple more kids in there. A small board was placed over the sink to allow Jessica or Josh to sleep on that. The dining table could be put down and that made another little bed. A little couch pulled out for a bed for Mom and Dad."

These were memorable vacations, where Jerry often referred to his lists to make sure nothing was left out of the bus. There were plans for emergencies and lists for provisions. As Jennifer recalled: "Dad had his lists to follow. Everything had its place; everything was packed in a certain order and unpacked in a certain order." The bus, however, frequently broke down, and Jennifer recalls one memorable vacation where they all spent three days in Wyoming with the bus in the garage. After this, Jerry decided to get a small motorcycle to save him hitch hiking for help when the bus gave out again. On the next vacation, the bus gave up, and Jerry confidently unpacked his bike. It would not start, much to his frustration, because according to his lists he had checked it prior to departing home. So he still had to hitch hike for help, leaving the bike behind. As Jennifer recalled: "It drove him nuts, I'm sure, because he just had to know the way things went. Even in his garage everything was set up nicely so he knew where he could put his hands on things." Several of Jerry's children recall vividly the image of Jerry staring at the bike, baffled because he checked it again and again.

Jessica: "Dad was the man with a plan. Mom basically let us run wild in that she gave us so much freedom to be creative and to be whoever we wanted to be. I always felt that Dad wanted us to be more disciplined and have a plan. I wasn't close to Dad. During those years, I think I was uneasy because I did not understand what was going on. When we got away from home, things were better. I do remember family trips to Arkansas. I enjoyed those. We were out in the woods and it was just us. We had to communicate, we had to have fun with just each other. Dad seemed to be happier because he did not have to be an astronaut and tough guy. He didn't have to please anyone but himself, and he could just be the down-to-earth guy I think he really is."

Then there was the time when they vacationed in the Texas hill country. JoAnn noticed the Blanco River water level had risen since the previous day, and mentioned it to Jerry. When they returned from their walk the water was closer to the bus, which, at first, would not start. By the time the bus was started the river was washing debris around the vehicle, which had all four wheels in the water. Luckily Jerry managed to drive the bus out of the rising water and up to higher, safer land… another adventure on a Carr vacation successfully completed.

The family always used a system for assigning chores. Jamee recalled a huge wheel and on each spoke there was a job to be done by one of the six children whose names were placed around the rim. It would be turned every day and whatever spoke fell by their name, that was their job. Each member of the family cleaned their own room, and took turns to do other chores. There was a tall jar with which they could opt out by placing in their assigned chore of the day and taking one of their brothers' or sisters' chores from the jar instead. They all took turns to cook, do the laundry, set the table and do the dirty dishes. John, who disliked kitchen chores, always traded dish washing with his siblings. John: "I had to trade with my brothers or sisters for work in the garage or anything outside which was great. If I got the kitchen I would not want to do it, and I would pay my brothers and sisters to do it for me.

For all the family, the realization of what Jerry actually did for a living came in November 1973, and the start of his long awaited mission into space.

A very early contractor model, identified as "Apollo Applications Cluster Configuration". The *Apollo Applications Program* spawned the Skylab program.

Chapter 7

Skylab 4 - The First Month

"The technology developed for Apollo is certain to be used even after its main purpose. One plan envisions a space station assembled from Apollo modules and spent Saturn V upper stages which could keep from six to nine men in orbit for at least a year, with crew members and supplies being orbited every month or so by ferry vehicles." **Wernher von Braun, History of Rocketry and Space Travel, 1966.**

Eight years after von Braun wrote these words, the "space station assembled from Apollo modules and [a] Saturn V upper stage" was in orbit. It was called Skylab. Less grand than originally planned, Skylab had nonetheless survived setbacks encountered on the day of launch, and had hosted two three-man crews for 28 and 59 days. It was about to become the home of a third and final crew for up to three months.

Forty years after von Braun had predicted operating a six- to nine-man space station for up to a year, the concept still remains to be demonstrated. The Soviet Mir and the International Space Station have proven that routine space station operations are possible, and re-supply by ferry vehicles has been accomplished. But station crews have only comprised more than two or three for a few days, or a couple of weeks at most – usually during crew exchanges or visiting missions. Larger station crews are expected aboard ISS over the coming years, and the first step towards this was achieved in 1973 aboard Skylab. The first and second manned missions proved that man could live and operate effectively in space for between four and eight weeks, but a space station 'tour' in the future was being planned for between three and four months. One of Skylab 4's objectives was to see whether this really was feasible.

Between 1961 and 1972, most manned missions into space had been recorded in days, not weeks. Only three manned missions had exceeded 14-21 days and only two (the first two Skylab missions) had logged a month or more in orbit. The success of Skylab in increasing manned spaceflight endurance from three weeks in 1971 to three months in 1973/4 is often overlooked. At a time when spending a few days in space was still a challenge, the prospect of spending weeks working and truly *living* in space was a massive undertaking. Jerry and his crew were about to create space history, writing a new chapter in the annals of long term space exploration.

A new comet and a longer mission

Right from the beginning, in 1968, the third manned mission to Skylab was planned to last for 56 days, as with the second. However, two months prior to the launch of the unmanned workshop, celestial forces were to amend this long-standing plan. On March 7, 1973, Czechoslovakian astronomer Dr. Lubos Kohoutek was working at his Hamburg Observatory when he discovered a new comet that would subsequently bear his name. In the United States, Skylab officials at NASA became aware that there might be an opportunity to observe the comet towards the end of the year.

Using information supplied from other spacecraft studying the new comet, it was determined that the perihelion would occur around December 28, when the comet would fly closest to the Sun at a distance of only 13.4 million miles. NASA realized that a range of experiments and equipment aboard Skylab could be used to study the comet. However, according to the original flight plan for Skylab, the OWS would have been vacated well before the comet's closest approach. Without a crew aboard, this data would be lost. As a result, on August 16, it was decided to extend the third mission by an additional three weeks. This would take the flight into the new year and would give Jerry and his crew the chance to observe the passage of the comet. A tentative launch date was therefore set for November 9. This soon moved to November 11 and then back to November 10. Jerry was very pleased about this latter date, as it was the official birthday of the Marine Corps. The Commandant of the Marine Corps and other staff members would have been at the launch. Much to Jerry's disappointment, however, other events transpired that would delay the launch to November 16, missing the birthday of his parent service.

Cracked fins

In the 53 days between the undocking of the second resident crew and the arrival of Jerry, Bill and Ed, Skylab orbited the Earth unoccupied for the third time since its launch. Though there was no one aboard, Skylab was not totally inactive. A range of experiments continued to gather data. There were over 4,000

Skylab 4 is rolled out to Pad B LC39 on August 14, 1973.

SL4 crew pose in their pressure suits in early November in front of the Saturn 1B that will rocket them into space later that month

photos taken with the white-light and X-ray telescope, and over 560 hours of UV activity were completed. The Neutron Analysis experiment passively collected data on the impact and passage of low energy neutrons through the workshop.

Processing of the Skylab 4 hardware proceeded without incident until early November. (At the planning stage, there were at least four missions under Skylab. Skylab 1 was the unmanned workshop, which would be followed by the three manned missions, designated Skylab 2, 3, and 4 (Skylab 5 was proposed but cancelled). The astronaut crews were known as the first, second and third manned missions and the 'three' on Jerry's mission emblem reflects the third *manned* mission, not the third Skylab mission.) The launch vehicle for the mission was designated Saturn 1B number 208. The second stage (S-IVB) for the launcher had been at the Cape since November 1971, with both the Instrument Unit (IU) and the first stage (S-1B) arriving on dock in June 1973. When the leaking RCS system was discovered on the Skylab 3 CSM, the intended Skylab 4 launcher was prepared as a potential rescue vehicle. By August 14, the combined Apollo/Saturn 1B was on the Pad 39B in a holding position, in case it should be needed to bring home the Skylab 3 crew early. In the event this was not required, but to save time the Skylab 4 stack remained on the pad. It was put through the sequence of pre-flight tests and checks in preparation for the start of the formal launch countdown on November 8, and the launch which was due two days later.

Two days prior to the start of the countdown, a pre-launch management meeting at KSC was held. At the same time, final inspections were being held at the pad to verify the hardware and prepare to authorize the launch. However, reports from the pad revealed that cracks had been discovered in the aft attachment fittings on the fins at the base of the first stage. Similar cracks had been discovered on the SA-209 Saturn 1B in the Vehicle Assembly Building, which had led to the inspections on the launch vehicle intended for Skylab

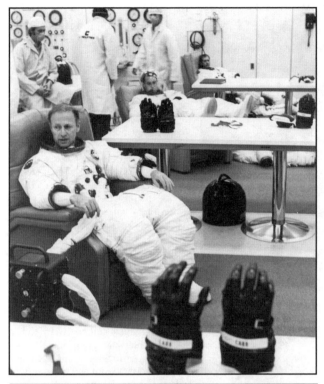

These two photos show Jerry and his crew completing pressure garment fit and pressure checks at KSC during early November 1973.

4. Cracks were found on all eight fins, with the largest measuring 1.5 inches. The decision was taken to replace all the fins, which delayed the launch to November 15, then to November 16.

On October 20, Jerry, Bill and Ed had begun their 21-day pre-flight quarantine and dietary intake control period, in preparation for a planned November 10 launch. With the postponement of the launch, this control period actually lasted for 27 days. From the start of the control period, the crew resided in a special set of trailers within a large hangar near the astronauts' gymnasium at JSC. Each crewmember had their own room inside the trailer, as well as a kitchen and a dietician who prepared their meals within the strict guidelines of the medical experiment program. Meetings with their families were allowed, but they were kept at a distance to avoid impinging the quarantine restrictions. The day prior to leaving Houston in their T-38s to fly to the Cape, news came through of the discovery of cracks on the launch vehicle fins, and that a delay of about a week would be required to fix the problem.

JC: "We just had to settle back and continue our Command Module training just to keep our skills warmed up. We went back to training and waited the extra time. The delay did not bother us at all that much, but it was a little irksome. We even began referring to the booster as 'Old Humpty Dumpty'."

The crew finally flew down to the Cape, arriving at Patrick Air Force Base on the afternoon of November 13, some twenty-four hours after the countdown for the final Skylab mission had begun. The following day, the astronauts completed proficiency flights in their T-38s. In those days, there were concerns over an astronaut's equilibrium, so they continued a series of flights that they had begun out of Ellington AFB near JSC, building up resistance by flying aerobatic maneuvers.

JC: "Whenever I was at the controls flying the acrobatics there was no problem. But when I was in

the back seat and Bill Pogue was flying it was a different matter, because I did not feel well. I turned green."

Their time was also occupied with final training sessions in the CM simulator at the Cape. This was mainly launch day activities and the launch sequencing.

Jerry and his crew also went out to the pad to take a look at their launch vehicle and spacecraft, while the final items (mainly film and data tapes) were being placed into CSM-118. There was also a package of 12 science demonstrations and a new camera to photograph the comet Kohoutek. The S-201 Far Ultraviolet Electrographic Camera was the former back up unit to the Far UV camera flown on Apollo 16. There was also a US Department of Agriculture payload (which comprised two vials each holding 500 gypsy moth eggs), plus an Airlock Module primary coolant servicing kit and a quantity of additional food supplies

A reflective look is on Jerry's face as he completes pressure checks of his suit and pre-breathing, prior to walking out to the transfer van which will take him and his crew to the Saturn 1B launch vehicle and their long-awaited launch into space

The management decision to extend the mission left the final duration open-ended, allowing for a

On their way. Jerry leads his crew to the transfer van as they depart the Manned Spacecraft Operations Building en route to the launch pad on November 16, 1973. Ed Gibson and Bill Pogue follow him.

At the 320-foot level, Jerry, assisted by a white room technician, prepares to enter the CM atop the mobile launcher on Pad 39B. Bill Pogue is on the right. Ed Gibson is already inside the CM.

Skylab 4 is launched at 09:01:23 am EST, Friday November 16, 1973. After serving as an astronaut for over 7 years, Jerry is finally on his way to space.

69-day mission if everything was nominal, but with a maximum duration of 84 days if possible. This meant that an additional 159 lb of food had to be squeezed into the CM, including over 59lbs of 'high density food' to supplement the supplies already on the station. In his 2000 oral history, science pilot Ed Gibson passed comment on this additional food supply:

EG: "The difficulty in staying up that long was that we had only enough food for 56 days, and we had too many experiments to take up in the Command Module. It was overloaded, so we agreed that every third day we would eat nothing but food bars. That was probably one of the most supreme sacrifices ever made for the space program. We had four of these little guys to munch. Breakfast consisted of four or five mouthfuls and that's all. But the idea worked and we stayed with it. It had all the minerals and calories you needed. It's not an ideal way to make it work, but it worked... but I still have a tough time looking a food bar in the face now."

On November 15 at Cape Canaveral, the crew were accompanied on a tour of the launch complex by their wives (who had arrived in time for the launch the following day). Then the three astronauts retired to quarters for their final night on Earth for three months.

Ascent to orbit

The crew had started to shift their Circadian clock several days prior to launch day, so that on November 15 they went to bed at about six in the evening to wake up at two or three in the morning.

JC: "I went to bed early that night, knowing full well I wouldn't sleep worth a hoot."

At 03:50 am CST on Friday, November 16, 1973, Skylab 4 crew training coordinator Elmer L. Taylor walked briskly through the corridors of the astronaut quarantine building, located on a security site of the NASA Florida launch complex. He opened a door, knocking gently as he put his

The spent S-IVB stage that placed Jerry and his crew into orbit is photographed from the Skylab 4 CM shortly after the start of the mission's orbital phase on November 16, 1973

head round the door: "Jerry, it's time to get up now. Your spacecraft is ready and waiting for you." Jerry had managed to drop off to sleep, but woke with a start and got up.

After seven and a half years wait, it was time. As he rose, Jerry felt an air of excitement. Today was the day he would be launched into space. Since turning in for the night, he had indeed had little sleep, waking three or four times wondering whether it was time to go. Now that time had come and he was ready.

Following a quick shower, Jerry put on a pair of slacks and a T-shirt and took a quick stroll to the medical examination room for a last minute check up. Here Jerry, Ed and Bill were given the once-over and pronounced fit for flight – a welcome relief to all three. One part of the physical was microbiology scrubs:

JC: "They took swabs from many parts of our bodies to find out what kind of flora and fauna were living on us, and they catalogued that. It was part of a long term experiment. They wanted to know how much of it we would leave on the spacecraft and if we would pick up anything from crews ahead of us." (At the end of the mission, when the crew got back, another micro scrub was conducted and found that they had, indeed, brought back some new micro organisms. They had been carried up by the second crew.)

Passing the final medical is something of a formality – if anything serious were to have shown up, they would have been taken off the flight well before now. Training accidents and injuries had cost several astronauts and cosmonauts their seats over the years, however, and it had fallen upon the back-up crew to replace a single astronaut, or in some cases, whole crews.

The most well-known incident of this type had been the case of Ken Mattingly in 1970, who lost a seat on the ill-fated Apollo 13 mission to the Moon to Jack Swigert 72 hours before launch. He had been exposed to German Measles and it was deemed too risky for him to fly in case he developed the disease during the flight (he never did catch the bug). The launch day medical may be a formality, but passing it was an important and often overlooked milestone on the road to space.

After the medical, Jerry and his colleagues went across the hall for breakfast. Astronaut breakfast was the first of a series of traditional pre-flight rituals. With the crew were Deke Slayton and Alan Shepard, along with Bill Schneider, the Skylab Program Director from NASA Headquarters in Washington, and Ken Kleinknecht, from the Skylab Program Office at JSC.

The astronaut breakfast of steak and scrambled eggs had been a tradition since the early days of the Mercury astronauts, no matter what the time of day. In those days, the dieticians thought that the best thing for astronauts was a large portion of steak and eggs, but in subsequent years Jerry modified his diet to reduce or eliminate both of these items, mainly due to their cholesterol and fat content.

Jerry and his crew were at their own breakfast table, idly chatting about anything but the mission ahead – they were all talked out on that subject! Following breakfast, they moved to the suiting room to get dressed

Home in sight. The forward docking port on the Multiple Docking Adapter on the Skylab OWS is visible as Jerry brings in the Apollo CSM for docking. Also visible is the thermal shielding installed during the first two missions, the sole remaining solar wing and the lower part of the ATM.

for their big day. It was 05:05 am as Jerry settled down in a comfortable recliner to be assisted into his spacesuit by technician Troy Stewart. Each of the crew had their own suit technician, and Troy was one of the best, keeping Jerry very relaxed as they worked. Overseeing the operation was leading suit technician, Joe Schmidt, who had participated in the suiting of every American to leave the launch pad since Alan Shepard in 1961. Shepard himself was also in the room, along with Deke Slayton, both quietly observing the suiting process.

The Skylab Pressure Garment Assembly (SPGA) was a development from the Apollo lunar suits worn since 1968, and was in effect a mini spacecraft in itself. The major differences were in the need for mobility and comfort in the planned EVAs during the mission rather than for walking on the Moon, and that the Skylab suits were connected to umbilicals, rather than backpacks.

To begin with, biomedical sensors were attached to shaved areas of the astronaut's body, to record medical data during launch. This was followed by the Liquid Coolant Garment (LCG) – basically a suit of "long handled underwear" – with fine tubing enclosed in mesh like a spider web all over the body. Warm or cold water could be pumped across the body to provide warmth or cooling and then returned to the spacecraft's systems for temperature regulation. For launch, each man carried a portable recirculation system, which would help keep him cool until he was connected to the spacecraft's systems on the pad.

Attached to the astronaut under the LCG was the Urine Collection and Transfer Assembly (UCTA), for the astronaut to relieve themselves while wearing the suit. Long delays on the launch pad in the early days of the program caused some concern if the astronaut needed to urinate. When this happened, with the astronaut prone on his back in launch position, urine would pool in the small of the back, seep through the suit and possibly short circuit some of the instrumentation. Indeed, his need to 'go' during one such delay earned America's first astronaut Alan Shepard the title of 'Wet Back' as he waited for launch. For the Shuttle, the astronauts are only allowed to remain on their backs for up to three hours during a launch delay. Any longer, and they are taken out of the vehicle again. This is for general comfort as well as to prevent any modern day 'wet backs.' For all male crews in the early days of the program, a condom-like sheath was used to transfer liquid into the UCTA, but several astronauts found that they were not elastic enough, were either too tight or too loose, and had a tendency to slip off!

On top of the LCG, the astronauts wore an Apollo A7LB Pressure Garment Assembly (PGA), which was a made-to-measure one-piece garment made from multiple layers of fabric. It was designed to provide an oxygen environment for breathing, suit ventilation and pressurization. Electricity connections were provided for biomedical instrumentation and communications. There were fewer layers to the suit than for Apollo, as less thermal protection was required in Earth orbit than on the Moon.

The suit had gloves locked into rings on the wrists, and a clear Plexiglas helmet that locked to a neck ring, allowing a flow of oxygen around the wearer's head. Each astronaut also wore a soft communications

The three 'crewmembers' await the new residents of Skylab. They were left by the departing second crew. Their poses reproduce the 'see no evil, hear no evil and speak no evil' saying.

A Skylab 4 image of a full moon from Earth orbit, a target Jerry should have reached with Apollo 19.

cap, called a 'Snoopy Cap', containing earphones and microphones and resembling a World War I flying cap. There was no need for the heavy lunar boot on Skylab, so the boot assembly had a rigid sole. It was designed to interface with the foot restraint devices outside the spacecraft in order to provide a firmer grip. These suits were to be used on each of their EVAs and would be located inside the station between spacewalks for the duration of their stay.

For many astronauts, their spaceflight starts when they snap on the helmet of their suit and, inside their own cocoon, are separated from the rest of the world. All that you can hear is the rush of air being pumped around the suit, your own breathing and, some have said, your own heartbeat. As excited as he was, Jerry still managed to drop off to sleep for a while during the long periods of waiting while suiting up.

The improved suit also included more provision for mobility and easier access into and out of the suit, since they were to be worn several times during the mission. As the final elements of the suits were fitted, the technicians added small personal mementoes which each astronaut would carry in suit pockets, such as pens, flashlights, scissors and so forth. Jerry also wore an extra watch, around his ankle, for the launch – a self-winding Movado mechanical watch. He was curious to see if the weightless environment would have any effect on the working of the watch (which, of course, it didn't, since it was wound by the motion of the body and limbs).

Jerry and his crew were very impressed with the way the suiting team completed their tasks – "That's the way to send a crew out to the pad," he commented later. Once suited, the astronauts had to sit and pre-breathe pure oxygen for 30 minutes to cleanse nitrogen from the blood and prevent the 'bends.' To do this, the astronauts lie quietly in the suits for 45 minutes or so with the lights turned down, to allow them to catch a few minutes sleep if they wished. Thankfully, there was no piped music coming through the communication link!

At last it was time to go, and after kicking the recliner leg support down, the three were given a helping hand to stand up. The suits were not pressurized, so it was fairly easy to walk down the corridor to the transfer van. As the astronauts moved down the hall, they waved at several members of staff who came out to see them off, continuing pre-launch traditions. Among them was Silvia Salinas who had been Jerry's Astronaut Office secretary and the Skylab office secretary at JSC for the past 8 years. As they walked past, each man gave her a brief hug. Jerry recalled that this was one of the more moving experiences of his time at NASA, seeing all those people genuinely wishing them well and to have a safe mission. Carrying their portable oxygen generators, they emerged into the glare of press lights and walked straight into the van with nothing but a wave to the crowd. This was the penultimate crew to go through this ritual for some time and the atmosphere was charged with nostalgia. At least they did not have to follow the Soviet cosmonaut tradition of urinating against the wheels of the transfer bus before they boarded it!

Inside the van with the crew was Charlie Buckley, Chief of Security at KSC. He was originally from Boston and a great friend of the astronauts. The ride to the pad three and a half miles away was long and eerie. It seemed painfully slow, especially as all three astronauts had driven a little faster than they would have done on a normal highway when visiting the pad during training. The launch vehicle was visible all the way, and at last Jerry started to believe they were actually going to do this for real! The lack of activity around the vehicle struck all three astronauts, who were used to seeing people everywhere working on the vehicle. This time, there was almost no one – just the closeout crew, the three astronauts and those on the van.

The waiting launch vehicle was poised atop the 'milk-stool' launch pedestal, which enabled the smaller Saturn 1B to be serviced and accessed at the same upper levels as the larger Saturn V used in the Moon program. Crew access to the structure surrounding the capsule, called the White Room, was at the end of the swing arm, where technicians were fussing over the final elements of the checklist prior to the arrival of the crew.

As the three men exited the transfer van and headed up in the elevator some 320 feet (97m), they passed the two stages of the launch vehicle that was to boost them into orbit. The first stage of the Saturn 1B was 21 feet (6.6m) in diameter, 80 feet (24.4m) long and, when fully fuelled with 66,000 gallons (249,810 liters) of Liquid Oxygen (LO) and 41,500 gallons (157,077 liters) of Kerosene (RP-1), weighed 997,000 lb (452,240 kg). The resulting thrust from the eight H-1 engines was 1,640,000 lbs.

Stage two was the S-IVB, which was used as stage three of the Saturn V Moon rocket. This was again 21 feet (6.6m) in diameter, but was only 58 feet (17.8m) long. With 20,000 gallons (75,700 liters) of LO and 64,000 gallons (242,240 liters) of Liquid Hydrogen (LH), the stage weighed 253,000 lb (114,750 kg). The single J-2 engine gave a thrust of 225,000 lbs. This was the same type of stage as the one that had been adapted from the lunar program to form the Skylab orbital workshop.

On top of this stage was the instrument unit which controlled the Saturn V during the boosted phase. Measuring 21 feet (6.6 m) in diameter, it was 3 feet (0.9 m) high and supported the Spacecraft Lunar Adapter (SLA) which had housed the Lunar Module (LM) on Apollo missions, but which remained empty on Saturn 1B missions. It tapered from 21 feet (6.6 m) at the base to 12.5 feet (3.8 m) at the top and was 28 feet (8.5 m) long. It weighed 4,300 lb (1,952 kg) and consisted of four petal-like panels that would spring free to separate the spacecraft from the spent stage in orbit.

Above the SLA was the Apollo spacecraft, the first element being the Service Module (SM) that housed the major propulsion and consumable elements for the flight to and from the Skylab station. It was 24.5 feet (7.5 m) high and 13 feet (4 m) in diameter, weighing 51,243 lbs (23,243 kg). On top of this was the Command Module in which the three astronauts would ride. This measured 11.5 feet (3.5 m) high and 13 feet (4 m) across the base. Aside from the three crew positions, the module contained controls for maneuvering the vehicle in space, control jets, the docking equipment to link to the Skylab station, and the recovery heat shield and parachutes used at the end of the mission. The 33 foot (10 m) tall, 4 foot (1.2 m) diameter escape tower topped the stack and would be used in the event of a pad abort or during the first stages of the launch. It weighed 8,910 lbs (4,041 kg). The whole stack topped 224 feet (68.3 m) and fully fuelled weighed 1,300,000 lb (589,680 kg).

As the crew exited the elevator and walked across the swing arm to the White Room, there was an obvious sense of calm around the pad, but with an ever-increasing feeling of detachment. Now, there were just the three of them and the launch closeout crew in the White Room. Everyone else was at least 3 miles (4.8 km) away. Ed Gibson recalled the twenty minute wait on the gantry walkway as Jerry and Bill were strapped into the CM. "I was the last one in the vehicle because I had the center seat. So I had a chance just to stand outside and looked at the vehicle. At that time it was still being fueled, and was creaking and groaning because of the cold fuel and the shrinking of the metal, and because of the weight. It started to come alive; the electrical side of it was working, unlike what we had seen before, where it was just a passive hunk of metal. Now it seemed like it had a life of its own. It was a very exciting time. Most of the time you're busy and you don't have time to reflect. But I had around twenty minutes there where I could just sit back and watch. To this day I really just felt lucky."

Continuing the traditions of launch, Jerry carried a bag of American Sour Balls (tart sweets) for the Pad Leader Guenter Wendt, which Jerry passed to him in the White Room. Guenter Wendt, nicknamed the 'fuehrer' of the launch pad, was a former employee of McDonnell Douglas, the contractors for the Mercury and Gemini spacecraft, and of Rockwell International, the contractors of the Apollo spacecraft. Wendt had served as Pad Leader since Glenn's flight on Mercury. One of the reasons for Guenter's nickname was his reputation of running the pad crew with an iron hand. In 1968, as the Saturn 1B carrying Apollo 7 lifted off on the first manned Apollo test flight in Earth orbit, astronaut Donn Eisele quipped; 'I vunder vere Guenter Vendt?' placing the Pad Leader in the history books as one of the characters of the Apollo era. He was still there to bid farewell to the Skylab 4 crew. As the three astronauts moved into the White Room to prepare for the mission, they saw a picture of the latest *Playboy Playmate* taped to the wall, with the inscription 'Keep This in Mind.'

Support crewmember and astronaut Hank Hartsfield had been in the spacecraft for some hours configuring the switches in the CM for the launch profile, and would assist the crew into their couches. To get into the spacecraft required help from both inside and outside the CM. Jerry entered the CM first. Grabbing a handrail above the hatch, he hoisted his feet into the spacecraft and, guided by Hartsfield in the lower equipment bay, eased himself into the left hand couch. Then Pogue slid into the right hand seat, followed by Gibson into the center. Checking that there was no snagging of the pressure garments, the three astronauts were strapped securely into their couches. The final restraint was achieved by one of the ground crew placing a foot on the astronaut's shoulder and yanking the strap tight. Connections to the spacecraft's environmental control system pumped oxygen through the suits and allowed the crew to disconnect from the hand held carriers they had used since the suiting room.

Then followed a period of spacecraft and communication checks between the astronauts, Launch Control at the Cape, and Mission Control in Houston. Communications between the crew prior to launch were handled by fellow astronaut Bob Crippen acting as 'Stoney' in the 'blockhouse' at the Cape. In the early days of ballistic missile development, those who launched the vehicles were stationed in concrete blockhouses close to the launch pad. In the event of a pad explosion, these stone and concrete bunkers afforded protection to those inside. These were used in the manned program up to the mid 1960s and a fellow astronaut would be stationed inside to communicate to the flight crew on the pad. By the time Jerry and his crew were ready to leave the launch pad, the 'blockhouses' had evolved into more sophisticated computer-filled launch control centers some distance from the pad. The names derived from the 'steely eyed missile man' era stuck, however, as did the identification of the person who talked with the crew from a stone blockhouse.

It was now time for Hartsfield to leave the spacecraft and for the ground crew to close the hatch. The normal way for the support crewman to leave the spacecraft once all three astronauts were in place was under the central seat (Gibson's) and up between the headrest and the lip of the hatch. But on this mission, there was so much extra gear stowed that Hartsfield had to crawl over the top of Gibson – who made several glib remarks as he clawed over him and out of the spacecraft. This was not a big problem, although there was some concern over the possibility of throwing or breaking a switch as he passed the control panels.

Following more checks and farewells, Guenter Wendt finally shook hands with all three men before the hatch was closed and locked. The crew later commented on the hatch labeling. The words 'LATCH' and 'UNLATCH' for both the gearbox and handle confused them since, from their position, this had to be viewed up above them. They suggested that perhaps in future a simpler OPEN/CLOSE would suffice. The hatch had been redesigned following the fatal Apollo 1 pad fire in January 1967 to open much quicker than on the original spacecraft. The original design had consisted of two hatches and took 90 seconds to open. The new one was a single hatch capable of opening in 30 seconds.

There was still an hour or so before the pad crew vacated the pad as they continued with further checks and rechecks. As the Skylab 4 crew waited for the final minutes, they looked at each other and "grinned like Cheshire cats." Jerry managed to lean forward a little against the harness and looked over to his friends and colleagues. He recalled that all three were "giggling like schoolgirls." They had been waiting years for this moment and they were determined to enjoy it.

JC: "I remember saying something like, 'I can't believe it's really going to happen'."

During the launch delay caused by the cracked fins, when the crew started calling the vehicle "old Humpty Dumpy," they didn't realize that it would come back to haunt them. They were joking, but the press had heard about it and reported the crew's comments. To those working on the vehicle around the clock to try to meet the launch day, the comment was not well received, but was not mentioned almost up to the launch. At 20 minutes prior to launch, however, the crew received a message from the launch team: "Good luck and God speed from all the king's horses and all the king's men!"

After all the training and simulations, this time the vehicle felt more alive, as they could now hear the pumps and systems working as countdown proceeded. As the White Room crew departed and the White Room was swung back, Jerry noticed a beautiful blue patch of sky out of the side window as dawn approached the Cape. With an amazing sense of calm and a little feeling of detachment, they began to realize that they were alone on top of the 'stack' waiting to be blasted into space. All three still found it hard to believe that they were finally about to do the job for which they had prepared for what seemed like a lifetime.

Bill Pogue (BP): "Ah! Doggoned shoulder harness ."
(The harness restricts Bill's movement in his couch, being so tight.)
JC: "OK, we've got nothing to do except pick our noses. But the helmet's in the way."

As ground control passed on best wishes and God speed for a good flight, the crew continued their program of switch setting and display checks. In front of each man was a myriad of computer technology that was state of the art in the early 1970s. In front of Jerry on the left hand side were the flight control panels. In the central position in front of Gibson were the caution and warning controls, the environmental control systems and cryogenics, and in front of Pogue on the right were the controls for the Service Propulsion System in the back of the SM, electrical power and more environmental controls. Each man had prime responsibility for the panels in front of him, but they were also cross-trained and the other panels were within easy reach. It was very much a team effort to fly an Apollo CM, but all of them were trained to fly solo if called upon to do so. The crew noted that most instruments were accessible, but in a suited, pressurized condition, others were not. Some displays were difficult to read due to the suits or were out of direct line of sight. It was suggested that in future crew station designs these areas would be addressed.

In the seconds before launch, Jerry felt the SM engine bell "swing its tail" when the gimbal motors worked. The sounds of pressurization and the movement of fluids attracted their attention. It was 09:01:23 am EST on November 16, 1973, and the mission of Skylab 4 was about to begin.

Ed Gibson (EG): "Clock, start."
Jerry Carr (JC): "Roger, Tower clear. We're getting a roll program (and a) smooth ride."
Capcom Dick Truly (CC): "Roger, Jerry, and thrust looks good on all engines. Skylab, Houston, your feet wet, and everything is looking real good."

(Feet wet meant that they were over water, and in the event of an abort they would land in the Atlantic Ocean.)

With Jerry's hand on the ABORT handle, "just in case," the crew felt rather than heard the ignition of the eight first stage engines, but were told of "Ignition" by ground control. As the vehicle built up thrust, it was restrained by hold-down clamps until the appointed moment, then released to leap from the pad. Unlike the impression of astronauts who had ridden on the Saturn V, which felt like a slow lumbering ride, the Skylab 4 crew felt that the Saturn 1B "really hauled the mail," as it climbed and burned its fuel. The immediate impression was the "eyeballs-in" feeling of being pressed against the seat as they rose and gathered speed. "Here we go," all three shouted in unison. All through the ascent, the crew was busy setting switches in preparation for updating the controls in the event of an abort, as the computers flew the ascent after clearing the launch tower. The ascent smoothed out a little but the whole ride was, "as rough as a cob," to Jerry and his crew. The Saturn 1B was a "rock and roll ride," similar to the Titan II that propelled the Gemini spacecraft, as opposed to the heavier, slower Saturn V. Jerry's heart rate reached 104 during the ascent to orbit.

Ed Gibson recalls the event as like being inside a very familiar room (the CM) on top of a very tall building (the Saturn) when suddenly the bottom of the building explodes. "Intellectually you know what's going on, but those images flash through your mind. You're thinking, 'Oh my God, this building I am in is shaking and rumbling.' It's like being in an earthquake where the world underneath you is no long stable. But we know, of course, what's going on."

With no time to savor the ride, they were aware of the raw, rough power beneath them as they climbed higher. Described by many crews as a "great train wreck," the time for staging approached as the propellants of the first stage depleted and the vehicle began carrying excess weight. Dumping the unwanted and empty stage and igniting the upper stage was a design feature of the multi-stage rocket, so that it carried only the usable portion into space and the unwanted stage fell to destruction in the atmosphere or into the ocean. Though it could not be recovered for reuse, this was the best technology available at the time. The crew did think that the call from the CAPCOM of 'GO' or 'NO GO' for staging was somewhat unnecessary, as it would happen automatically whether they were ready or not. [NB: CAPCOM: meaning capsule communicator. In the early days of the program, the astronauts flew in CAPsules (although the astronauts disliked that term – to them, they were pilots flying a spacecraft as they would an aircraft). Those who talked to the astronauts from Mission Control (always fellow astronauts) were COMmunicators. Capsule communicators became CAPCOMS, a term still used in the Shuttle program today].

The crew could hear the clattering and banging behind them as the stage began the separation sequence. The first stage engines shut off as commanded. and with the separation command, the crew was slung forward in their seats into the harness straps. The single J2 engine of the S-IVB stage subsequently ignited and they were slammed into their seats again to continue the powered ride. As Jerry put it in his diary, "… staging… HOLY COW!"

> CC: "Skylab, Houston. You're looking good. You're GO for staging. The thrust looks good on the S-IVB."
> JC: "Roger. It looks good here too. Tower Jet [tower jettison] smooth as glass Houston."
> CC: "Roger, we had a beautiful picture on the TV all the way. It looks real good. You're GO at 5 minutes. You're right on the money."

The second stage was a much smoother ride for the crew. Then, as the escape tower jettisoned (it was now not needed as the crew could 'abort to orbit' or, if required, complete a ballistic entry in an emergency), a bright light entered the windows on Jerry's side. It temporarily blinded him and obscured his view of his instrument panel. However, on Pogue's side, the capsule was in complete darkness. As Jerry raised his arm to shade his eyes, he felt it weighed about 40 lbs (18 kg). Glancing at his crew mates, he saw that Gibson was just "wide eyed with excitement," and Pogue was "all teeth," grinning from ear to ear.

> CC: "Skylab, Houston. We're predicting cut-off at 9 [min] plus 37 [sec]. We confirm you're GO for orbit. You're looking real good."
> JC: "Roger, cut off at 37. OK, Houston, we're looking at a 120.7 [mile] by 83.3 [mile] orbit. This is really great Dick."
> CC: "I envy you."

The trip from selection on April 4, 1966 to the launch pad on November 16, 1973 had taken Jerry seven years, seven months and 12 days. The trip from the pad to orbit had taken 9 minutes and 37 seconds. As Skylab 4 started its first orbit around Earth, the unmanned workshop was on its 2,685th revolution around the planet. When the second stage shut down at 11 minutes and 20 seconds, it was not immediately separated from the CSM. They entered zero-g, and promptly found lots of items floating in the cabin – screws, washers, pencils. This was the signal that they were really in space at last. Twenty-four minutes and fifty-nine seconds into the mission, the CSM separated from the stage, and Jerry maneuvered the CSM to observe the spent stage. Panels open wide like a metal flower, it soon arced to a re-entry and destruction over the ocean.

> JC: "OK Houston, we're ready for SEP, and the bird's (the S-IVB) just coming into sight. The SLA panels are all open, the deployment looks normal."

Views from the family

In the days prior to launch, Jerry received news from the family that lots of friends were calling home wishing the crew luck. Though he was in quarantine, Jerry kept track of what the family was doing as they left their home in Houston to fly down to the Cape for the launch. Once they had checked into the Holiday Inn, they also kept him informed of the activities that were going on – well, some of them at least.

JC: "A large group of family friends from Houston rented a bus, drove it to the Cape and had a fine time. The kids had a lot of fun there. They told me stories about some of the shenanigans they pulled at restaurants and things which they knew they couldn't get away with when Mom and Dad were around, but were able to when somebody else was with them."

JoAnn recalled the activities from the family side: "Jerry had to leave the house a month before his flight day. I could go out to JSC and see him but the kids couldn't. We stopped by there on Sundays after church, and he could come to the door (of the quarantine quarters) and we would all wave (from the car). Then Jerry's launch slipped. Once, we even got into limousines to be taken to the airport to fly to the Cape, but we were brought back home again because there were cracks in the fins, which was disappointing. We had a lot of friends who were primed to go to the launch who could not do it because they had already made plans they could not change." Some friends stayed longer than planned down at the Cape to be there with JoAnn when the launch occurred, however. JoAnn agreed she should have had a checklist with her; "Jerry's the one who always has a checklist. I sort of freewheel." When the family arrived at the Cape, a NASA representative was on hand to monitor their needs and safety. The Carr's were looked after by Protocol man, "Robby" Robinson. After a few days, JoAnn asked Robinson what he thought when he learned he would be looking after her and six children. Tactfully, he replied, "Well I heard you were difficult." JoAnn explained that she was not 'difficult' but simply outspoken and said what she thought. The children were well behaved and caused him no problems, and he finished the assignment very pleased that it had gone so well.

The family was assigned a limousine for use at the Cape, but JoAnn would not ride in it and instead obtained a convertible so that they could ride around and see the sights. The children got to visit the pad and went up to the White Room to see inside the capsule in which their dad was about to be launched into space. This was an overwhelming experience for JoAnn, "It was something we had waited for so long for, and I was really on a cloud of air. I wasn't worried; I didn't feel concerned. I was just euphoric that this was finally going to happen... a sense of relief. Down at the Cape, Robinson and other NASA security staff took the children (as well as the Gibson's and the Pogue's children) down to Disneyland when the flight was delayed. With extended family there too, they had a great time on the beach and in the pool. Jennifer Carr laughingly recalled the experience: "We got to Disneyland and then the ten kids went in all different directions. The NASA security guys were saying, 'Uh, oh we've lost them already!'"

The evening before launch, JoAnn was beginning to feel the tension. That evening, there was a Manned Spaceflight Awareness Party at which selected hard-working people from the factories that built the spacecraft and launch vehicle were honored. These honorees, had a chance to meet with workers at the Cape and the wives of the prime crew. Annie and John Glenn were guests of the Carr's for the launch and Annie accompanied JoAnn to that party, and also to a smaller one close to the Holiday Inn (with Jerry's fraternity brothers, colleagues from the Marine Corps, relatives and close friends). Film star John Wayne, whom they had met at a USC alumni function in Houston, was invited, but was unable to attend the launch. It was a fun party, followed by a trip to see the launch vehicle all lit up at night, a couple of miles from the pad. JoAnn recalled the experience years later with vivid clarity: "To see the vehicle at night when it's all lit with spotlights, it's just incredible... emotional. A beautifully pristine piece of machinery that is ready to go. It's an overwhelming sight. It renders you speechless. It was out there in all its glory. It was like it was carved out of ice. When we got back to the hotel, the kids were all in bed asleep and I could relax. So I sat out by the pool with a friend until 3 or 4 in the morning. Finally, I tried to get some sleep, but we had to get up early to get on the buses to get out there (for the launch)."

"So I got to see him launched and that [experience] is really hard to even tell you. All that I can tell you is that we were standing there across a marshy area, and before you hear anything you see the grass start to blow down and the birds start to fly up. And then the sound reaches you. It's the sound that gets you because it resonates and it moves you, literally. I was holding on to the top of my son's head because I thought I was going to go off the ground with them if I didn't. And then it happened so fast I thought, 'Oh my God, I want to see them do it again'. It was too fast. It was over. This little thing we had waited for so long for was over so quickly. The kids came back to Houston with my sister that day, but I stayed down at the Cape for another night because I did not want to be in the air when they docked. I wanted to be where I could hear all that was going on." Later, back at the hotel, JoAnn was sitting out by the pool with friends when the Glenns came along and invited her over to hear the docking. The next day, after another sleepless night, JoAnn returned home to Houston.

Some of the older Carr children recalled other aspects of seeing their dad launched into space:

Jennifer: "It seemed like there was a party every night, but NASA had stuff for the kids to do. My mom was pretty busy, spending a lot of time with the other astronaut wives fingernail biting, but NASA took care of us kids. During one particular meal at the Howard Johnson Restaurant, along with the kids from our church, we began building up the silverware on the table into a tower. The waitress came over and it was obvious we were astronaut kids, so she says 'Whose kids are you?' and in unison we all looked up and said 'Gibson's'. For the launch we all got into a big bus and drove down to the launch site. There was a motorcycle cop with us and there was one car that would not move out of the way. So the cop rode up next to him, flipping his lights and kicking the side of the car to move over, which he eventually did. At the time of launch I had a little brownie camera with me. When the flight went up I had tears streaming down my face, but I was taking pictures. I didn't look directly at the rocket. Well, I did a little bit, but I got a sequence of pictures. But it was the sound… I mean you see it through the smoke and then you hear it. That day was kind of a surreal day. I was in a fog that day. I was aware of who was there; I remember taking pictures… but… Dad was there, then he was gone. Click, click, click and then there was nothing. He got out of sight and he was gone. It was a bit of a relief almost because that's the scariest part over."

Jeff Carr: "I had a football game the Friday night before the launch, so I was unable to go down with Mom and the rest of the family. I had to catch a different flight, a private jet that was owned by the President of Texas Instruments. As a member of our high school football team I had to stay for the game. Having an astronaut for a Dad might get you some things, but…" Jeff told his brothers and sisters that he had told people on the plane he was a track star. In fact, he helped the hostess serve a group of VIPs vodka tonics all the way to Florida. "I remember the launch vividly, especially the drive in the limo to the Cape. I kept looking for some high tech space stuff, past the VAB to the astronaut family viewing area. I was very aware of a lot of people. I used the binoculars to look across the marsh at the Saturn − while listening to the air-to-ground commentary − zeroing in on this little capsule on top of this monster where my Dad was.

"When we counted down, I put the binoculars down and watched. I saw a flash, steam and smoke. I started hearing a low level rumble, and I saw the shock wave come across the marsh in front of us. It hit me in the face and blew back my hair. Then the sound increased to a crescendo, to a deafening roar, and I watched the lift off. It [the Saturn] rolled and it pitched, and just as I was thinking, 'My God, it's going to fall over!' it began accelerating. I remember feeling that I was being pounded because of the shock waves; being pounded like somebody was hitting me in the chest and face. I remember my mother (who had John Glenn by one hand and the minister of our church by the other hand) jumping up and down. I don't remember if I yelled anything. I was thinking 'GO! GO! GO!' but I don't think I breathed. It just struck me as absolutely awesome because this huge amount of energy was being released… and my Dad was sitting on top of it.

"I was much more sensitive to it. It was clearly a more powerful experience to me than was Apollo 12. No comparison. I was left absolutely shaken after that. As things went on I began to hear my father calling back, responding to mission control. And I heard my father in his element. I heard the tone of his voice. He was doing everything he was trained to do. He was on his game. He was calling out the marks. He was sharp, and he was on top of it. I felt suddenly all my fears had gone, and I felt very comforted, more confident about it, because I could hear that tone in Dad's voice that I knew, that I recognized from my life, from around the house. I recognized his 'matter-of-fact, ticking things off, getting things done' tone and I knew it was all going to be OK. That's how he always made me feel when I heard that tone of voice. He was in charge, and if he was in charge everything would be OK. So I was on a huge high after that."

John Carr: "As soon as the rocket launched, that's when it really hit me hard that Dad wasn't going to be here. That's when I started to get a little concerned." The feeling didn't really subside until he knew Jerry was safe aboard Skylab.

Jessica Carr: "The launch…that was a trip! I remember a lot about that. I even remember what I wore! It was a powder blue wool pant suit. People were everywhere and in our faces constantly. I never wondered where Dad was because as far as I knew he was at work. I remember the rocket taking off and the noise coming over me in a wave. It made my chest rumble and the air electric. It was so exhilarating at first, and then as it continued a kind of panic set in. Tears were in my eyes, and I heard my Mom yelling something like, 'Go, Go!' Then it hit me. My Dad was in that thing and it scared the heck out of me. I think I lost it at that point, and I felt bewildered."

EXTRACTS FROM THE DIARY.

Mission Day 1: Friday November 16, 1973. Inside the CSM during second orbit:
"The blue and white of the Earth is absolutely breathtaking. Space is stark and dark. The horizon is many hued blue and green. At sunrise and sunset, orange and blue. Must take a picture of it for a keepsake. Totally disorientating geographically. Didn't see anything I recognized for hours. Then all of a sudden saw Spain, then the Med, Sardinia, Italy, Greece, Suez. Rush, rush, rush …no time to enjoy the view. There'll be plenty of time later."

CC: **"How is everybody liking zero-g today?"**
JC: **"I'll tell you Dick (Truly). It's really neat."**
EG: **"We can hardly take our eyes away from the window. There are nose smears all over the windows. All I can say is it's worth eight years of waiting."**

It was not until Jerry spotted the familiar 'boot' shape of Italy that he finally worked out were he was. He was delighted to see that Italy did indeed look like a boot. Despite the smudged windows from the RCS exhaust gases and escape tower jettison earlier, all three were delighted by the view. It was now time to chase Skylab.

The journey to the Workshop took around eight hours, during which time the crew took turns doffing their pressure suits and donning flight overalls. The path to the space station, which was several miles ahead of them in a higher, 235 nautical mile orbit, was mainly sequenced by the on-board computers. Jerry and his crew completed a systems check of the CSM and gathered the latest tracking data from the ground. Bill Pogue also performed a star sighting using the on-board sextant in the lower equipment bay under the seats. This provided an accurate update to the Inertial Measurement Unit and determined exactly where they were in space. This was the same system that had been used for the Apollo lunar missions.

For the rendezvous phase, all input into the computer was checked by the ground control team and verified. The sequence to load the computer with the update program was then passed to the crew, who then manually input the command sequence before a second check by the ground team. In the propulsion phase, the computer informed the crew of the attitude, and the time and duration of the burn. Jerry had the option to manually control the burn, but decided to let the computer do the work. His hands remained close to the controls and he would closely follow the elapsed time clock during the burn, just in case he had to manually stop it. After the ground team checked and confirmed the numbers and angles, the computer counted down to the burn and then asked the crew to confirm whether they wanted to proceed. Jerry pressed a button on the control panel, and the large SPS engine ignited to adjust their orbit and begin the chase of the space station.

Maneuvering the CSM was, to Jerry, like "floating on oil." The astronauts had been told by the previous crews that the firing of the SPS engine was "a good kick in the pants," which resembled the afterburner of an aircraft. For Jerry's crew, their first SPS separation burn was a definite thump, like a catapult shot off the deck of a carrier. The severity of the acceleration took the astronauts by surprise, with all three laughing and shouting "yippee." To all of them, this was certainly more than "just a kick in the pants." There were four such burns initiated. Skylab 4 was now in a circular path 10 nautical miles below the workshop.

JC: **"IGNITION."**
BP: **"Holy Moses! Lookee!"**
EG: **"Oh Man, it really pushes you around. You know, once that thing [the SPS] settles down, it feels very comfortable."**

Maneuvering a vehicle in space is completely different from driving a car on Earth. If you are driving a car and wish to catch up to the vehicle in front, you put your foot on the accelerator, the engine speeds up and you decrease the distance to the other vehicle. If you wish to slow down, you touch the brake and this slows the vehicle. In Earth orbit this is the exact opposite to what you want to do. As a vehicle circles the Earth, the higher its orbit, the longer it takes to complete. Any vehicle in a lower orbit is closer to the Earth and therefore has a shorter path to travel to complete an orbit. Thus it will complete its orbit faster than the vehicle that is further out.

For the astronauts in the CSM to catch up with the Skylab, which was in a higher orbit, they had to apply engine thrust *against* the direction of flight (effectively, braking). This slowed them down and placed them in a lower, shorter orbital path, which allowed them to catch up their relative position to the Skylab. As they passed underneath the station in its higher orbit, they turned their spacecraft around and fired the engines *with* the direction of flight (effectively, accelerating). This boosted their capsule to a higher (slower) orbit to match the target vehicle. Using a combination of the large engine and the four sets of small thrusters around the outer surface of the CSM, Jerry gradually caught up with the unmanned station in the same orbit.

> **JC: "Ok, we see it. She looks pretty as a picture. We're about a quarter of a mile out..."**
> **EG: "We're about 100 feet out now... We're looking over the vehicle. The MDA has got a fair amount of discoloration to it. It appears to be light gold on the Sun side. One part of the OWS which originally was gold had been discolored quite a bit. We can see red and green also on it. There's a fair amount of discoloration in that cord they used to put up the last solar shield. Other than that it's looking a good home."**

As the crew closed in on the station, they decided to have a snack, as all three were feeling fine. The crew saw the station on the fifth revolution (GET 7 hrs 25 min] as they flew over the Pacific between Australia and Guam. As they approached from behind and beneath the station, Pogue handled the photography. Gibson supported Carr in the approach by supplying calculations using a hand-held keypad, reading off the approach distances and angles as Carr maneuvered the CSM toward the front docking port of the station.

EXTRACTS FROM THE DIARY.

> **Mission Day 1: Friday November 16, 1973. Inside the CSM, fifth orbit, approaching Skylab:** "The workshop... stark and beautiful in the black sky... white, gold and black. Very sharp, angular and sterile looking. TACS [Thruster Attitude Control System] puffed back at us every time we maneuvered. Thought of a bull fight – a snorting bull."

The docking system used on Skylab was the same as the one used on the Apollo lunar program. The astronaut lined up a T-shaped target on the station near the docking cone with crosshairs on the windscreen of the CSM and with an optical sighting device. By aligning both crosshairs with the T-shape, the actual docking mechanism at the front of the CSM was aligned with the drogue on the station, which was situated to the right of the astronaut out of the line of sight. As the extended probe enters the drogue, it is guided by the conical docking sleeve to the central receptacle, where the probe's three capture latches trigger to perform a 'soft dock.' Retracting the probe draws the two vehicles together and lines up the twelve docking latches around the circumference of the docking mechanism. When activated, the latches form a perfect attachment and airtight seal, whereupon an indicator panel in the CSM signals that docking has been achieved. Removing the docking probe mechanism from the CSM gave access to the station through the docking tunnel, in the same way that astronauts moved from the Apollo CSM to the lunar landing vehicle.

As Jerry closed in with the CM, he approached too gently. When he soft docked and got the 'barber pole signal', the crew thought they had docked with the station and informed the ground of their apparent success. But when they felt side-to-side movement, they realized that they had not achieved the soft docking. The workshop moved away from them a little. Instead of re-cocking the probe and trying again, Jerry imparted a little thrust and went back into the drogue from about one foot away and slightly off center. He bounced off, then reversed the CSM to try again. This time, after a reminder from the ground to re-cock the latches, he imparted a little power to the approach. By "goosing it," as he later described it, he hit hard and the workshop bounced back. But this time they heard the familiar shot gun blast sound of the latches firing, securing the CSM to the OWS. They had arrived at their home for the next three months. Once safely docked, Pogue checked the latches to confirm the secure docking.

> **JC: "Roger, Houston, we moved in, made contact, got our barber pole and did not capture. We tried to shove it one more time and it didn't work. So we backed out [to station-keeping again]. We gave it a pretty good shove."**

JC: "We were rejoicing that we were captured, and then we were dismayed to see ourselves drifting off. The second time there was no doubt we were captured because we got a little bit of backlash. When we hit it, the workshop moved away and then bounced back and hit us again. We knew were at the end of the pole"

The plan after docking was for the crew to rest. After all, it had been a long day. Entry to Skylab was planned for the next day. However, some days prior to launch, they had requested entry into the OWS shortly after docking. After several days of consideration and evaluation on the ground, it was recommended that the crew stay in the CM for the first night. This was mainly to avoid prolonging their day even further in the event of some unforeseen problem.

After an 18-hour day, all three were to eat a meal and sleep in the cramped confines of the CM. They would transfer to the workshop the next day to begin several days of reactivation before getting down to routine science work. All three were still amazed that they had actually arrived in orbit:

JC "Ed is still wagging his head and is wondering how he got here... Actually, we all are."

The crew were to turn in early and have a restful night, but they worked fairly late and then decided to take dinner. Bill Pogue told his colleagues he was not feeling too well. They talked this over and decided that he should eat something. They hoped that it might make him feel better.

About an hour after docking and just after eating, Pogue became ill and vomited. He had felt nauseous during the rendezvous maneuvers and had taken medication to relieve the discomfort. But his condition worsened. His stewed tomatoes went down and came up again into a vomitus bag. He felt miserable, and was afraid to move should he vomit again. One of Pogue's responsibilities was the configuration of the communication system, and unfortunately, he left the switch on that was recording the conversations in the CM.

After vomiting, Bill Pogue stated he was going to slow down for a few minutes. His colleagues helped him take off his jacket and snoopy communications cap to help him cool down, as he was feeling very warm. They encouraged him to rest, stating that if Houston called there wasn't anything that the other two couldn't handle:

JC: "Good. Things are quiet, and we don't need you so just hang loose. Why don't you take the rest of the day off? [laughter]"

The crew was very aware of the importance of this mission and, not wanting to risk an early return by informing the ground, they discussed the incident in the CM, off the 'hot mike' connection to the ground. Unfortunately, they forgot about the on-board tape recorder that was usually left on to record casual crew comments during the mission and which was automatically dumped by the ground during sleep periods.

As Jerry and Ed discussed whether to let the ground know, there was some surprise that it was Pogue that had become sick. Known in the astronaut office as 'Iron Bill' or 'the astronaut with cement in his ear,' he was the former Thunderbird pilot who made most astronauts flying with him ill when he was piloting the T-38 in aerobatics.

Having forgotten about the tape recorder, they thought that, since it was such a mild attack, it was not worth reporting and perhaps jeopardizing their mission so early. Gibson thought the doctors might over react, but since everything had to be reported about food intake, his not feeling well might just be passed off as adjustment to the new environment, since they would have to hand in the bag containing the vomit. However, Jerry commented that they could also "just toss it into the trash airlock and forget the whole thing."

In the evening report, Jerry commented on the water intake of the crew, and that he had eaten Pogue's strawberries by mistake for lunch, so he had not eaten his strawberry dessert for the evening meal. He had also elected not to try the turkey and rice soup and not to take two coffee drinks. He also reported that Pogue had not eaten his evening meal.

CC (Bill Thornton): "Don't feel bad about not eating the food... You can always blame the cat. We're LOS (loss of signal) in about a minute. We'll be seeing you tomorrow. Looks like everything got off to a good start there today."

One of the major concerns at the time was the still unknown phenomenon of space sickness. The crew had been put under tremendous pressure to record everything that went into their bodies and everything that came out, by whatever means. Every item of body waste, including vomit, had to be bagged, recorded and stored for post-flight analysis. This did not initially bother the crew, as they knew they had volunteered to be human guinea pigs for the duration of the flight. They had also been told that they were probably the last crew that would have to endure this type of regimented recording of everything they did, said or felt. The day before leaving JSC for the Cape, the medical specialists asked the crew to take medication during launch and rendezvous to counteract any signs of space sickness. They had tested all three on which type of medication would suit each astronaut. Jerry's medication was scopalomine-dexetrin. He observed that if he took any medication to the degree he was being asked to, he would not be allowed to fly an airplane or even drive a car. Now he was being advised to pilot a multi-million dollar spacecraft while under the influence of scop-dex. "I said I'd take the sickness rather than the side effects, so I decided not to take it. Bill wanted to be a good patient, however, and he took it, as he was not going to be piloting the vehicle."

Ed Gibson recalled discussions about the Space Shuttle and of a sick crew trying to bring home the vehicle after only a few days. NASA was trying to convince the politicians that the space sickness problem could be solved. So the crew was concerned that a full report on the illness of Pogue might potentially lead to early termination of the mission – due to political pressure rather than operational reasons.

The next day, after sleeping in the CM (Jerry slept for about an hour at a time interspersed with 15 to 30 minutes awake), the CAPCOM indicated that Chief Astronaut Alan Shepard wanted to speak to them. Apparently the incident of Bill's sickness had been revealed when the tape was dumped, and Shepard was not a happy man. He gave the crew a mild but very public reprimand over the intercom:

> **Shepard: "I just want to tell you that on the matter of your status reports, we think you made a fairly serious error in judgment here in the reporting of your condition."**
> **JC: "OK Al, I agree with you. It was a dumb decision."**

There was some discussion afterwards, but it was dropped. However, it did sow the seeds of frustration over the next 4 to 6 weeks as the crew was under pressure to emulate the performance of the second crew. There had been a feeling that they would be unable to discuss their thoughts across the very public air-to-ground channel (A channel). Even the more restricted B channel recorded every success and failure, and this was also released to the public. Their concern was understandable, given that their shortcomings would make headline news on the news stands the next day. Some astronauts had even used the so-called private channel for operational discussions, so nothing was actually private and confidential in the crews' eyes. For now, however, they would be occupied with reactivating the station, a task that would take several days and would allow Pogue to recover.

During the unmanned period after the Skylab 3 crew had left, the atmospheric pressure had been maintained at about one half pound per square inch to conserve consumables and inhibit microbiological growth prior to the new crew arriving. Following pressure checks on both sides of the hatches, they cracked the hatches and floated through to begin reactivation and orientation. Gibson and Jerry were the first to enter Skylab, floating through the Multiple Docking Adapter into the Airlock Module and then the main workshop. The previous crew had stuffed some spare clothing, making dummies. One was on the exercise bicycle with Bill Pogue's name tag on it. Ed Gibson's tag was on a manikin lashed down in front of the ATM console, while Jerry's name-tagged manikin was near the storage lockers in the upper part of the workshop. This greatly amused the three of them. They were so busy in the first few days that they could not find time to take the 'other crewmembers' down, and Ed Gibson recalled that it was slightly distracting catching a glimpse of one of their 'colleagues' out of the corner of his eye.

Due to his illness, Pogue initially remained in the CM couch, swapping jobs with Jerry to allow him to recover. As they activated the workshop, the CM was put into dormant mode – a passenger or parasite that would not be used until it was time to come home. They did use an umbilical to keep a flow of air into the vehicle, and they later found that it was cold and had a tendency to freeze up somewhat. The CM was also used to make personal family phone calls on VHF, away from the hustle and bustle of the rest of the station.

While Pogue was the only one who was actually ill, Jerry recalled "a feeling in my stomach that was kind of a big knot, but I wasn't sick. Ed just didn't have any problems at all." Both Jerry and Bill have always though it a marvel that Gibson, the science pilot with the least flying hours, was the crewmember who felt the least discomfort in adapting to spaceflight.

Over the next few days, the crew slowly adapted to their new home and completed a paced but busy schedule, activating the station, unloading the CM and overcoming the adaptation to microgravity conditions. The completion of their first week in space would see the first EVA during the mission, and the commencement of the varied and packed program of scientific investigations. The crew had only been in space a few days, and the pace at which they were working given what lay ahead was going to tell.

The official status bulletins issued by NASA briefly mentioned the adaptation of Pogue to microgravity:

Mission Status Report. Day 1.
The SL4 crew was launched today. As scheduled, all boost events and rendezvous events occurred in a nominal manner. A hard docking was achieved at two attempts. The pilot reported slight nausea and did not eat the evening meal.

MD2 Status Report.
Major crew tasks today (were) workshop activation and CSM reconfiguration. This activity ran slightly behind schedule due to periodic nausea symptoms. The crew is continuing with anti-motion sickness medication.

MD3 Status Report.
Crew continue to adapt well. Commander and pilot (have minor) stomach awareness. The science pilot reports no symptoms of motion sickness. The crew did perform exercises and feel that this improved their subjective state a great deal. More equipment was transferred from the Apollo CM.

MD4 Status Report.
The crew had another very busy day. Activation of the orbital workshop with no further reports of motion sickness. A fire drill was completed and the crew successfully re-serviced the primary Airlock Module Coolant Loop

MD5 Status Report.
The crew is adapting well to the new environment. Troubleshooting the ATM console coolant loop, which kept the control and display panel at a comfortable touch temperature. The major medical activities commenced this day with runs on the LBNP and Metabolic Activity experiments by both Jerry and Ed.

MD6 Status Report.
Motion sickness is no longer a problem. The crew completed the activation of the workshop and transfer of equipment from the CM.

EXTRACTS FROM THE DIARY

In his personal diary, Jerry recalled the first few days' adaptation to the new environment of space:

Mission Day 2. November 17: "Slept better than I expected, but it was still a poor night's sleep. Bill stayed strapped to his couch. Ed slept with his feet in the tunnel and head on his footrest. I stayed under the couches. Got Bill to take some breakfast... he felt better, thank God. Ground got it [the vomit reporting discussion] on tape, and has us dead to rights. Shepard called and slapped our wrists. It was a dumb stunt for us to pull. I wish I'd had more sense than to try it. All we did was discredit ourselves. Didn't do the shuttle program any good at all. Bill is improving... kind of drifts around, looks a bit disoriented."

Mission Day 3. November 18: "What a terrible day! We got so far behind and made so many mistakes that it's ridiculous. We told them before we left that we wanted to take our time, but they've just kept after us and crowded us until we tried to hurry and then we screwed up. We were so tired last night. I've had a headache for two days now because of the head congestion. Looked at myself in the mirror. My face is all fat and my forehead and neck veins are standing out. Eyes bloodshot. Ugly, ratty looking beard. Hope things ease up pretty soon. This ain't no fun at all."

Mission Day 4. November 19: "Still behind and dog tired. We've got to catch up and get a routine going! So far space flight isn't much fun. Don't even have time for a look outside. They had us scheduled for a fire drill today. What a laugh! It's been a Chinese fire drill ever since we got here."

Mission Day 6: November 21: "We still seem behind, but we're making headway. A couple more days and we should be on top of things. We're still goofing up a lot too. That sure doesn't do much for your ego – making mistakes while everyone is watching. They seem to enjoy telling us when we've done it. Yesterday I had my first LBNP run. Was really easy!! Sooner or later I guess it'll get me, but so far it's no sweat. We're starting to exercise, and that makes you feel better. Pulls the blood out of your head and into your muscles so the congestion goes away for a while. Still not much time for looking out at the world. We see Siberia and China every night after dinner. Sure looks bleak and dreary. Makes you appreciate our green country. We stayed up late last night getting ready for tomorrow's EVA. Kind of wish I had scheduled myself out [on the] first [EVA] after all. Oh well, I'll probably get out the day after tomorrow to do the S193 antenna job if they don't get to it."

Meanwhile, back on the ground

When JoAnn returned to Houston after the launch she received a call from Alan Shepard telling her that he had had to slap the crew's wrists over Bill's sickness. Shepard said it was no big deal, but they [NASA] had to make an issue out of it. "Knowing Jerry the way I do," JoAnn recalled, "he has a real need to do things right and in the proper order. I knew this was something that was really hurtful and really devastating to him, although he would suck it up and would not let you know it had hurt him. I had two squawk boxes, one in my bedroom one in the living room, so I could listen to the air-to-ground communication plus some press conferences. I became aware really early on – I think it was about day three or four – that there was really something badly wrong because I could tell by the sound of Jerry's voice. I could tell by his tone he was backed up against the wall and this worried me."

JoAnn called Kenny Kleinknecht, the Skylab program manager, who was also a family friend. He told JoAnn: "We've got them over-scheduled and they're just running around trying to get everything done." JoAnn was relieved that NASA knew what the problem was, but the problem did not go away. This puzzled her: "I thought, 'The head of the program knows what the problem is, so why is there still a problem?' I rarely left the house after that, except to go to the grocery store and take the kids to appointments. A transcript of the air-to-ground conversation was delivered to the house every morning. I had a tracking map out all of the time, listening critically to all the air-to-ground I could. I didn't like what I was hearing at all. Every third day, each spouse was allowed a family telephone call to their husband…"

NASA requested that all family conversations be taped, but not listened to, just in case there was something of interest after the flight. JoAnn was fine with this, so long as she could have copies. "I also asked Jerry to keep a diary for me and he was pretty diligent. I could really track exactly what was going on. I was becoming more and more distressed about the way I felt the press was handling things. I became very sensitized to it. Perhaps too sensitive, because I have gone back to it recently and looked at those clippings and they don't look nearly as negative to me from this distance [almost 30 years later] as they did then. But it seemed to me they [the press] never missed an opportunity to point out that the crew were rookies when any little thing went wrong. Mission Control did not address the problem and neither did the crew. Jerry's mind set was always, 'Here are my orders. Here is what I have to do. Don't question it, just do it'."

The responsibility of command fell squarely on Jerry shoulders. Though not unfamiliar with the role, with Ed busy at the ATM and Bill increasingly frustrated at unpacking and finding equipment, it is clear from Jerry's diary and reviewing the air-to-ground commentary that he was not comfortable with the situation that was developing. His only direct contact with the ground was with the CAPCOM on duty via an open mike, or the medical doctor on a closed channel. Neither system was particularly helpful in explaining the situation privately without raising further concern on the ground. Neither could Jerry fully explain his predicament to JoAnn during the short family conversations without alarming her unduly, although JoAnn was already beginning to sense something was not quite right.

In those days, before emails and before regular long flights, the psychology of space was unknown territory. This flight was proving a steep and difficult learning curve for both the ground the crew and the families. As strong willed as she was, and as concerned for Jerry, JoAnn decided she was not going to be the one to tell NASA that she was hearing things that led her to believe Jerry and his crew were not receiving the support they should be. "I'm trying to support this guy, not trying to tear him down while he is up there," she later commented. Jerry's colleague, friend and fellow Marine, Jack Lousma (pilot on Skylab 3), was off on post flight tours, so JoAnn could not turn to a fellow Skylab astronaut for support. Afterwards, Jack's wife Gratia told JoAnn that they felt bad about not being there to support her and the family, or to put a word in for Jerry during the mission. JoAnn tried not to tell the children about the problems she knew Jerry was facing.

Jeff Carr remained attentive as the mission progressed. "I was checking with Mom and asking her questions and I was looking at the flight plan. It was clear which ground station was being used when they called. I could hear him and we could talk. But Mom was not sharing things she thought we could not handle. I was worried about the challenges they faced early in the mission. I was sophisticated enough to read between the lines of what was said in the newspapers and to take what I read with a pinch of salt, so I would read the articles, and ask my Mom questions. She helped me understand.

"Nobody had to tell me that my father and his crew were going to be encountering things they had never seen before. Sure I was proud of the fact that they were putting their best foot forward and challenging their situation. I knew my Dad knew what he was doing, because I'd seen the way he dealt with problems, with frustrations. I worked with him in the garage a few times before the flight, so I knew exactly what he was up against. I knew from my experiences with him that he would overcome this. And frankly, I think I always felt that when I read this stuff. I always chuckled a little bit about it. 'These guys don't know my Dad; he'll show them.' I was so blindly confident in my Dad and his crew, that they would be able to work through the typical early stages of the project. I'd seen it from him before; evaluating it and getting his arms around it, and then conquering it the way he always has…"

EVA 1 (Thanksgiving Day) November 22.

The Mission Status Report for Day 7 stated that Jerry, Bill and Ed were adapting well to their new environment, and that their sleep patterns were improving. They were expected to be well rested in order to perform the planned EVA.

There were four EVAs planned for Skylab 4, with the first scheduled for the seventh day of the mission. During the previous mission (Skylab 3), the first EVA had to be postponed from Flight Day (FD) 4 to FD10 due to system problems and the health of the crew. After the illness Bill Pogue suffered during the first day in space, he had recovered quickly and was well enough to join Ed Gibson on the excursion outside the workshop. Most of the sixth day was taken up in preparing for the EVA to the Apollo Telescope Mount. Originally, the first EVA of the mission was to include the retrieval of a meteoroid collector and partial installation of ATM film during a two hour activity. By the time the mission was launched, this had increased in complexity and duration to almost six hours. The two astronauts were now scheduled to retrieve film cassettes from the X-ray telescopes, from one of the H-alpha instruments, and from the XUV/spectroheliography and Chromospheric XUV Spectrograph telescopes. Additional tasks for Bill and Ed included setting up the Trans-Uranic Cosmic Ray detector, deploying the Particle Collection and Thermal Control Coatings experiment and attaching the collecting foil sheets of the Magnetospheric Particle Composition experiment to the truss of the ATM. They were also to obtain 40 exposures using the Coronagraph Contamination Camera (experiment T025). This was to have been used from the science airlock, but it was now blocked by the deployment pole of the parasol solar shield. Finally, the two astronauts would also pin open the viewing aperture protective door of the H-alpha 2 telescope

As if all this wasn't enough, the EVA planning team had devised complex repair procedures for the S193 Microwave Radiometer/Scatterometer/Altimeter (part of the Earth Resources Experiment Package) that was located on the opposite side of the Airlock Module. This experiment was used to obtain thermal measurements of the land and ocean across the planet. Its field of coverage was accomplished by gimbaling the antenna 48° forward and 48° to either side of the ground track over which the Skylab was flying. During the 29th Earth resources run on the second mission, the gimbal capability of the hardware was lost, restricting the use of the experiment. Therefore, a difficult EVA was planned for SL4 to restore the unit to full operational use. No EVA operations had been planned for this area prior to the launch of the OWS, so there were no hand rails, foot restraints or restraint attachments in the area, and it was assumed that a second EVA might be needed to complete the work.

With Jerry helping his colleagues to suit up, the EVA preparations included refilling the Airlock Module (AM) suit cooling system with water. Skylab EVAs accessed the outside via a former Gemini program hatch in the side of the Airlock Module, with the internal hatches sealed at both ends of the Airlock Module. The third crewmember (in this case Jerry) remained in the Multiple Docking Adapter, so that he could still access the Apollo CM if the EVA crew could not re-enter the workshop (because the AM could not be re-pressurized or the EVA hatch sealed, for example). The EVA proceeded smoothly, except that only five of the planned 40 exposures were obtained using the Coronagraph camera. Control knob problems halted the operation early. All other operations progressed smoothly, including the repair to the S193 antenna. At the close of the EVA after 6 hours 33 minutes, all scheduled activities had been completed. The EVA was compared in difficulty to the deployment of the stuck solar wing during the first Skylab mission. Despite the lack of restraints near the antenna worksite, Pogue and Gibson's success clearly demonstrated the value of having a crew available to repair scientific apparatus that had failed in space.

The crew had trained underwater to complete the antenna repair and found that attaching foot restraints on the truss provided some stability. Bill Pogue got into the foot restraints and then Ed Gibson got on Bill's lap and stood up to do the work on the antenna while being held by Bill. This type of ad lib worked very well for this particular EVA.

Problems encountered included torque effects from suit venting, which caused them to use up an additional 528 lb/sec of nitrogen gas from the Thruster Attitude Control System (TACS) to compensate. There was also difficulty in keeping the two umbilical oxygen supply lines apart as the astronauts moved across the worksites.

EXTRACTS FROM THE DIARY:

Mission Day 7, November 22. EVA day: "We were still behind, but finally got out the hatch about an hour and a half later. Ed and Bill really did a good job and even got the EREP antenna repaired, so there's no EVA tomorrow. I was hoping to get outside tomorrow, but I'm glad the work is done. I guess my chance comes on Christmas Day. They stayed out 6 hours 33 minutes. [Apparently] that's a new record for an EVA. I thought Lousma and Garriott stayed out for close to 7 hours. We asked to verify that. Ed was really euphoric about the thrill of standing up on the sun end of the ATM and looking at the world.

That night, while the crew were asleep, Control Moment Gyro Number 1 began rapidly overheating, requiring an immediate shutdown of the CMG. Operating in a 2-CMG control mode required more use of the TACS supply, which would have serious impact on future activities. If another CMG should fail, limited control of the station's attitude would still be possible, but this would seriously hamper all activities. Without accurate attitude control, all ATM/sun activities would be impossible. Under mission rules, the flight would probably be terminated and the crew brought home. Because of this failure, some of the early maneuvers planned for turning the station towards Kohoutek for better viewing and photography angles were cancelled.

Despite this setback, the crew were able to conduct some hand held photography of the comet out of the Command Module window. Over the next ten days, similar photos were taken at 12-hour intervals. Though some activation work was being completed on the ATM, full-time science research was still some days off, so the crew completed unloading of the CM and transferring items to stowage positions in the workshop.

EXTRACTS FROM THE DIARY

Mission Day 8, November 23: "[We were] told tonight that 6 hours and 33 minutes is a new record. They checked the records and confirmed it. That's a good record! I think we have another record too – the most mistakes made in a seven-day period. If the pace ever slows up and we catch our breath, it's going to be a great day. Today's schedule was changed so many times I can hardly read it – mainly because we are so far behind. They just cancelled a bunch of items and tried to give us some catch up time, but it wasn't really enough. We're all beginning to feel reasonably well now. I wonder if the EVA had anything to do with it. Maybe it was a morale milestone. Got to talk to JoAnn today. My first private conversation. I had tears in my eyes the whole time. I guess I've spent so much time feeling sorry for myself these last few days that I had a lot of emotions stored up in me, and talking to family released the tears. I really feel happy to be able to talk to them. I think we'll be able to get on a twice-a-week schedule for calls. That'll be great. Had some disturbing news this morning. When we got up they told us that while we were asleep CMG #1 failed. It apparently just froze up – the bearing went bad I guess. It's no danger to us – unless it comes apart. The two remaining gyros seem to be in good shape. If another one fails, that's all for the mission. We'll have to use the CM for stabilization until we can secure things and go home. God, after all the suffering we've gone though so far, it would be a real heart breaker to have to quit short of our goal."

The following day was the first formal rest day, and apart from some general housekeeping duties, the crew could at last take a breather from the fast pace at which the first few days in space had required them to work. The Mission Status Report for the day explained that the first rest day had been moved up from Day 13 in order to give the crew some down time and a chance to complete activation tasks. Generally, the crew's health remained good, and the first results from using the treadmill indicated that it was effective in stressing the lower extremity muscles.

The next day was mainly occupied with exchanging failed equipment on the ATM console and associated equipment in the AM, and preparing for the first run of the EREP, scheduled for FD11. However, by the evening of FD10, this had been cancelled by bad weather. That allowed the crew to complete the checkout of the ATM and to take some photos of the Atlas Mountains, bedrock in the Sahara Desert, river cities in the Congo, shield volcanoes in Hawaii and drought areas in Mali. But Jerry was becoming increasingly concerned about the amount of time it was taking to complete specific tasks. He asked the ground for more time to complete one task, commenting that they were still trying to determine the profile for optimum activity onboard the station. On FD11, the first major activity using the S201 Far UV electrographic Camera with the comet Kohoutek was accomplished. On FD12, an excess of TACS gas was used to maneuver the station from, and back to, solar alignment to observe and record the effects of a barium metal material cloud that had been released from a Black Brant IV rocket at an altitude of 348 nm. This also resulted in the cancellation of the planned EREP run, but an extra three hours were spent at the ATM console instead. Unfortunately, a filter wheel in the X-ray Spectrographic telescope had become stuck between two positions and would require manual adjustment in a forthcoming EVA.

In his 2000 NASA oral history, Jerry recalled these first few days aboard Skylab:

JC: "It took three or four days to just get things moved in and set up before we even really began doing our experiments. It was a lot of set up time, but we got all of that done and then began doing our experiments. That's when the schedule caught up with us. We found that we had allowed ourselves to be scheduled on the daily schedule. The schedule was so dense that if you missed something, or if you made a mistake and had to go back and do it again, or if you were slow in doing something, you'd end up racing the clock and making mistakes, screwing up an experiment or not doing a procedure correctly."

In his 2000 oral history Ed Gibson added:

EG: "That's where the miscommunication started, because all of a sudden it put the ground on one side of the table and us on the other. Then we got behind. Bill Pogue was working at half efficiency, and the poor guy was struggling. Then he'd make a mistake, feel worse about it, and then struggle some more. We were, of course, just getting used to that environment, trying to find things and where they were. The more we got behind the more detailed the messages sent up to help us. But we didn't perceive them that way. And there were no private communications. You couldn't just call them up and say, 'Hey guys, lets talk this out,' because everything had to be open for the world. We thought, 'Okay we'll work through it,' but it didn't work. It just got worse as we got further behind."

Bill Pogue highlighted some of the early problems during the first few days onboard Skylab in his 2000 oral history:

BP: "Trying to get all the right stowage books – to find the right things to use – seemed to be the immediate problem. We worked until about 10.30 pm Houston time the first day trying to catch up. After the initial bout of nausea I had no more problems, but it was humiliating. It was tough. It was a surprise to me and to everybody. What I felt worse about was that the doctors didn't know their ass from third base. They had the medicine, they had all the theories, and none of them worked. I had an awful time finding equipment in the lockers. But one of the most frustrating things was transferring equipment from the Command Module and stowing it in drawers. Everything was stored loose in large floppy laundry bags, so I opened them up and of course everything just floated out of there. I had to identify each one of them. It took me a lot longer than it should have to transfer items."

On FD13, the crew completed over four hours work at the ATM console, along with medical experiments on Jerry and Bill. The 200th day in orbit for the workshop (FD14) was also the first day the crew were finally able to complete an EREP pass. Over the next few days, more EREP passes were completed as the science program gradually built up. There were sometimes two runs per day. Observations of Kohoutek continued, as did preparation for extensive work at the ATM console and further medical experiments and examinations. Kohoutek turned out to have a period of 2000 years, so the last time it was seen by humans on Earth was around the time of the birth of Christ. Some theologians speculated that what was called the "star of Bethlehem" may have really been Kohoutek. There was considerable international interest and speculation in this possibility for a while.

EXTRACTS FROM THE DIARY

Mission Day 9, November 24: "Our first day off. We slept in and it really did us a world of good. We're beginning to get adjusted physically to spaceflight. The congestion is still with us and I guess it will stay to some degree throughout the mission. I still get tired pretty quickly – especially when I get hungry – and then I begin to feel badly. Our biggest problem right now is mental attitude. We're so damn mad at the guys on the ground that are scheduling this mission. It looks like they're trying to get everything finished early or something. They seem to be more interested in filling squares than in what they're doing to us. We've started passing the word down on tape when they schedule us into a box. If we could give them enough complaints maybe they will get the picture and ease up. Started taking some pictures of the comet today. Can't even see it yet. We're just pointing the camera in the right direction and snapping. We did some funny TV today. Bill made some large cardboard swim fins, and paddles for his hands, and I televised him in his crazy get-up trying to paddle from one end of the forward compartment to the other. He put lightning bolts on the helmet. I laughed so much I could hardly hold the camera. I made up the dialog to go with it – called him "William Pogue Aerospace Pioneer." Hope the folks on the ground get a kick out of it."

Mission Day 11, November 26: "Had another LBNP run today and still no problems. I wonder when the axe is going to fall. I thought we'd have some trouble sometime during this 'period of adjustment'. Had my first run in the rotating chair. No problem there either. Next time we will run at 30 rpm and see if that has any effect. Eleven days into the mission and we're still not up to speed yet. We've been getting better, but just about every day we get hit by something new that throws the schedule all out of kilter. Then we scramble for the rest of the day. Not making any mistakes now, but still need to do a lot of improving. The ground still catches us napping."

Mission Day 12, November 27: "Things are looking up. The ground says that our gyros are running great, and we're showing progress in our battle to hang on to the timetable. I got to do some TV today. My beard still looks a little ratty but most of the scene was on my feet anyway. I demonstrated the treadmill. We called it 'Thornton's Delight' at first, and then when we found out how much work it is we renamed it 'Thornton's Revenge'. It really works the legs over. Sure hope it keeps our calves in good shape for recovery. Today was our first day for EREP, but it got cancelled because of the weather. Nuts. I was looking forward to a change of pace. Bill and I practiced hard on EREP, so I think we can do a really good job on it."

Mission Day 13, November 28: "Pretty much of a nothing day. Bill and I did LBNPs on each other. Got to look out of the window a little bit and saw a lot of Africa, Mongolia, Siberia and some of North China. Got a look at Peking. Couldn't see the Great Wall. Couldn't find it. Looks like all of China is washing into the sea. They must have recently gotten hard rains or something because the sea shores are all stained with yellow and tan silt for miles out to sea. Sure is an ugly looking country. No vegetation to speak of. The Yellow and Yangtze rivers are both ugly and dirty looking even from here. Korea and Japan both look green and healthy. We did TV of our supper tonight, and Ed wanted to narrate it, so we turned him loose on it. He rattled on about all the things we have done and all the great things we are going to do. Then he talked a little about how we eat, and Bill and I demonstrated our salt dispensers and such. When it was just about all over the ground called and said they saw we were recording TV but the audio switch wasn't on. So we gave them silent TV with Ed standing there beating his gums. He was really mad!"

Mission Day 14, November 29: "Busy, busy day today. We had our first EREP and carried it off really well. Made us feel pretty good to do something perfectly for a change.

Made a run up through Central America and Texas. Bill was on the VTS [Viewfinder Tracking System] and I was on the C&D panel [Controls and Displays]. He had a few cloud problems but we got our data. Looked for the laser beam today but couldn't see it. Don't think I was looking in the right place. It's located between Washington and Baltimore at Goddard, but I couldn't find it. Did the French experiment, S183, and when I finished and took the film carousel out one of the frames was hanging out and a piece of glass floated by. Looks like that one is ruined. It's been plagued ever since SL-2 launched. Haven't gotten any data yet. The food is pretty good. I've been waiting for my taste to change like the 'old heads' said it would, but it hasn't. I can still smell pretty well in spite of this damned head congestion. Talked with JoAnn tonight. She doesn't like the way I sound. Thinks I have a cold. They haven't seen the treadmill on TV yet. Maybe Monday. I'll be interested to hear how it came out."

Mission Day 15, November 30: "Had another EREP today and did a good job again. If all we had to do was EREP we'd have it made. Everyone had been pretty worried about these EREP maneuvers with only two CMGs, but it looks like ground has a good scheme worked out for getting away with it. We spend a little TACS, but it doesn't amount to much. At this rate we can go the whole mission and have plenty left over. Did some more TV today. Nothing spectacular or particularly interesting. Just had it on while we set up the French stellar experiment in the anti-solar airlock. Beginning to pick up on the exercise now. Pumping 4,000 watt min. on the bike plus various isotonic exercises with the Mark I and the springs. Walk two minutes per day on the treadmill and run one minute. We passed below the tip of South America this morning and I spotted a huge ice island with clouds all around it. It was beautiful – even whiter than the clouds. That's how I noticed it – it was so pure white. Got a picture of it. Kind of made me think of the booster on the pad with the spotlights on it at night."

Into December

As the mission moved into December, the first few days were spent completing EREP passes (on FD16 though 23), as well as continually manning the ATM experiments. On FD19, Ed Gibson had requested more sleep time – a desire for a full 8 hours. He had only managed 6.5 hours a night, partially due to the constant need to sustain requests from the ground. On Apollo, the crews followed a definite flight plan with little in-flight experimentation during the journey to and from the moon, so their rest time was assured. On Skylab, things were beginning to become more strained due to the added requests from the ground. By the third week of the mission (FD21), it had been noted that, despite the added work and difficulties in maintaining the flight plan, the crew had still accumulated 38 hours and 31 minutes on the ATM, 106 hours on medical experiments and 68 hours on EREP equipment. Their planned 90 minutes a day on exercise had actually only reached 45 minutes, below the pre-flight matrix. However, the crew was averaging 5-8,000 watt/min energy expenditure on the ergometer, which was the same as the second mission and far above the 2-3,000 watt/min attained by the first manned mission.

Each day, Mission Control issued a status bulletin on the progress of the flight, experiment operations, observations and on other activities accomplished during the flight day. Generally, these were up-beat and positive in their reporting of the crews' activities and accomplishments in accordance with the flight plan. It was noted that the general health of the crew remained good, although Bill Pogue reported feeling a fullness of his left ear for several days. This gradually improved. The rather dull reports gave no indication of the rate at which the crew was working to try to get everything done.

During the night of FD23/24, it was noted that the filters for the multi-spectral photographic facility had been left off after the EREP checkout during FD9. The nine EREP data runs made with this experiment since were degraded, but the crew would be able to collect good data by flying over the task sites again before the end of the mission. On FD24, a procedural error uplinked from the ground resulted in more TACS gas being expended than was planned. As the first month on Skylab came to a close, the experiment program was gearing up, and the stress on the crew was causing concern. Something had to be done to improve the relations between the ground and the astronauts.

EXTRACTS FROM THE DIARY

Mission Day 17, December 2: "For a day off we sure were busy today. Back-to-back EREP passes really ate up the time, so we ended with only about 2 ½ hours of leisure time. That's not quite what I had in mind for Sunday so I sent a message down on tape for them to try to keep away from the double EREPs. We really need to be left alone to charge our batteries. Ed seems to feel he's obligated to spend every waking hour at the ATM panel. He came down from the MDA this evening all nervous, upset and tired. Went right to bed. Too tired to even take a shower. I'm going to have to watch him to keep him from driving himself into the ground. Had a nice long call with the family today. They gave each of us three full phases that amounted to about 20 minutes apiece. Mine started at the Vanguard [tracking ship] off South America, handed off to the Canaries [tracking station] off Africa, then finished [at] Madrid. How amazing it is to be flying across about ¼ of the world while talking on the phone to your family. Twenty years ago no-one would have believed it. It's dazzling sometimes how fast our world is changing. Sounds like everyone at home is getting along OK. Sure hope things go smoothly for everyone there – no breakdown of cars or appliances or the like."

Mission Day 18, December 3: "Not much of interest today. Flew two EREPs and I finally got to be the VTS operator. My site was Lake Sommerville in East Texas. Really an easy site and I found the lake OK, but I had trouble locating the proper inlet, so my tracking wasn't smooth at all. Felt very dumb after I finished. Hope I get some more VTS jobs so I can get my skills up again. Rest of the day was slow – even managed to stay up with the schedule for a change. That's quite an improvement for us. Did some window looking today. Saw the Philippines through the clouds. Even saw Clark AFB and Cubi Point. Haven't located Okinawa yet. I remember well those long over-water flights from Atsugi and Okinawa to Cubi Point. It even looks like a long way from up here. The Philippines and Indochina and the East Indies are always covered with clouds. We want some photos of Sumatra, Malaysia and the Philippines but haven't seen enough to photograph yet. Was scheduled to get some volcano photos in southern Japan, but it was cloudy there too."

Mission Day 19, December 4: "Had a busy day today. But more fun than usual. Was VTS operator for EREP and my targets were some volcanoes in Honduras. Enjoyed looking for them, but there were too many clouds so I missed them. They tied a handicap on me too. We tried having me operate the ETC [Earth Terrain Camera – Experiment S190B] in the EA SAC [Scientific Air Lock] so Ed could stay at the ATM panel and monitor the ELV [Earth - Local Vertical] maneuver. Turned out to be a Chinese fire drill. I couldn't get down to the ETC and back to the VTS soon enough to set up a good track on my targets. The folks on the ground are pretty worried about our CMGs so they've got us watching the attitude control system like a hawk during maneuvers. Rode the M131 Chair today at 30 rpm and didn't feel a bit of a twinge of dizziness. Sure do get desensitized in zero-g."

Mission Day 21, December 6: "Weather over Houston was clear this morning. Ed and I watched Houston go by as we ate breakfast. We could see Clear Lake. Couldn't quite make out Taylor Lake. Next chance I get I'm going to use binoculars to see if I can see El Lago. There's a real thrill at seeing home from here. It flashes familiar feelings and things through my mind. It happens so fast I can't elaborate or even specify what the feelings are. I guess it's like the tape recorder has been backed up for a bit and then run fast forward to catch up with now. It certainly is stirring. Had my turn on the LBNP today. Was scared that I'd conk out again at 50 mm. Drank a lot of water before the test and got some air with it. Suffered the belches and gags through the whole damned run, but made it. I hate to faint. It's like watching your life ebb away. Talked to JoAnn tonight. Sure do enjoy those calls. Wish they were longer. She sounded cheerful and happy. Hope things aren't too tough around home. That's a big burden for one person. Glad she's having so much fun with the MG. I may have to arm wrestle her for it when I get back."

Mission Day 22, December 7: "Got up about an hour earlier today for EREP. Was sure it would wipe us out, but we took it in stride. Had a pretty successful day. Sound EREP, good JOP-13 [Joint Operating Procedure] – no errors. Looks like they took my request in earnest and are going to free up our PSA times. Had time tonight to look at the stars and to watch the ground from the Caspian Sea and all way through India, Burma, Malaysia and Australia to New Zealand. I'll never tire of looking out the window. Taking pictures of the ground is the most interesting and exciting part of the mission for me!! Bad news tonight. Another of our CMG's has been glitching so the ground says. Hope it can hang on. If it goes we'll have to give up and come home early. Oh, I wouldn't mind coming home, but would hate to miss the chance to set an endurance record."

Mission Day 23, December 8: "Today was a barn burner! We did a little of everything. Solar viewing, exercise and medicals in the first half of the day and EREP the rest. Didn't finish until 9:00 tonight. Had very little opportunity to look out the window. Saw JSC this morning from over New Orleans. Amazing! Could still see Houston from Tampa, Florida. Spent some time trying to find Kohoutek out the STS [Structural Transition Section – a 10 ft portion of the AM that attached to the MDA. It included four viewing ports at 90° intervals, all with removable covers operated from inside]. We lost it from the CM window. Looks like we'll have to wait a few days before it comes out from behind our solar panel. Today Bill uncovered a mistake that he made on day 4 or 5 which degrades all the EREP data we have gotten so far. The guys on the ground contributed, but it doesn't make him feel any better. He was really shattered when he realized that he had forgotten to put the filer in the cameras. I really felt for him because he had to swallow his pride and send word down that he had screwed up royally. It brought back the chaos of the first week we spent here and all the unhappy and frustrated feelings we had because we couldn't keep up and were making so many mistakes. I really feel bad, and I knew he felt worse. My beard is beginning to look more presentable now. It has filled in quite well and is itching less. Soon I'll have to trim my hair and blend my sideburns into the beard. Then it will look natural. No new developments on our CMG. Guess we'll just wait it out. That's all we *can* do. They told us our coolant loop is hanging quite well. That's good news. We'll have a good EVA on Christmas Day."

Mission Day 24, December 9: "Today was an easier day. No EREP and a relaxed pace. Worked on a very complicated ATM experiment with Ed this morning. I have always felt that I was inclined to choke up in very complicated situations that require that you be right first time, and I felt that it was a serious shortcoming. But I am beginning to realize that its pretty normal behavior. Ed had it bad this morning, and luckily I was able to stay loose, so we muddled through in good shape. Had another LBNP run again today and made it fine. That darned device has got us buffaloed. If we start flunking out too frequently they'll have to terminate the mission, and the ironic part is that we're all feeling great. We're all doing much more exercise here than on the ground and I'd guess we're in better shape. The only problem, I guess, is blood pressure. We keep our exercise up for fear of bombing out on the LBNP. Talked to the family tonight. Enjoyed telling them about eating peanuts like goldfish do. Josh really enjoyed that – I figured he would. John is sure proud of the antlers that Kenny Kleinknecht gave him. I have to remember to comment on them next time. Jennifer sounds like she likes her job. Sure hope she can convert it to something permanent. Tomorrow's our second day off. Looks like a nice one too! Not too many responsibilities at all. We'll spend some time playing with TV and writing a goodbye message to Dee [O'Hara, the astronaut's nurse who was retiring from NASA]."

Mission Day 25, December 10: "Today was kind of nice. They left us alone until 08:00, so we caught up on our sleep. Had some light work until noon and then nothing for the rest of the day. Exercised before lunch and then took a shower. A shower costs 45 minutes but it's really worth it. It's great to feel clean. A sponge bath with a cloth just really doesn't do it. If I can work it in I'm going to try to get one every 3 or 4 days rather than one a week. Spent my free time today looking out of the window, reading, listening to my music, and just floating around. Did some TV too. We have most of that music on old records at home. I remembered how much we enjoyed them and had a few flashbacks of our early married years. I remembered our Fredericksburg and Kingsville homes, but not much of Pensacola. I wonder why? I don't specifically remember playing the music anywhere, but it sure called up those two little hunks of our lives. We're not seeing the US on our day passes any more. Mostly it's Asia, Australia and New Zealand, and the oceans… not interesting. Got a good look at Vietnam. Sure glad I didn't have to spend any time there. Looks rugged. Our TV was a bit on the 3 dummies. Ed had been itching to get on TV and philosophize. Then we showed how we suit up for EVA. Hope we can think up something more interesting next week. Not very inspired."

Mission Day 27, December 12: "Yesterday was a bear! It was a moderate Go-Go day, but we had to stay up to 01:00 to make some UV stellar observations and to photograph a rocket as it launched. At the last minute they cancelled the launch, so the night was wasted. We got the stellar photos but that possibly could have been gotten earlier. Got to sleep till 09:00 this morning, then had a pretty routine day. Had to lean on Ed today. He wants to be an eager beaver solar physicist and spend all his time watching the sun. Feels obligated. On our second day off we did a double EREP pass and he did four ATM periods, which boils down to essentially no day off for him. He was exhausted that night and wouldn't take a shower – just went straight to bed. Tried to go for 6 ATM passes this day off but I insisted he limit to 4, so he got ½ day off and a shower. Then he started bitching to the ground that they weren't scheduling him enough for ATM, and that he would take after hours stuff and 6 passes on days off. I kept thinking 'Sunday'. I had to step in and stop that, and he didn't like it at all, so I had to remind him of his first day off. He gets so uptight when he's working at the ATM, that it's really not very relaxing. So do I for that matter. Anyhow, I've got him going at a reasonable pace for now. Hope we can keep it that way. Talked to JoAnn and the kids tonight. Jeff's voice is really deepening. Darn guy is growing up too fast! Glad to hear JoAnn cancelled the new car that could have been a real burden if we had trouble selling the old one. Taped a goodbye note to Dee [O'Hara] tonight. Sure sorry to see her go. Glad on the other hand though. She needs a change."

Mission Day 28, December 13: "We're $\frac{1}{3}$ of the way through our mission now. Hard to believe. The first 10 days were slow and painful. Now they're going faster. We've equaled the SL-2 time and we feel like we're just getting started. We're not working as hard and beginning to enjoy it more. Did some window looking today. Just don't see how Al Bean could say he'd go crazy if he had to look out of the window for [more than] 60 days. There's always something to see day or night – be it land, sea, clouds, stars, lights, fires. I first try to imagine who's living down there. There's so much desolate country down there, but you can still see the marks of man. The US is so green and neat compared to Australia, South America, Africa and China. Soon we'll be able to see Europe by day. That should be interesting. Bill saw the lights of London today. I'm going to find time tomorrow to look at Europe in the evening. Today we started thinking about Christmas – trying to figure out a tree. Got some good ornament ideas. Green plugs from urine bags. Red food containers. Paper chains."

Chapter 8

Christmas in Space - The Second Month

As Jerry and his crew surpassed the Skylab 2 duration and completed their first month in space, program managers gave the "go" for a 60-day mission. This would be reviewed weekly to determine if a full 85-day mission was possible and worthwhile. Flight planners on the ground were becoming concerned that the man-hours of science performed each day was slightly reduced from the previous mission. Jerry and his crew were attaining 25-27 man-hours per day, whereas Bean's Skylab 3 crew had reached 29 man-hours by the end of their first month in space. But several factors had to be taken into consideration. The six day launch delay cancelled some of the planned EREP runs (and the bad weather delayed others in flight). The loss of CMG-1 on FD8 inhibited maneuvering the workshop, and of course there was Pogue's illness. There had also been a substantial amount of stowage changes prior to the mission, for which the crew had not had sufficient time to train.

To illustrate this, in the final four weeks prior to the launch of Skylab 2 there had been 6,000 stowage changes. For Skylab 3, 8,000 changes were made. In the final month prior to the launch of Skylab 4, the records reveal that over 17,000 changes had been made to the stowage, as everybody tried to get as much onboard the final mission to the station as they could.

Bill Pogue became frustrated on the very first day inside Skylab thanks to the stowage inside the CM: "I have to get all this stuff out of the way so I can open panels. Man, it's a mess, this is really too much. I have to move 50 jillion things to get to one place." Ed Gibson added his own observations in communications with the ground: "Oh it's going pretty slowly. We're doing a lot of on the job training and stowage is a little bit difficult, what with everything being helter-skelter in the Command Module. We're having to sort though a lot of piecework."

Jerry commented on time utilization via the B tapes early in the mission (FD9), reflecting the hectic first week onboard the station. "I think, looking back, the best word I can use to describe it is frantic. I think it's too bad that we have put ourselves in this position where we are just breaking our necks trying to get a lot of stuff done. It would have been a whole lot better, I think, at least for our morale if for nothing else, if what took a week could have been spread into two weeks, because the biggest time consumer we had was finding things. We were not really familiar with locating everything."

Despite this, the crew began picking up speed over the first month on Skylab and worked hard to achieve their pre-flight optimum work profile. They were beginning to enter a routine, although up to Christmas this was still hurried and continued to build up the barriers between the crew and the ground. Once the air was cleared, they began to become more productive, and in fact surpassed many levels attained by the first two crews. As his son Jeff commented, Jerry was 'dealing with the problem', but it would still be a few more days before the 'problem' was finally addressed.

The period from Mission Day 29 through to Christmas was one of working the major science objectives of the program, as well as learning to live and work in space for a crew who had never been there before. The primary objectives of Skylab were in solar science, Earth observations, biomedical investigations, and astronomical, technical and student experiments. As the crew entered the second month of orbital operations, they had already accumulated 160 hours 24 minutes at the ATM, 4 hours 48 minutes with the EREP, and 157 hours 26 minutes on the medical experiments. In addition, Jerry and his colleagues had spent 58 hours 22 minutes on 22 corollary experiments and 32 hours 39 minutes on corollary and Kohoutek observations. There had been 2 hours 50 minutes spent on student experiments and 14 hours 41 minutes on various other scheduled tasks.

Combined, this totaled over 430 hours science work, but this was still 150 science man-hours less than the Skylab 3 crew had achieved at the same point in their mission. It is therefore a tribute to Jerry and his colleagues that they turned around these averages in the last two months to achieve the results they did by the end of the mission. One of the unspoken objectives of Skylab was to learn how to balance living and working during those early pioneering missions in space. Skylab 4 became the first mission to really utilize the in-flight experiences of rookie space travelers on a long mission. They returned some of the most important data gathered on any manned spaceflight up to that point, and for many years to come.

SKYLAB - A HOME IN SPACE

One of the significant differences (other than the mission duration) in flying a Skylab mission, compared with anything that had preceded it, was the sheer size of the spacecraft. With the Skylab 4 CM docked to the station, the combination measured over 115 feet (35 m). The total mass was over 200,000 lb (90,000 kg), with an internal habitable volume of 12,500 cubic feet (354 cu. m). The description favored by the media was that Skylab was like a small, three-bedroom 'house in space', which was not far from the truth. Compared to the tight confines of the Mercury, Gemini and Apollo spacecraft, Skylab was positively immense.

Learning to live and work in this cavernous volume was a 'space adaptation' experiment in its own. As with Skylab 2, it was one of the secondary objectives of Skylab 4, and was listed under Habitability Studies. The sometimes frank impressions of hardware, facilities, procedures and experiments recorded by the crew on the restricted access B Tapes were later down-linked to the ground. Their purpose was to assist in post-flight debriefings and evaluations of the mission in order to improve future flights. These "restricted access" B Tape transcripts, which the crew hadn't realized were being released to the media, were often misinterpreted by the press and taken out of the context for which they were intended. Had the crew known that the B Tapes were being quoted by the press, they probably would have been more circumspect in their down-linked criticism and would have saved the more frank discussions for the closed debriefing on the ground. Skylab was to have been the first of a series of orbital workshops, and the experience gathered from the first series of missions could have been used to upgrade and improve conditions on later versions of the Saturn OWS. A similar step-by-step program was adopted by the Soviets for their Salyut/Almaz/Mir program. When the other Saturn workshops were cancelled however, the three manned missions to Skylab afforded America the *only* opportunity to fully explore long duration manned space flights for some considerable time. In fact, it was to be twenty years before Americans returned to long duration residence in space. But it wasn't only the Americans who benefited from Skylab 4. In 2001, during a visit to the UK, Jerry Carr met Valeri Polyakov, the Russian physician-cosmonaut who had spent over 22 months in space on two missions to Mir (8 months and 14 months). Polyakov revealed that the post-flight documentation and research results from Skylab, especially from Jerry's mission, were instrumental in preparing for long flights on Mir.

Learning to work efficiently on the Skylab required quickly learning to orient oneself inside the OWS in a way not possible in the 1-g trainers on Earth – by floating through in zero-g. One of the most enjoyable pastimes for all the Skylab astronauts was to try to float through the internal length of the workshop by pushing off from the 'floor' of the CM and reaching the other end of the workshop without touching a wall.

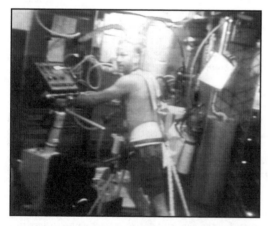

A TV picture of Jerry exercising on a treadmill-like device in the OWS. Jerry is held down by bungee cords attached to a hardness onto a Teflon coated aluminum sheet bolted to the triangular grid floor of the workshop This downward pressure creates a force to allow leg and muscle exercise while in zero-g.

Traveling through the forward hatch of the CM they would first enter the Multiple Docking Adapter (MDA). This was cylindrical and measured 17 feet (5.2 m) in length and 10 feet (3 m) in diameter, with a habitable volume of 1,130 cubic feet (32 cu. m) This compartment housed the second docking port (at 90° to the first) and the control panels for both the Apollo Telescope Mount and the Earth Resources Experiment Package. Also located in the MDA was a furnace chamber for the materials processing experiments, as well as various tools and spare parts for use by the crews as required.

Permanently attached to the rear of the MDA was the Airlock Module (AM) that measured 17.5 feet (5.3 m) in length and had a habitable volume of 615 cubic feet (17.4 cu. m) It had a hatch at each end to allow it to be isolated from the rest of the spacecraft and was the only way into the main workshop, via the Structural Transition Section to the forward dome. An EVA hatch (a spare Gemini hatch) was located here to allow EVA egress and ingress without depressurizing the whole vehicle.

The Orbital Workshop (OWS) was the modified S-IVB stage, equipped on the ground and used as the primary living and working area during a resident mission. With the engine removed and replaced by a thermal heat rejection unit, it consisted only of a hydrogen tank and an oxygen tank. The 10,300 cubic-foot (292 cu. m) hydrogen tank provided the habitable volume, divided into two sections by an aluminum triangular open-grid floor. It was thermally protected by the micrometeoroid shield and, after the loss of part of that shield, by the parasol and solar sail devices installed on SL2 and SL3. Ten water tanks around the upper dome each held about 600 lb (272 kg) of water. There were also 25 storage lockers, containing the majority of supplies needed by all three missions. The Forward Experiment area (forward of the triangular grid floor) contained food lockers, freezers, two scientific airlocks (one of which was unusable because of the deployed parasol assembly) and experiment hardware. A central opening in the grid floor provided access to the lower deck and wardroom. Here were the 'living quarters', with an individual sleep compartment for each astronaut, the collapsible shower assembly, and the water management and hygiene facility (euphemism for bathroom). The wardroom, or dining room, consisted of a meal table/food preparation area, food and equipment storage lockers, a small freezer and a large double-paned 1.5-foot (0.46 m) window, which was heated to prevent fogging. An environmental control system (ECS) provided a temperature range of between 16.6 and 32.2°C (62-90°F). An oxygen/nitrogen atmosphere was used with an internal pressure of 5 psi (3.45N/cm²). Centrally located in the lower floor was a trash airlock to transfer filled garbage bags into the former oxygen tank of the Saturn stage. By the time Jerry's crew arrived on the station, however, the airlock's mechanism was wearing out and wouldn't latch down easily. The crew solved the problem by one of them bracing themselves against the 'ceiling' to 'stand' on the lid, forcing it closed while the second person latched it. It was another disruption they could have done without.

Located outside the OWS were solar arrays for electrical power, and the Apollo Telescope Mount (ATM) structure for solar observations, with its own 4-panel solar array system. The two OWS solar arrays should have generated 10,500 watts of power at 55°C, but the loss of one of them during the launch phase seriously threatened the success of the program early in the mission.

Living On Skylab

There is no such thing as a 'typical' day in space, as there is always something different going on. This was certainly true of Skylab, which was an experimental spacecraft in itself. However, by overviewing the daily activities of Skylab 4, one can appreciate what it was like for Jerry, Bill and Ed to live in space and call Skylab home for three months:

JC: "The people on the ground woke us up every morning at six [except on days off], and it was always pleasant because you never knew what they were going to do. We had some really strange wakeups. One time it was Brigitte Bardot. I think they played something from a tape recording she had made. Another time, they played a very funny Jerry Reed song entitled, 'When you're hot, you're hot. When you're not, you're not.' Still another time, they played 'On top of the world'."

Sleeping: Jerry describes the individual crew sleep stations (bedrooms) as being about the size of a telephone booth. The 'bed' hung on the wall and was like a sleeping bag laced to a tubular steel frame that was fastened to the wall. The inner layer of the sleeping bag was netting and the outer layer was fabric. If it was warm, Jerry would unzip the outer fabric, leave it open and just sleep under the netting layer, or unzip the netting layer part way. At about thigh height was a wide (about 8 inches, or 20 cm) elastic strap that attached across the rack for the purpose of pushing the thighs back against the bed. A second wide strap could be secured across the chest to push the upper body against the bed:

JC: "The reason for this was to give the sleeper the feeling of pressure. It had been learned that when you try to sleep while you are weightless, quite frequently your subconscious would say, 'Hey, I'm falling out of bed!' and would wake you up with a start. It was like the American Indian women who wrapped up their babies tightly to give then the sense of security when being carried while asleep. At the top of the bag was a turtle neck and the idea was to zip the bag down to get in and zip it up with your head sticking out the top. We had a pillow which was a piece of firm sponge, with a hairnet attached to it. You could pull the net down either under your nose or under your chin, and this would prevent your head from flopping around while you were sleeping."

The crew normally slept at the same time, and the three sleep compartments were located adjacent to each other in the lower wardroom area. Jerry never felt the need for eye masks or ear phones to shut off light or noise while he tried to sleep. Skylab was pretty quiet, considerably more so than the Shuttle, Mir or International Space Station.

A couple of times an alarm went off during a sleep period and Jerry recalled during post-flight debriefing that the design of the sleep restraint was not conducive to responding to an emergency situation: "There was a lot of struggling and banging and thumping going on while getting out of that thing. It's too bad that we didn't have a system that you could whip out of quicker."

Directly across from the bed on the opposite wall of the little cubicle called the 'Sleep Station' was a stack of drawers and cabinets. At chest level was a door that unlocked at the top and folded down, much like a desk. This is where Jerry kept the diary he wrote while he was in space, plus photographs of his family, spare clothing, writing utensils and similar items. Air flowed from a set of louvers in the floor to another set in the ceiling. The compartment was designed so that the occupant could close a flexible door and turn out the light. Zipped up in the sleeping bag, Jerry felt warm and comfortable and generally slept well.

Daily Mass Measurement: Immediately after wake up, one of the first things all three astronauts had to do was to 'weigh' themselves as part of the medical experiment program. The Body Mass Measurement Device (BMMD), labeled Experiment (M172) was developed by fellow astronaut Bill Thornton. Located in the work area of the workshop, it featured a small 'rocking chair' mounted on an oscillating framework. Once strapped tightly into the device, the astronaut released a latch assembly and the chair would rock back and forth, creating a frequency of oscillation different from that when the chair was empty. A small computer calculated this difference and converted it into the mass of the astronaut. On Earth, mass is equal to weight, but weight changes with the force of gravity while mass does not. In Earth orbit, an astronaut had no 'weight' but possessed the same mass as on Earth. Consequently, the computer could calculate the Earth weight of the subject to within one tenth of a pound. It turns out that Jerry gained a total of two pounds (approximately 1 kg) during the mission, while Bill and Ed lost a little.

Morning Chores: While one of the crew used the BMMD the other two went about completing other routine chores that were required each day. These rather less glamorous duties of an astronaut career included changing the urine bag in the waste management compartment. "Your urine bag needed to be changed every morning. Right after the first urination, you would change it, put a clean bag in and process the used bag. We had to shake it up to mix it well and then attach a small, soft, 100 cc plastic cube to it via a small soft aluminum tube and fill the cube full of urine. That was the urine sample for the day. We then crimped the tube, removed the cube, and what was left over was thrown into a trash bag for disposal. The cubes of urine were placed in a freezer for Earth return and were aptly named 'urine-cicles.' So the morning chores consisted of weighing ourselves, doing the urine bag change out, and normal hygiene. Frequently, there was some kind of experiment that needed to be started because of where we were in our trajectory as we orbited the world." The crew would stagger their morning activities so that one would be shaving or washing, while another was being measured for mass, and the third was off doing something else.

Scientist Astronaut William E. Thornton's feet demonstrate to the press in a pre-flight photo the operation of what became known as 'Thornton's Revenge' during Skylab 4. Thornton developed the device for Jerry's crew after reviewing the medical results from the first two missions.

Shaving: There were three choices for shaving. One was a wind-up German mechanical shaver, another was brushless shaving cream with a safety razor, and the third was not to shave at all. Neither Jerry nor Bill were impressed with the shaving equipment and gave up shaving early in the mission, though Ed carried on with it. "We tried shaving the first couple of days of the flight, but it was such an unsatisfactory experience that we just gave up. That's why we grew the beards." Jerry found the wind-up razor did not work that well and when he tried the safety razor, it became clogged up after only a few strokes. That meant it needed cleaning, and with no running water, it required wiping the blade constantly, which also dulled the cutting edge.

Washing: The washing facility was a little squirt valve, with which water could be added to a face cloth to give a sponge-bath type of wash. Another squirt device delivered the liquid soap on to the wash cloth. Normal bar soap was also available if desired. Jerry would mix both to lather up. After body washing, the wash cloth was placed in the circular wash cloth squeezer, which opened like a potato ricer. By closing and depressing the lever, the water could be squeezed out. At the bottom of the container was a hole with a filter in it through which the excess water was drained into a waste water containment bag. The wash cloth could then be used again either by adding more water or leaving it damp as it was. During Jerry's mission, the crew used this system for nine days. On the tenth day, "it was our holiday, and we took a shower."

The shower featured a 42-inch (107 cm) metal ring (or hoop), which was normally latched to the floor of the crew quarters work area. To take a shower, Jerry would step inside, unlatch the ring and pull it up around him, snapping it to fasteners on the upper grid in the wardroom ceiling. Attached to the ring was a cylindrical beta-cloth shower curtain, sealing off the 'wet' area inside when in use. A flexible hose with a push button trigger on a spray head allowed approximately a gallon (3.5 liters) of water per shower. A vacuum system was used to remove water from the body and to draw the water back out of the shower. Wash cloths and shampoo were provided, along with liquid or bar soap. The drawback was that once the curtain was lowered, the occupant felt the change of ambient temperature and very low humidity. Evaporation caused a real chill, and they had to race for the towels to dry their body and get dressed as soon as possible. The design was good, but it took about 45 minutes to set up the screen, take the shower and then clean the unit and re-stow it after use. On one of the downloaded habitability debriefings on the B channel, Jerry mentioned that the shower, though time consuming, was a welcome event. However, he commented that the liquid soap was not very pleasant smelling and compared it to dog shampoo. He wondered why a more pleasant smelling soap could not have been used. Shortly after mission completion, he reported that the Nutragena company had sent him a complimentary package of beauty soaps, with a card saying that they, "hoped [he] would not relegate them to the family pet."

Waste Management: The bathroom was designated the Waste Management Compartment (WMC). The toilet on Skylab was on the wall, with the seat vertical to the floor. As Jerry recalled, "It was kind of a strange thing. I felt like a bat hanging on the wall. The seat also had a seat belt. You'd get up on the seat and strap yourself tightly so that your buttocks were up against the seat of the toilet. Then you turn on a switch that opens a valve and causes air to start circulating all around inside the toilet bowl and backwards toward a pre-installed fecal bag. In drawers just below the toilet each one of us had our own urine collector. So to use the toilet, I found I just had to disrobe from the waist down in order to be reasonably comfortable. You had to have your urine container nearby very quickly because you normally urinate at the same time as you defecate. You turn on another switch that starts air flow into the base of the urine collector. The next problem was to get yourself off the toilet and clean up… getting the paper and everything thrown into the fecal matter with the fan running so it would stay in the bag and not drift out on you. When you were all cleaned up you would seal the bag and turn off the fan. Then you'd weigh it on a Specimen Mass Measurement Device (SMMD), which was a miniature of the BMMD. The bag was marked with the name, date and weight and then placed in a small heated oven that was vented to the vacuum outside. Over a 10-hour period, all the moisture from the fecal matter would be evaporated, making a little brown brick. Those bricks were then double bagged and brought back to Earth to be reconstituted for medical evaluation.

As part of a metabolic study back on the ground, doctors would add distilled water to each sample, bringing it up to its original weight. By taking small specimen samples from the fecal and urine samples, they could determine how food, vitamins and minerals were processed by the astronaut during the course of the mission. Only one urine sample was taken per day but all fecal matter was collected, even if an astronaut 'went' more than once a day. Any vomit was also collected and stored in fecal bags for return to Earth as part of the metabolic study.

Personal Hygiene: Hair cutting did not cause the crew any problems. It just meant having the vacuum cleaner handy to suck up loose cuttings as they were snipped off. Washing hair was fine with the wash cloth or shower. The drive to keep personally clean was still there, even with the restricted equipment and time, as Jerry recalled: "We found it easier to stay clean up there because we did not sweat as much. We found that one full body wash per day and one shower [every tenth day or so] was quite adequate. In fact, you could get along without the shower, if you kept up with the body wash and did a good job with that. But there is no substitute for running water all over your body and getting it in your hair. A shower is a very refreshing thing, but again it's very time consuming." Jerry also pointed out during post-flight debriefing that a crewmember should choose his own deodorant, shaving cream and toothbrush. "I think the toothbrush was too soft. We told the folks before we left and still feel that way. I was very grateful to get back to my good old hard toothbrush at home." Better deodorants were also suggested, as were biocide wipes and 'air fresheners.' The toothpaste was ingestible, so each person had the option to either swallow it or spit it into a tissue.

Clothing: Because the missions were of some considerable length (4, 8 and 12 weeks) compared to those which had preceded them (up to 14 days), and with the reliability of the workshop pressurization system, the crew were not required to wear their full pressure garments. Depending on the temperature inside the vehicle (averaging 74 to 78°F), the crews of Skylab could wear a pair of fire retardant trousers, resembling those of a flight suit but with a cloth belt and larger pockets. On occasion, when the relative positions of the sun, Earth and Skylab meant the station experienced direct sunlight for longer, the temperature would rise. At one time, the temperatures rose to 83 degrees for three days because of this condition, referred to as "high Beta Angle." When that happened, the crew could unzip and remove the lower portion of their trouser legs, from mid-thigh down. The crews also had cotton T-shirts and a tan beta-cloth 'jacket', but mostly wore T-shirt and shorts. Their shoes were similar to tennis shoes, but with a specially designed plate on the sole fitted with a triangular cleat which could be fastened to the triangular floor grid of Skylab. During the Skylab 4 mission, there was some concern that the supplies of clothing might run out before the crew came home. Fortunately, the crew didn't have to take up one joker's suggestion that they should rotate clothing to solve the problem; i.e., Jerry would pass his used clothes to Bill, who would pass his to Ed, who in turn would pass his own clothing to Jerry!

Meals: There were approximately 71 items on the menu, and the dieticians had worked out precisely what vitamins, minerals and calories each contained. A six-day menu cycle was followed, with the menu repeating itself from day 7 for another six days, and so on. Prior to the mission, Jerry and his crew sat down with the dieticians to select their breakfast, lunch and dinner menus for the six-day cycles. A target of about 2,900 calories per man per day was set. Once selected, each meal was analyzed for calorie consistency and set levels of calcium, potassium, magnesium and other elements. Any shortfall was made up by dietary supplement pills, which kept the menu stable and consistent. If for some reason a meal was skipped or not finished, residual food was precisely weighed on an SMMD located in the wardroom – a different one, of course, from the one in the Waste Management Compartment. This data was reported to the medics on the ground every evening. They were keeping records on food and caloric intake as part of the metabolic analysis program. Meal variation was not allowed without reporting, because the doctors required knowledge of what each man should have been eating and drinking and when. Accurate reporting was essential to validate the medical experiments. Every sip of water was ingested via a personal squirt gun with a measuring device on it, which tallied each one-half ounce of water taken by each crewman:

JC: "They had a very, very precise record of everything we ate and of everything coming out the other end. We were being analyzed just about like you would analyze an internal combustion engine. You measure the fuel and air that goes into that engine, the brake horse power that it delivers, and you measure the exhaust gasses to see how efficiently your engine was burning. That's basically what they did with us.

Jerry, strapped to the triangular grid floor by his right ankle is photographed for a medical experiment to record changes in the human body during an extended spaceflight.

"It got a little boring after three months but it was manageable. We had very good foods… three kinds: frozen, freeze-dried and thermal stabilized. Nothing fresh. The frozen items included lobster newburg, filet mignon, roast pork and dressing and ice cream. We had freeze-dried chili con carne, spaghetti, soup, a couple of kinds of vegetables and strawberries, plus tea with lemon, lemonade, orange and strawberry Tang and coffee. Thermal stabilized items, another NASA euphemism for canned items, were mostly fruits and things. There were also snacks like peanuts and sugar cookies. These items were augmented with supplementary food bars in order to stretch the food items to fit the longer mission time. The diet people in Mission Control modified our daily menus with the food bars while maintaining the caloric, vitamin and mineral levels established pre-mission."

For Skylab, a wardroom dinner table was developed to allow the crew to eat their meals together. The table resembled a three-position snack bar, with each position facing the other two. Jerry recalled that it worked very well. The top was slightly lower, and thigh restraints, combined with the grid cleats, locked the crewmen in a crouching position around the table. However, the thigh restraints proved uncomfortable and they soon stopped using them. They were pleased to find, however, that the body's natural zero-g posture held them at a comfortable crouching

Jerry floats in the "neutral body posture" in zero-g to demonstrate the body position, as part of a medical experiment.

position for eating. This also meant that more things were within reach when leaning over toward a cupboard, or to the back of the meal pedestal, without unlocking their shoes from the grid. "Each of us had a food tray which had holes in it that matched the diameters of the large and small food 'pop top' cans. The tray itself was about 3.5 to 4 inches (8-10 cm) thick and had some thermal heaters, some heating elements and controls. The tray itself was very highly insulated so that you could have 'cooked' food in one area and you could have ice cream nearby, which would not melt. It was really well done."

Food was retained in place mostly by its consistency. The ones that gave the most trouble were the soups and cereals. "The cereal had dry milk mixed into it, so when we added the proper amount of water and shook it, the milk became liquid. The problem with soups and cereal was that the liquids were thin and were easily shaken loose by a careless eater. Thicker sauces had more surface tension and tended to hold the food in the container." When one meal was finished and cleared away, the next was loaded into the tray. The containers would be opened and reconstituted, adding water to freeze-dried items. The timer would be set to come on and heat the food at the proper time, so that the astronauts had a warm meal ready at meal time. Despite being busy, Jerry found that he and his crew were able to make most meal times as planned. Lunch was a little more casual. "Sometimes, if you were running behind and you needed a little more time, you would grab your lunch and take it with you to where you were working. At dinner every night, we tried to eat as a crew."

Personal items: For relaxation, the astronauts had a deck of cards, a Velcro dart game and checkers. However Jerry's crew never bothered with any of it. "Each of us had about 7 inches (18 cm) of shelf space to put in whatever paperback books we wanted, and these were put in fire-resistant covers. Jerry read Arthur Haley's *Wheels,* then *Man the Manipulator,* followed by John Gardner's *Self Renewal* and Al Bean's copy of *Childhood's End* by Arthur C. Clarke. There were about 20-30 music cassettes specially cut for the astronauts, which they listened to during exercise. Jerry liked listening to his music at 'night' before going off to sleep:

JC: "The most enjoyable entertainment up there was looking out the window. We spent more time doing this than any other form of entertainment. We took photos of all we saw, talking into a tape recorder about the photographs we had just taken. We'd note the time and the subject of the photography and why we had taken it. We dumped all those tapes down to the ground and brought all the film back. I also wrote my own diary.

"Running around the wall lockers of Skylab (near the upper dome) was kind of fun. It was not part of the [formal] exercise routine. You could start by crawling, and then as you picked up velocity you could stand up and run using centrifugal force to hold you on the lockers. To stop, you kind of do a tumble or just slow down and grab something as you float free of the lockers."

Newton's Second Law came into play and as the astronauts ran or tumbled around the lockers, the spacecraft wanted to rotate in the opposite direction.

Jerry's crew was restricted from running the lockers early in the mission. With one of the gyros down, it was feared that the remaining two might not be able prevent counter rotation.

Housekeeping: Keeping the workshop clean was beneficial to both health and safety. The vacuum was used quite a bit on Skylab. Apart from collecting hair during a hair cut, the most frequent job for the vacuum was to clean the air conditioning filter intake system:

JC: "You would go up there, and you'd find that those screens were just covered with this grey looking crud. So we started asking 'What is this stuff?'. The guys on the ground replied, 'It's skin.' We looked at it very closely and realized they were right. That, and lots and lots of body hair. That's just natural sloughing.

[On the ground] we are all continuously shedding skin and hair and not realizing it. It all falls to the floor or collects in your clothes and is vacuumed up eventually."

Any lost item of equipment usually floated in the gentle, but steady, air flow towards the filter screen. This is where most items were found after a few days. "We referred to it as the 'Lost and Found'." The major cleaning areas for the crew were the bathroom (WMC), small spills and dribbles around the urinals, and the grid ceiling of the wardroom above the table where the thin liquids collected.

Trash was stored in small white cloth bags. They were filled with used clothing and trash, and after gathering several, they were disposed of in the trash airlock. To operate the trash air lock, the upper hatch was opened and trash bags were pushed in. Then the hatch was closed, the volume depressurized and the lower trap door was opened allowing the filled bags to be levered into the 22-foot (6.7 m) diameter oxygen tank. By the time Jerry's crew arrived, the first two crews had deposited nearly 90 days worth of rubbish in there. Towards the end of their 84-day mission, Jerry's crew had to brace themselves for the aroma back flow when they opened the upper hatch. The airlock smell was alleviated somewhat by using splashes of precious 'Old Spice' aftershave. Old Spice and Joy liquid detergent (for de-fogging the inside of space helmets) were the only two sources of any pleasant fragrance in the station, so they were used sparingly.

For the most part, Skylab had little or no noticeable aroma, mainly because the crews were meticulous about cleaning. As well, the crew felt that the designers did a very good job in habitability design. They hoped that their regular critiques on the downlink would serve to ensure that any future space station design would be even more compatible with the human occupants.

A Place to Work

There were many different work stations inside the orbital workshop, as well as outside at the sun end of the ATM. There were the ATM and EREP control stations, the medical experiments, and scientific airlocks. Each one was a different design, but this was more a minor irritation than a significant problem. "When we got back, we mentioned in the habitability questionnaire [during debriefing] that if the workstations had been more uniform in nature, we probably would have been more efficient workers. But I think it is hard to do that when you have a bunch of experiments with different goals and methods of operation. You just can't make all of the workstations the same."

At almost every workstation, there was a floor grid in which to lock their feet. "That was one of the beauties of Skylab. There was always a good place to lock your feet in. We tried to get the designers of the Shuttle and space station to go with the grid again, but we just could not get them to do it. They were all taken with the idea that it was much better to wear soft 'slippers' and just have foot loops for restraint. We didn't find the foot loops to be as satisfactory as positively being able to lock yourself to the floor. Sometimes you could feel the strain of leaning to reach some of the equipment, but most of the time, like at the ATM panel, you could tilt the floor grid so you didn't have to lean into it. We would set the floor grid so that we were comfortable standing in front of the workstation. As we relaxed, our bodies drew down into somewhat of a crouch, and we could just sit there and comfortably crouch in front of the instrument panel."

Initially there was a seat at the ATM panel, but Pete Conrad's crew soon removed it. According to Jerry, "the seat was more of a hindrance than help. It restricted your motion and forced you into an unnatural body position. The zero-g neutral body position is not the same as when seated. Actually, it's a lot like the position you would be in if you were in a reclining chair or if you were floating relaxed in the water. Your knees are slightly bent and your back slightly curved. When there is little gravity, your muscles just pull you to that normal position. And of course, the longer you stay up, the more your muscles atrophy, so that normal body position is going to change slightly. My natural body posture at the beginning of the mission was slightly different to that at the end of the mission." This change in body posture was another ad lib measurement made during the flight. "We would strip to our underwear and take photographs of each other floating in neutral body positions."

The plan was to conduct all 90 minutes of exercise in one go. Jerry's crew discussed trying to split this up to improve the work regime, but soon realized that this would mean getting sweaty twice and therefore waste more time by having to clean up twice. "So we stuck to one hour and a half every day, and in reality this was an hour and ten minutes of exercise and twenty minutes of clean up". The exercise equipment available included the bicycle ergometer, inertia-reel springs, isometric straps and the sheet of Teflon-covered aluminum on the floor, called 'Thornton's Revenge'. Jerry jokingly reasoned that "because he [Bill Thornton] didn't get to fly on Skylab, he got even with those that did by developing a rudimentary treadmill."

The ATM control and display console located in the Multiple Docking Adapter was the primary control station for the ATM. This gave the observer/operator information on which experiments were currently active, which programs were being used, and precise pointing capability for each instrument. The sun could be viewed via any two of five TV cameras at once. From here, the astronaut could also control the ATM experiment pointing system, instrumentation, communications and power system. The EREP control and display equipment was located in another sector of the MDA and was used for operating the suite of Earth resources experiments.

The solar physics experiments included UV X-ray solar photography, a white light coronagraph, an X-ray spectrographic telescope, a UV spectrometer, a dual X-ray telescope, a UV coronal spectroheliograph, a UV spectrograph, and a trans-uranic cosmic ray passive detector. There were also multilayered foil collector sheets mounted externally and retrieved during EVAs. These sheets gathered magnetospheric particles, and their composition could be studied upon their return to Earth.

The Earth Resources Experiment Package included a multi-spectral photography facility, an Earth terrain camera, an IR spectrometer, a multi-spectral scanner, a radiometer/ scatterometer and altimeter, and an L-band radiometer. There were also studies involving the atmospheric absorption of heat, and of volcanoes. Skylab also carried 13 astrophysical experiments, 17 medical experiments, several bioscience experiments, technology experiments and a number of student science demonstrations.

Jerry disagrees that it was the crew's inexperience of spaceflight that caused the difficulties in maintaining the work pace early in the mission: "It was the workload. The inexperience of the crew had nothing to do with it. The daily mission planners started us out at the same pace that the crew before us had finished. We had no time to warm up, to get up to speed. They even put in two experiments that we had never seen before, so we started having trouble right from the beginning. We were in such a rush that we made mistakes."

Generally, the ground would read up the news to the crew while they had breakfast, but they were left alone during other meal times. Though the communications were not turned off, the astronauts were left to get on with their meals, early morning hygiene, or post-sleep activities as required. However, the ground did not read up *all* the news. During the Skylab 4 mission, news of the Watergate scandal broke and the ground never read anything up to the crew about that.

JC: "I think they were afraid we might say something derisive on an open channel." (In fact, the crew learned about this via private family conversations, but knew enough to keep quiet. Since nothing was coming up on the open channel, they suspected that they shouldn't mention the situation.)

Skylab was packed on the ground with all the logistics required for the three manned missions because the Apollo CM could only accommodate a small amount of extra equipment. When Jerry docked his densely packed CM to the front port of Skylab, they carried additional supplies and equipment sufficient to support a three-month mission. The first two crews had used a lot of the supplies on-board the station, and the early difficulties with the launch and thermal shielding of the OWS had ruined some film, food and other supplies. The extra supplies their CM carried, along with extra experiments, were the source of many of the problems that the third crew suffered during the first few days of their mission. Everything had to be transferred from the CM and stowed in the OWS:

JC: "One of the things about Skylab was that there was a place for everything, and it was important that you put everything back in its place. Unfortunately, some of the crewmembers before us didn't always do that, and we ended up not being able to find certain equipment. There was a written inventory in a book on the ground that reflected the location of everything stowed on the spacecraft. From that, the people in Mission Control were supposed to know where everything was. The rule that all the crews tried to follow was that if you took something out of storage and used it, then decided that there was a better place to store it, you would report that to the people on the ground. That way, they could change the storage book and be able to tell you later on when you forgot where it was you put it."

Unfortunately, busy crewmen forgot to update storage locations, and the book became inaccurate. Consequently, time was lost looking for items, and productivity decreased. It's interesting to note that inventory control turned out to be a significant problem on the Mir space station and is becoming one on the International Space Station.

If an item of equipment broke down, the crew would contact the ground for assistance in repairing the item. In some areas, none of the crew was trained for detailed repair of certain items of equipment, so referral to the ground support teams was usually the only way to get the item back up and running again. The engineers in Mission Control became adept at sending up text and simple diagrams on the narrow tapes of the teletype system.

EXTRACTS FROM THE DIARY

As the mission progressed through the second month, the efficiency of the crew improved. But they were still being pressed by the ground, who were scheduling catch-up work on what were supposed to be their rest days. With two EVAs scheduled around the Christmas period, including observations of Comet Kohoutek, the month was already tightly packed with activities, and the strain of trying to please everyone finally took its toll on the crew. They began to voice their complaints to the ground; a move which shook a few people but certainly changed the mood and approach of the ground/space team working relationship.

Mission Day 29, December 14: "Really had a thrilling sight tonight! We were setting up for an EREP pass over Indonesia and Australia and finished the set up early just as we were coming up on the coast of Spain [midnight Europe time]. Bill and I opened the EREP window and Europe was nearly cloud free. We could see Madrid, Barcelona, Paris, and London all at once. It looked like one of those black velvet paintings – like somebody had painted spider webs in silver and white on black velvet. God! What a sight! I had heard that Europe is small [from space], but tonight's viewing brought it all into focus. We went from Spain to Russia in about 4 minutes. Bill and I sounded like two kids on Christmas morning with the entire 'Hey look at that!' 'Wow!' etc. I think that's about the most enjoyable out-the-window session I have ever had. The more I see of these places around the world, the more I want to go to see them on the ground. I sure hope we can wangle a good post mission tour."

Mission Day 31, December 16: "Didn't even think to write something last night. Went to bed bushed. Talked with JoAnn and the big kids last night. They were at [a family friend's] for a Christmas Party. They all hollered 'Merry Christmas' to me. Sure are a good bunch of kids. Had another LBNP run today. They're getting easier again – I'd better not get too complacent. I figured I had it knocked before I conked out last time. Today we got our first solar flare. A hot spot started acting up this morning while I was on the console. Got a burst that didn't amount to much. Late this afternoon while Bill was

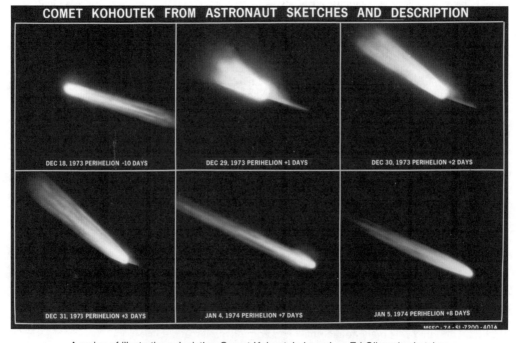

A series of illustrations depicting Comet Kohoutek, based on Ed Gibson's sketches and crew descriptions during the third mission.

on, it really let go with a nice flare. He caught it as it started up, then it faded back for a few moments and then went right on up to a peak. By then all three of us were at the console – all eyeballs. We got the entire event on film, so there ought to be some happy scientists when we bring it home. After it was over Ed was all down in the mouth because he wasn't lucky enough to have been at the console when it popped. He has worked like a dog at the ATM. I hope the old sun gives him a good flare pretty soon."

Mission Day 32, December 17: "Today was a 'day off' except there was a lot to do. We've about decided that all 'day off' means is shower day. I got about two hours of free time aside from my shower time. I guess if they [the ground] do that to us again I'll have to holler. Tonight we had another EREP pass and saw the whole Southeast US. It was really beautiful; even better than Europe. We could see the entire Gulf Coast from Brownsville, Texas to Key West in Florida. I wonder if JoAnn and the kids were listening to us. Everyone on the ground seemed to enjoy our description of what we saw. The days are really going by fast now. Last week was a breeze! Hope they all go as fast. We're approaching the half way mark."

Soyuz 13 is launched

On December 18, 1973, the Soviets launched the two-man Soyuz 13 spacecraft. Upon entering orbit, the mission created a 'space first' by having American astronauts (Jerry, Ed and Bill on Skylab) in space at the same time as Russian cosmonauts (Commander Major Pyotr Klimuk, Soviet Air Force, and civilian Flight Engineer Mr. Valentin Lebedev). The two cosmonauts were to spend a week in orbit aboard Soyuz 13 before returning to Earth on December 26. This was the second successful manned Soyuz mission in the past three months (Soyuz 12 had flown a planned two-day test flight of a Soyuz ferry craft in September 1973), vindicating the changes made to the spacecraft in the wake of the 1971 Soyuz 11 tragedy. Neither crew reported sighting each other despite being in similar orbits, and no communication between the two spacecraft was attempted (although Jerry recalls telling Mission Control to invite the cosmonauts to "stop in for coffee!")

A color photo of Comet Kohoutek, taken from the University of Arizona Catalina Observatory with a 35 mm camera on January 11, 1974, supplementing the observations taken aboard Skylab.

Soyuz 13 was a spacecraft specially adapted for independent flight at a time when most production line Soyuz craft were manifested for flights to either Salyut or Almaz space stations. At the time, the US and Russia were developing the joint Apollo-Soyuz Test Project international docking mission planned for July 1975, and this flight was a valuable demonstration of the improved capabilities of the Soyuz to the Americans. The RSS-2 spectrograph on-board Soyuz 13 actually used NASA-supplied film for stellar studies. During the mission approximately 10,000 spectrograms of over 100 stars were obtained, while the crew also performed a range of medical and biological experiments, navigational exercises, and observations of the Earth and its atmosphere. The mission was also probably timed to study Comet Kohoutek.

EXTRACTS FROM THE DIARY

Mission Day 33, December 18: "Today started at 04:00, and I was ready for bed again by 11:00. Had trouble getting to sleep again last night so I only had four hours. Thought I was going to die until I had some exercise and some lunch, then life wasn't so bad. The only time I looked out the window was the EREP pass. I saw Peru, and that's all. Must be more alert next time. Talked to JoAnn today. Nice to talk to her alone for a change. The guys from [The] "GO" [Show] [a CBS TV children's program] sound pretty interesting. Sure glad to hear of their interest in Jessica, Josh and John. Hope they do some filming.

"We're finally getting some Christmas things going up here. We've got our tree nearly finished and will start decorating it. Looks better than expected. Hope we can find enough different colors around here to splash it up a lot. I can think of green, yellow, red, blue, silver and gold. We turned on some Christmas music this evening and worked on the tree. It suddenly dawned on me that I'm not going to hear Rogge's [Rogge Marsh, an ebullient family friend] rendition of 'Deck the Halls with Boston Charlie.' That's kind of sad. I've got the tapes of all the good stuff we've done at the church, but forgot Rogge's 'piéce de résistance'. When I heard 'Rudolf the Red Nosed Reindeer', I could feel Josh's excitement. I bet he's on his horse now. No doubt the piano is ringing with Christmas Carols in his own style. We don't have another day off until the day after Christmas [EVA day], so we'll have to ask for some special TV on Christmas Eve so we can show our tree and say our thing. I've got to start putting some thoughts together for Christmas too. Ed's all hung up on a Santa Claus bit so I'm going to turn him loose on that and I'll stick to world peace and brotherhood.

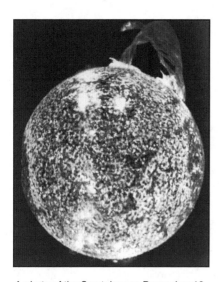

A photo of the Sun taken on December 19, 1973, revealing one of the most dramatic flares ever recorded. It spanned over 588,000 km (365,000 miles) across the solar surface. The photo was taken by the extreme UV spectroheliograph instrument of the US Naval Research Laboratory aboard the ATM solar observatory.

John Carr - TV star

The 'GO SHOW' was a product of the NBC network and was filmed towards the end of the flight in the Carr household, which added its own stress to the situation. JoAnn became particularly annoyed when the TV network wanted to film one of the telephone conversations between JoAnn and Jerry. A NASA official at headquarters in Washington got to hear about it and phoned JoAnn, stating that they did not like that idea and told JoAnn she could not do it. JoAnn responded by stating that no one could tell her what to do in their own house and that if they wanted to remove the 'squawk box' equipment that was fine with her. She would call her own press conference with all the six children, and simply state that NASA wouldn't let her or the children speak to the astronauts, and that NASA had removed the equipment. The official was left speechless but JoAnn was mad over a silly phone call. She did not work for NASA and would not be dictated to by them. Deke Slayton called a short time later, understanding the frustration JoAnn was feeling. He assured her that everything would be fine. The filming of the teleconference did not happen, but the pettiness irritated JoAnn for some time afterwards.

NBC wanted to do a special on one of the children of the Carr family, and John was the one on which they wanted to focus. "I think I was Dennis the Menace three times over," he recalled, "always getting into trouble. I was fearless. I didn't have the fear when I was younger that I have today. I was off the school bus and changing boots and running back out again – non-stop. I was the all-American kid I guess, playing football in the street. I had my animals in the back yard and many diverse areas of interest, so they picked me to do it. My older brother Jeff was fully occupied with things in high school and Josh was heavily into music and plays. Jessica and Jennifer were into horses, and Jamee was into fashion, so I was their main focus. I was either playing football with friends, all over the wood pile behind our house or in our tunnel system. I don't think our parents ever had to worry about me – their main worry was to keep us busy. I don't want to ever detract from my brothers and sisters because I wasn't aiming to get the main part. It just happened that way."

John remembers the presence of a security man sitting in the classroom while the children were at school. "I never knew why. I thought it was to protect us just in case something happened. Mom had so much on her hands that when Dad was in space I spent a lot of time with a neighbor. His name was Doc Stulken. He just mesmerized me. I slept over there on weekends and he showed me the outdoors. He was the one who got me interested in animals. He had ducks, pheasants, quail and so on. While Dad was up in

Jerry works at the materials processing facility in the Multiple Docking Adapter of Skylab

Jerry (right) and Bill Pogue work at the Earth Resources Experiment Package control panel area in the MDA.

space, we built a boat out of wood. I would spend days over there, Doc and his wife were interim parents while Mom was dealing with Dad's flight. My brothers and sisters were out doing their own activities, but I don't think I would be the person I am right now without him. My father is my father and had done everything you'd expect a father to do, but [Doc] set me in the right direction with a love for the outdoors."

On the 'GO SHOW', some of these outdoor interests were featured, along with the normal views of a young boy being difficult when asked to get out of bed. The TV show reflected John's day in comparison to Jerry's day in space. Life in space was shown as was life on Earth for the young John. TV pictures from space were edited with segments of John inside the Skylab mock-up at JSC. One great shot is of John in the Skylab sleep compartment, with John clearly stating his preference for his own bed over the Skylab one. At school on one occasion, John was shown being stumped by questions about the space program, which clearly embarrassed him. But he clearly knew who was in space at that moment. A school drawing showed the family house, his mother and siblings on the ground and Dad in space aboard Skylab.

Being much younger than his brothers and sisters, Josh Carr did not pay too much attention to the details of the mission while the flight was progressing. He knew that dad was 'at work' and would be home in a few months. What he recalls clearly from those weeks, however, was being very upset about older brother John becoming the lead 'star' in the TV show.

During filming, over several sessions, John tried not to look and smile at the camera, but he was the center of attention and felt really special. However, as the mission continued, his 'fame' soon faded and instead of requests for photos of Jerry, or his own autograph as the 'son of astronaut Jerry Carr', John found himself getting signatures of best wishes from his friends for when his Dad came home. Though his daily life did not change much, the questions about and interest in what was happening to his Dad increased during the mission. John speculated that "all that space stuff would probably be over by the time I grow up." He didn't seem inclined to follow in his father's footsteps: "I wouldn't go up there for a million dollars," he commented towards the end of Jerry's mission.

EXTRACTS FROM THE DIARY

Mission Day 34, December 19: "Today was a medium day. No big mistakes. Some variety in jobs. Ed and I spent a couple of hours trying to spot the comet with the ATM instruments. We couldn't see it on the TV monitors, but we knew where it was. It's just not bright enough. Glad we told them to schedule two of us at a time for these [comet] experiments. They're so complex that one guy needs to just watch and catch mistakes. The looking outside isn't very interesting now. We must spend 80% of each day coasting over water. We'll soon be over the US again, and it'll be more interesting. Our Christmas tree is coming along well. Tomorrow we should finish decorating it."

Mission Day 35, December 20: "One more week and we're half way there. The last couple have gone pretty fast. Made it though another LBNP today in good shape. Looks like all the exercise is paying off. I'm exercising more (and enjoying it less) now than I have since my freshman year at USC. All my fat has gone. I wonder how long it will take me to get it back when I get home. Just after dark tonight we flew just east of Chicago and looking north we saw the Aurora Borealis. What a beautiful but eerie sight. The whole northern part of the world seemed to have glowing fuzz growing up from it. Sitting up here looking at things like that sure makes you think differently about the Earth. You get used to your own little niche – your own surroundings where you live – and you lose sight of the whole wondrous thing. It's so big, and there are so many desolate areas! There's so much water! Most of the world's population is all crowded up into a fairly small percentage of the surface. No wonder the population explosion is bothering the experts. There is all sorts of room to expand, but so little of it is viable. No one is going to volunteer to live out in the Hinterland to make more room for those in the temperate areas. Christmas is coming fast. We should make it with our tree, and I hope we can pull together some meaningful remarks to make on Christmas Eve. I guess we'll try to do our own thing around mid-day so that we can get started on the EVA prep and get to bed at a reasonable hour."

JC: "During the mission we lived on Houston time. Wake up was at 06:00, and bed time was at 22:00. As it turned out, eight hours of sleep per night was probably more than we needed. Most of the time we slept about seven hours and were well rested, except on those occasional nights when one's mind starts racing. That happens on the ground, too, so it's not all that unusual. I customarily spent the first hour of the rest period writing in my diary or just enjoying the view from one of the CM windows. An hour of quietness helped me get a good night's sleep."

Mission Day 36, December 21: "The Christmas and EVA excitement is picking up. The ground offered us some TV time on Christmas Eve or Christmas Day – about two hours after I taped a message down asking for some. I've done so much today I'm all muddled up. Hope I can sleep OK tonight. Had to give the ground hell again tonight for scheduling things during our so called 'free time'. Hope it does some good. Talked to Rusty Schweickart for a while today about the EVA. Besides the stuff we planned, I've got to try to fix up one of the filters in a solar telescope. Have never seen the mechanism before, so Rusty sent up a description of a plan of action. Should be interesting to see if we can pull it off."

A Christmas message from Skylab

Prior to his mission, knowing he would be in space over Christmas, Jerry wrote a message that he would broadcast to the world on 24 December – Christmas Eve. It would be five years after the crew of Apollo 8 transmitted a similar message from space. With the help of his Presbyterian Church minister, Dean Woodruff, Jerry mentioned that words played an important part in everyone's lives and are used to express aspiration and anxieties. These were especially meaningful at Christmas time. Jerry observed, "Our Earth seems very large to us. Yet those men who have flown Apollo to the moon say it's small. We see that there are vast areas of desolation and great masses of water, with man crowded into only the more temperate and hospitable areas. The men from Apollo perceived the Earth as a tiny blue island in the vast sea of space. The observation is humbling and emphasizes the need for man to get into harmony. The Christmas season serves to heighten our awareness of others and our concern for brotherhood of man. No matter the religion or country of origin, or what season it may be on Earth, one of the principles for the future must be to learn to live in peace and harmony with one another."

Ed reflected on the fact that from space there were no visible borders between the countries of Earth, while Bill continued the message of kindness between all people. All three men finished with personal

Christmas greetings to their families and a description of the Christmas tree made from the larger food can liners. Jerry made no reference to the Apollo 8 mission, the last time men were in space over Christmas, and a time when he was on duty as CAPCOM.

"So from the crew of Skylab, we wish all of you a very merry Christmas. And an extra special message to all the guys that are going to be away from their families tomorrow, and tonight, working with us on our EVA in order to try to get some good photos of the comet. To all you guys we thank you very much for your interest and dedication, and to you we wish a very special season's greeting. So long."

EXTRACTS FROM THE DIARY

Mission Day 39, December 24: "Today's been a unique day indeed. We started it off with our Christmas message. I wasn't sure that we had chosen the right tone for our message, and I was afraid to trust myself to say what I wanted to without reading it. I guess it went pretty well. The guys on the ground said it was good, and JoAnn liked it. She's my best critic, and I guess if she's bought it then it's acceptable. Looks like we're over the hump. The Christmas greetings started coming in. First one's [from NASA administration] typically carefully worded – slanted towards our adjustment to the environment – Shuttle in mind no doubt… but it sure makes us feel good to receive all those good wishes. This afternoon we began the prep for EVA… it's always a frantic affair. There are so many things to haul out and pre-position. Tomorrow is my opportunity to get outside and look around. I'm really excited about it. What a way to spend Christmas Day! I can hardly wait to get out to the sun end of the ATM and stand up like a wing-walker and watch the world go by."

Jerry floats into the lower half of his pressure garment, which was stowed in the upper work area of the workshop, in preparation for an EVA.

EVA-2, December 25, 1973

For his first EVA, Jerry accompanied Bill Pogue outside the workshop on Christmas Day, with Ed Gibson acting as the inside crewman. The tasks for this 7 hour 1 minute EVA included X-ray/UV solar photography, coronagraph contamination views, images of the Kohoutek comet using the Far UV camera attached to the ATM truss, another ATM film exchange and some solar telescope repairs. A total of 40 pictures of the comet were taken. They retrieved space exposure samples and Jerry managed to fix the filter he'd been talking to Schweickart about.

JC: "The problem was that a filter wheel in one of the solar telescopes had stuck so that the divider between two filters was directly in the center of the aperture of the instrument. My job was to stick a screwdriver into the filter wheel, destroy one of the filters and force the wheel to the next filter position, so that at least we could get data through that one. I had to use a dentist's mirror and a screwdriver to do the job, and it was really difficult to get into the tight space to see and do what was necessary."

As the two astronauts moved back to the airlock area to stow equipment, Ed Gibson used the TACS to move the workshop to a better attitude for taking UV images of the comet. Using the Far UV cameras, three 10-frame sequences were captured.

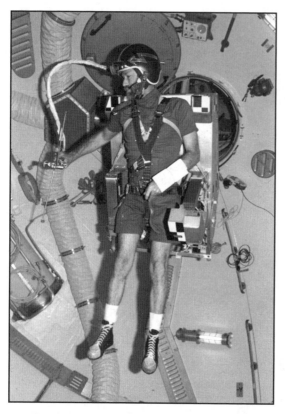

Test flying the M509 astronaut maneuvering unit experiment. Jerry flies near the upper dome of the workshop using the wand for control. The wand, like the foot controlled maneuvering unit, was considered unsatisfactory.

JC: "I was amazed when I got back in. I figured I'd have to go to the bathroom something fierce [because of the record duration], but I didn't. I decided that what happened was that I got rid of the waste water fluids though my pores. The system somehow compensated, and I sweated it out, because I was really sweaty. But I didn't really have to urinate. And I was just amazed that, after seven hours that I would not be pretty interested in getting to the urinal."

Following the EVA, the crew took off their suits and fixed them into the 'golden slippers' restraint on the upper grid floor. With the gloves and helmet reattached, air would be pumped through to re-pressurize and dry the suits. 'Joy' detergent was used to defog the inside of the helmets before an EVA, and this also made it smell pleasant. The liquid coolant garment was also hung up to dry. Prior to each EVA, the astronauts assigned to the spacewalk would lubricate all the moving rings and the rubber seals, clean the helmets and generally inspect the suit. The same suits were worn by the same astronaut for each EVA. There were a few spares on-board from the previous missions (such as an unused liquid coolant garment) but any serious suit malfunction would have prevented the astronaut from participating in any future EVAs. One problem they had to contend with was the garments becoming damp and moldy. Jerry found that splashing a little 'Old Spice' aftershave under the armpits of the garments made them bearable to wear. The crew were told to use Zephryn Chloride clear wipes initially followed by Betadine wipes to clean the mold, and a portable fan to dry the suits. Jerry observed, "The biocide is doing a real good job so I've got a lot of hope. It's not the kind [of cologne] you might wear on the ground. It might inconvenience you a bit… might lose a few friends, but I think [up here] it's looking pretty good."

Each IV (inside) astronaut would usually be located in the Multiple Docking Adapter (MDA). The latches from the airlock to the workshop and MDA were closed and sealed, so that if something went wrong, the third astronaut would still have access to the CM. Also in the MDA was the necessary equipment to communicate to his fellow astronauts outside, and access to TACS for OWS maneuvering if required.

EXTRACTS FROM THE DIARY

Mission Day 40, December 25: "Christmas Day - EVA 2 Day. Standing out on the ATM sun end and looking out over the world is an experience I'll probably never forget. It's like the thrill a little boy gets out of climbing as high as he can up a tree and being able to see the whole neighborhood – only this is *really* thinking big! As I stood there, the workshop was below my feet. The horizon started below and behind me and came around to above and in front, so I could see a bit more than half of the Earth at one time. As we came by Brazil, I could see the Straits of Magellan on one side and almost to Florida on the other. Saw the Virgin Islands, Puerto Rico, and Hispaniola. We flew right over the mouth of the Amazon. The awesome feeling I got is almost indescribable. I remember taking in a huge breath for some reason. I guess it was the only thing I could do under the circumstances. I went ahead and did my job out there, but I stopped every minute or so just to see what new thing had come into sight. Too bad there was nothing but water from Brazil on."

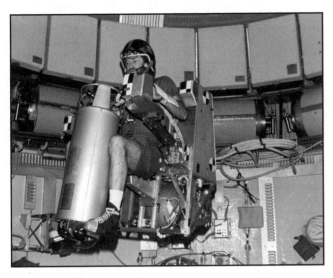

Flying a combination of the M509 with a tank in his lap, Jerry evaluates M509 performance with an external load for future EVA programs.

A humorous depiction of M509 AMU runs that "mysteriously" appeared on the check list.

Christmas on Skylab and on Earth

Jerry recalled his Christmas in space during his oral history: "We had built a Christmas tree. We had taken a bunch of food can liners from our kitchen and fashioned them into what looked like a little aluminum cedar tree. Then we had all kinds of orange and red and green decals and stuck them to the tree for decorations. We made a foil star with a trailer on it to put on the top, which was the comet Kohoutek. That was our tree."

The ground informed the astronauts they had hidden Christmas presents. They told the crew where to look and they soon scampered to find them. There was a beta-cloth Christmas tree, and Jerry's personal gift was a small silver Christian Icthus tie tack.

Jennifer Carr recalled Christmas time with the family – and the NASA security guard who was with them most of the time. The family also asked some friends to erect a big wooden flag pole out in the front garden, and they adorned it with 84 light bulbs. One was taken out as each day passed towards the end of the mission.

Jeff Carr also remembers that Christmas time in 1973. As well as visiting JSC to watch his Dad in space, he would occasionally participate in the family telephone calls. "Mom got to talk with them every few days, and about once a week she would let the kids participate. We had a chance to talk to Dad and tell him how we were doing, responding as if the conversation was like him being in the next room. If I had a really good football game we discussed it, and we talked about getting ready to take exams. At times it struck me as a little bizarre. When I realized he was aware of the fact that I was not doing well on history, it kind of struck me as a little strange. How did he know that? I guess that Mom was talking to him on the days we were not there.

"Skylab EVAs were great fun for me; especially the Christmas EVA. It was so exciting for me. I spent my

Jerry (lower right) and Bill Pogue handle
the garbage on Skylab.

Jerry in his personal crew quarters on the wardroom
area of the Skylab space station

Christmas Day glued to the squawk box. I wanted to hear every breath, and it was just spectacular. It was Christmas Day, and my Dad was probably doing the most exciting thing in his life. I remember sitting there thinking what a great Christmas present this is."

When Jerry was completing his EVA on Christmas Day, JoAnn was of course very busy at home. The six children were opening presents and she was preparing to have friends over for dinner.

JoAnn Carr: "Jerry started [the EVA] when I put the turkey in the oven and he was just finishing up when we were having dinner. It wasn't hard [that he was away for Christmas]. In fact, in some kind of strange way there was a closer bond between us when he was up there than when he was on the ground. I don't know why that is. I felt closer to him when he was on the flight than I ever did when he was home. Which is kind of sad, but at any rate it was not a sad time. We had a nice Christmas.

"I remember that John had a metal detector because he disliked junk lying around. I gave it to Jeff [now 15 years old] to put it together. He never told me that he didn't know how, bless his little heart, and so after Christmas I took it back to the store and said, 'This thing doesn't work.' They opened it up and it was put together with scotch tape. I thought, 'Oh God, poor Jeff is trying so hard to be the man of the house.' I also remember one of the conversations with Jerry, when I told him that I got Josh a chemistry set. Jerry said 'Oh no', but I reassured him that we would all still be here when he got back."

JoAnn managed to see Skylab go over Houston when Jerry was onboard just once in the three months. "It was in December, and I was sitting listening in the evening to the squawk box, and I heard them start over Mexico City. I heard them talking about seeing the lights of Mexico City, so I ran to the tracking map and saw that they would be coming over Houston. I ran outside and saw them and ran back in, and they were saying they could see Houston. That was amazing…it was magical."

EXTRACTS FROM THE DIARY

Mission Day 41, December 26: "Day off today. We paused in the EVA post procedure long enough to eat at about 8:00pm. We finished the EVA at 6.00, got out of our suits, put our food on to warm [thaw] and finally ate. The fruit cake was lousy, but a pleasant change from the routine, so Bill and I gobbled ours up. Ed doesn't like fruit cake, so Bill ate his tonight. I could only hack it once. The EVA really wasn't as tough as I had imagined. When we came in I was hungry but not too tired. [The ground] told us that we now have the world record for EVA [around] 7 hours. Ed and Bill broke the SL3 record on the first EVA and now we've broken our own. It was nice to get to talk to JoAnn again last night. I think this Christmas has gotten to me emotionally as much as if I'd been at home – although the emotions are different. I guess the thing I like most about Christmas at home is the Christmas Eve service and all the good feelings that go with it. I'm really thankful that in spite of all the hollering and turmoil that seems to characterize our family, we really do a lot together and love each other. I'll bet our kid's memories of these days will be happy ones. Anyway Christmas up here was different emotionally for me. I wonder how it was for JoAnn and the kids – family? (They're not all kids anymore). We tried for a couple of weeks to capture some anticipation for Christmas here, but it didn't come to us until a couple of the CAPCOMS indicated they wouldn't be talking to us until after the holidays. When the 24th came around we were deluged with good wishes by teleprinter and I got all mushy inside. I spent the rest of the afternoon with a lump in my throat. Then when I talked to everyone that night I had tears all over the place again. Darn things don't move away. They just fill up your eyes. Then last night I talked to JoAnn again and she played 'Deck the Halls with Boston Charlie' for me. That was great. We finally found our individual Christmas packages and opened them last night. I was really touched and pleased to get the fish (Icthus) symbol tie tack. I'll probably wear it more often as a lapel pin. At any rate, summing it up – I did have a happy Christmas just knowing that my family loves me, and that they were doing all the things that I like so much about Christmas. The EVA was a singular spectacular event in my life that I'll never forget, but I guess that I'll probably not connect it up too much with Christmas. It's just something special that happened at the same time. Maybe after I think about it more I'll feel differently. Anyway, today was a pretty nice day. The schedulers laid off us pretty well, so we got our showers and relaxation and can start another hard week tomorrow."

Mission Day 42, December 27: "Today went really well. We got off to a good start and stayed up with the flight plan all the way. Everyone went to bed feeling pretty satisfied I think. This is also the mid-point of our mission. From now on it's all down hill. I asked the guys on the ground to send us up a status report on how we stand, relative to mission objectives that were planned. The day after tomorrow we have another EVA. This time it will be Ed and me and should be a short one. All we plan to do is take some experiments outside and operate them. We also heard today that we are scheduled for a TV conference with Dr. Kohoutek. I suspect it's more of a PR play than anything else, but it'll be different. Ed gets really excited about the TV [stuff]… maybe Bill and I are just a couple of fuddy-duddies. Anyway, its too easy to make an ass of yourself, [up here] and most of the press find that to be more newsworthy than anything else we're doing. This thing tomorrow could be good. I wonder if the [TV] networks will bite. Saw Pensacola at dusk [today]. I still get nostalgic about Pensacola. We really liked it there. Every time I see all those airfields (Whiting, Corry, Saufley etc.) I get memories of flight training. Those were scary days."

Mission Day 43, December 28: "Got ahead of them again today, then they caught up with me tonight. Passed my LBNP again today – no sweat. Had an interesting experience today. Got to speak with Dr. Kohoutek about the comet. NASA set up a science conference with us on TV. I'm sure it was more of a PR effort on NASA's part, but it was fun, and we enjoyed talking with the man who discovered the comet. We hear that Dr. Kohoutek was pretty pleased about it too. Spent the afternoon getting ready for tomorrow's EVA… should be routine."

EVA 3: December 29, 1973

Jerry again stepped outside, this time with Ed Gibson, to complete a 3 hour 29 minute EVA. During the spacewalk they collected samples of the Airlock Module foil meteoroid collector to bring back to Earth for analysis (it had been exposed in the space environment for over seven months). The observations of the comet were also repeated. This time, Ed's suit leaked coolant water, causing ice to form on the front of his suit.

Jerry has described the comet as a disappointment in terms of brilliance. "Everybody, even on the ground, thought that it was going to be a beautiful brilliant comet, and it turned out to be very faint. We really had to work to find it. Once we found it, it was a gorgeous little thing, but was really small and faint. We took as many pictures as we could of it, but I don't think our film was sensitive enough to really get good pictures. I think the only decent one that was taken was with the coronagraph on the ATM. I think the people on the ground got better pictures of the comet than we did. So in order to describe what we saw, we did drawings." Ed Gibson did the drawings, followed by a TV report showing the drawings, describing the colors and explaining what was seen. "That was it. It was a disappointment, but it was really fun looking for the comet and finding it."

EXTRACTS FROM THE DIARY

Mission Day 44, December 29: "Did EVA-3 today. Never got more than about 10 feet away from the hatch. Spent the whole time photographing Kohoutek with different kinds of equipment. We never saw the comet on the last EVA. It just wasn't bright enough, and there was too much glare. This time it was much brighter, and we saw it with the naked eye. It was really a breathtaking sight! A bright ball with streamers in the stark blackness of empty space. Couldn't see enough of the tail yet to say it's graceful, but it won't be long before we can see a long one. The yellow and gold in the tail we could see though, and I'll bet that's significant to the experts. The people on the ground were so excited about what we saw that they scheduled an impromptu TV session… and we sketched it for them on live TV and described the colors. I guess we are pretty darned lucky to be able to see it when no one else on Earth can. Tonight we checked out some of our windows, and it's visible from one of them at every sunset, so we'll be doing some rubbernecking for the next couple of days."

Mission Day 45, December 30: "Tonight [the ground] teleprinted a message [in response to a request for a] status report on how we are doing compared to the mission plan as far as objectives are concerned. We got so far behind at the beginning of the mission and made so many mistakes that I figured the status was pretty grim. I had also asked for constraints we put on them that caused problems. [They] said, "You're running behind, and your exercise and free time constraints are killing us." That was what I expected, but then after dinner we spent four ground contact passes talking it out, and now I think we all feel better. (This conversation took place on the A Channel for all the world to hear. The crew had tried to get it done on the private medical report, but the press complained so they had no choice.) When we started talking they admitted that they had overscheduled the first 20 days or so of the mission. They were trying to get the same output from us at the beginning that they were getting from the SL-3 crew at the end of their mission. They allowed that the total of science man hours and exercise man-hours for the SL-3 and 4 crews were equal for like periods (MD 15-30). So since we had a requirement for 50% more exercise time it reduced our science time. The fact of the matter is we equaled the SL-3 pace. No wonder we were fit to be tied! On about Day 30, they finally wised up and quit trying to get 28 man-hours of science per day out of us and dropped it to 24. Then everything smoothed out. We felt the pressure come off immediately. Now we have some breathing room and are getting more efficient – making less mistakes. We've started shortening the time needed to do things, and we've decided to let them split our exercise periods in two as necessary to make scheduling more efficient. (As it turned out, that was a bad idea, and they quickly abandoned it because of the extra clean up time.) I wouldn't be surprised to see us equal the SL-3 science times now that we're approaching the problem more intelligently. We held out and got our quiet time in the evenings. They're going to pipe down at 9:00 pm instead of 10:00 pm, so now we'll be able to relax in space. Talked to JoAnn tonight… seems to be getting along okay. Sounds like people are helping to keep her occupied and it sounds like she's doing some pretty interesting things. I'm glad she's benefitting from this flight and staying involved."

JC: "People need rest. You need to be able to re-charge the batteries. I told [the ground] we needed to talk, so we ended up having a good session as we were making a US [communications] pass with people on the ground. We told them all the things on the first pass they were doing to make our lives miserable. They were doing things like scheduling heavy exercise right after a meal, and not giving us enough time between experiments to get from one to the other. We dumped all our grievances on that pass, and then on the next pass they dumped on us and told us how difficult we were making life for them down there. So we agreed that we needed a change in pace on the ground and a change in style up on Skylab. So the people on the ground thought about it a bit and so did we. We recommended that they change the schedule system so that we did not go through a day with every task on a time scale. We said look at only things that should be on a rigid schedule because of where we are in the orbital trajectory. Fix them in the schedule with precise start and stop times. In hindsight, this conversation turned out to be a pivotal moment in the mission.

"All the other stuff, you could say, 'We need you to do in the morning or the afternoon.' For housekeeping and other activities that are not time critical, we said, 'Send up a shopping list.' Each one of us from then on had a 'shopping list,' and it said, 'These are the things you need to do today sometime whenever you can.' These were 'change this, clean that, check those'. So, once we changed our schedule, our productivity went right up. It was amazing to see the difference in productivity because of going though life following a rigid timetable, and going through life where you are able to make choices about what you are going to do right now and what you are going to do ten minutes from now."

EXTRACTS FROM THE DIARY

Mission Day 46, December 31: "New Years Eve. Not much interesting went on. CAPCOM [Bob] Crippen was the bright spot in the day… he's a real gem! Sent us a New Year's card and an invitation to a Crimson [Flight Control shift] Team Egg-Nog party. What a bunch of characters. Story [Musgrave] came on in the evening and signed off with Guy Lombardo's "Auld Lang Syne." Don't think I'll even stay up 'till midnight Houston time… we have an early EREP in the morning."

Mission Day 47, January 1, 1974: "New Years Day. Ohio State 42, USC 21 – Bad Day at Black Rock."

Mission Day 48, January 2: "Day Off. Baloney! We got about two hours free and a shower. Spent most of the day quibbling over the details. We had our TV press conference today and that was rough! They asked some mean questions, mainly concerning Bill's sickness, our mistakes, our demands for time off etc. They also asked some rather philosophical questions about our view of the world, our inner selves, and the mission. We fielded them but felt bad because the critical questions indicated to us that we have been labeled as screw ups and slackers. We were also feeling like the managers weren't behind us either, otherwise the press wouldn't be asking so many critical questions. Talked to JoAnn tonight and she said we went over great. What a surprise! She also assured me that Deke [Slayton] and Kenny [Kleinknecht] are behind us all the way and have been fighting to keep us from being ploughed under by the schedulers. We're all very relieved now and feel much better. That's a morale boost really needed. Had just gotten to sleep when the Flare alert went off, so now I'm wide awake with only 4½ hours left to sleep. Got up a few minutes ago to watch Tokyo go by, then came sunset, so I went to the CM to watch Kohoutek. Still a beautiful sight. Hope it doesn't poop out before the ground gets to see it. Just as I was about to leave the CM I saw lights on the ground, and it turned out to be San Francisco. Further south came Los Angeles. It was clear from S.F. to L.A. so I had a field day. Saw the airport beacons at Long Beach and San Diego flashing. Could see the L.A. area from San Clemente to Ventura – solid lights. Absolutely breathtaking. Sure glad I bumbled in there."

JoAnn Carr: "It was our turn to have a telephone conversation that night of the press conference. The guys were very upset by this press conference and the first thing Jerry said to me was 'What's going on down there?' They [the crew] just got the feeling that they were being portrayed as a bunch of screw ups that were not doing their job. Of course I lied through my teeth and said, 'No that's not true, it's just the press trying to find something interesting. They are just nit picking.' I said I had talked with Kenny [Kleinknecht] and everybody's saying you're doing a great job. I assured him everything was fine and under control. Now when you put that against Jerry's diary entry for the day which read 'JoAnn is the only person I can trust and she says everything is OK,' I thought, 'Hell I'm lying through my teeth so he can't even trust me'.

"So I did what I should have done in the first place, in retrospect. I called Deke Slayton. The reason that I did not call him earlier was because I thought Deke wouldn't listen to a wife. He was a powerful advocate for 'his guys' and was totally on their side, but he had finally gotten assigned to a flight [Apollo-Soyuz] and he was in training for that, so he was not paying as close attention to what was going on with

Jerry's flight. I said I needed to talk with him, but did not want to come over to JSC as I did not want people to see me walking into his office [as it] would be all over NASA. So he came out to our house. and I was blown away because he sat there and listened to everything I had to say. He was clearly receptive, taking it in and listening to me; not taking me as one flighty little wife. I think he knew me well enough to know what I was talking about. So I told him everything that had happened, including the press conference with that medic and how the guys were feeling; how they felt alienated from the ground and were beginning to feel paranoid because they felt they could not trust anybody anymore. Bless his heart he took care of it. A couple of days later, he marched over to Mission Control, took the mike out of the CAPCOM's hands and talked to the crew. He told them they were doing a great job, surpassing records, and pointing out the good things to them.

"When things were settling down and when I could hear things going straight, I turned off the damned squawk box. I told Jerry that on the phone and said 'I'm not doing this anymore. I can't do eighty-four days, I'm through.' He said, 'Well, you've been flying this mission with us haven't you,' and I said yes. I didn't talk to the other wives because I figured they would call me if they realized there was a problem. I didn't talk to my best friend who was a widow. What was I going to complain about? My husband was having a tough spaceflight, but her husband was dead. I don't think so. All my best friends had gone, the Lovells, the Roosas. Other best friends who were civilians just would not get it. For me, there was no way to escape the intensity of what was happening on that flight. It was an intense experience for me, mentally, physically and emotionally. I was in touch with all those really intense feelings, and when Jerry came back I said please in your debriefings don't gloss over this because I think it's an important issue."

EXTRACTS FROM THE DIARY

Mission Day 49, January 3: "This was another of those super long days. We had to get up at 04:30 to get ready for an early AM EREP pass over central Africa, the Red Sea and Northern Iran. At 06:30 I had my exercise period, and by 08:00 I was ready for bed again. Felt like a zombie for the rest of the day. I'm sitting at the wardroom window watching the world go by... out over the Carolines and Marianas [in the Pacific] nothing but water. The islands are so small you really have to look for them. When you find one you're inclined to cling to it as if it were a link to reality or something. Seeing dry land seems to make me feel more secure or something. You don't realize how much of the Earth is covered by water until you fly over it. Then you appreciate every piece of ground."

Mission Day 50, January 4: "50 days! My gosh, it hadn't seemed that long. Our days go by so fast we lose track of time. Won't be long before we're in the seventies [mission days] and then we'll be the champs as far as total time in space is concerned. Hope all the systems hang together okay. Deke [Slayton] called today. Said we were doing an outstanding job in spite of having been sand-bagged in the early days. Told us we had just exceeded Pete Conrad's time in space so we couldn't be called rookies any more. Guess he feels pretty good about our conference last week on scheduling. He mentioned it and indicated that things would smooth out. In fact they might be hard pressed to keep us busy. I wonder if JoAnn called him after I spoke to her and clued him in on how we felt. If she did, I'm glad, because it made us feel pretty good to hear it from him. Before I talked with JoAnn I didn't know what to think. I couldn't believe that Kenny Kleinknecht and Deke would stand by letting these things happen to us. Was glad to hear they were behind us all the way. Now what I want to know is who had the whip in his hand and why couldn't he be stopped?

"I still can't get used to this beard. I keep surprising myself every time I look into a mirror. I like the look of it, but it still surprises me. I still haven't gotten over the 'grab and snatch' instinct when I 'drop' something I'm working with. I usually swing under it and knock it away in my frantic effort to catch it. I'm learning slowly, though it'll be a scream when I get back to 1G and drop things and just let them fall. Ed's still funny up here. We call him 'crash' Gibson. He's forever losing control and plowing into something. Hope we don't have to sew him up sometime."

Mission Day 51, January 5: "By Gosh I was right! JoAnn really did talk to Deke. I kind of had a hunch that she slipped him the word about what we were worried about, and that prompted him to call us yesterday. Sure glad the air is cleared up now. Maybe we can come out of this mission with a good feeling after all. Was beginning to think that we were going to come away with the shadow of our cover-up of Bill's illness and our own muddled first two weeks hanging over our heads and canceling out any good we might accomplish. Didn't get much time to look out the window today… Have tried for two days to get a picture of Iwo Jima but it's too cloudy. Would like to get one for the USMC."

Mission Day 52, January 6: "Looks like we're well on top of the schedule now. We're getting everything done on time or even earlier. We don't have to beat them back so that they leave us alone during the 'quiet hours.' We're all feeling much better about everything, and that'll sure make the rest of the mission go quicker and better. We're all in surprisingly good health too. No colds or upset stomachs, only an occasional tough time getting to sleep. Everyone is getting through the LBNP runs in good shape too. We're beginning to see the US during the day again. The EREP passes are picking up. The weather is pretty bad though, so were not getting to go for the targets we've trained on. The snow on the Rockies is really beautiful. Makes some terrific pictures. Sure looks cold – ice forming everywhere. I wonder how it is in Houston. JoAnn hasn't said anything so I guess everything is going well [at home]."

Mission Day 53, January 7: "Today was my day to mess things up. We did EREP again, and at the end I started us back to solar inertial an hour early (misread the PAD). Used a whole bunch of TACS and had to listen to it belching for the next half hour – how embarrassing! It was a dumb, dumb trick caused by lack of attention to what I was doing. It was a good lesson. Bet I never do that again. Mistakes are ten times worse with the whole world watching. Our spirits seem to be rising every day. We're working a whole lot better with the pressure off, and I think we're getting more done now than when we were scheduled to the eyeballs. Saw Japan today… also saw Clark AFB in the Philippines and remembered our long over-water flights from Japan. Also remembered that scary one from Ping Tung to Clark when I had fuel tank trouble about the time Josh and Jessica were being born. (On the way to the Philippines, Jerry's wing tanks stopped transferring fuel, so the only fuel left was in the fuselage tank. He had progressed too far to go back, so he calculated the time left to get to Clark AFB from where he was and the time he could remain airborne. The times came out equal, so he powered back to maximum endurance and let the air controllers know what the problem was. Then he started thinking about shark-infested waters and wondering if the Luzon natives were friendly to white folks who descended upon them from the sky. When he got to within gliding distance of Clark, Jerry powered back and descended for a straight in landing. The engine was still running when he touched down, but it quit as he rolled out, and he had to be towed to the parking area.) That was an interesting year, but I don't think I would want to do it again. It's ridiculous being away from your family that long. Well, actually three months is ridiculous too! And here I am. It's sure going to be nice to get back and get all the reports done so I can slow down and do more with JoAnn and the kids. I wonder if we'll do any official traveling. Everybody says that's terrible, but I'm still hoping we can get a good trip. That's got to be better than the 'soup and fish circuit' inside the US. It would be neat to be able to take JoAnn on a big trip even if we have to work for it.

Mission Day 54, January 8: "My mind's blank tonight. Not much worth remembering today. Best part was getting to talk to JoAnn tonight. Sure look forward to those phone calls."

Mission Day 55, January 9: "I'm lying here listening to Andy Williams sing some great songs while I think back over the week. I feel pretty good about things generally. We had a couple of tough days, but I guess the high points overshadowed the rough ones. We were really low until I talked with JoAnn on our last day off and was so pleasantly surprised to hear that our press conference wasn't a complete debacle. She's my best critic, and I didn't consider her opinion to be biased either. Then a couple of days later, after JoAnn had clued him in, we got a call from Deke reaffirming approval of our work and acknowledging how we had been 'sandbagged' early in the mission. His saying that on the air-to-ground loop right in the middle of Flight Control people made us feel that he was putting the word out where it would do the most good. From a schedule standpoint it was a good week, because we managed to stay even with, or ahead of them the whole time. You have such a different attitude when you're not being driven! We worked hard and enjoyed it because we could see progress. I think the last thirty days of this mission will go quickly and pleasantly if we can keep this pace. I understand that we're doing as much science with this plan of action as we were before. That's justification (and benediction) in itself."

Mission Day 56, January 10: "Now today was the kind of day off I've been harping for. We had all sorts of time for looking out the window or doing whatever we wanted. We slept in an extra 2½ hours, then we ate and heard the news. I then did an experiment that looked at Kohoutek, then had the rest of the day free until tonight when we did an EREP pass just south of Japan. What a pleasure that was. Ed spent most of the day fooling with the ATM and had a ball. Bill and I went camera crazy and took a bundle of pictures out the window of the US and Europe... what a magnificent view! I'm gonna steal every moment I can for looking out the window. This has to be by far the most fun of the whole mission (except for the EVA, maybe)! Bill and I are like a couple of kids under the Christmas tree – yelping and 'oohing' and 'aahing'. The Purple [flight control shift] Gang gave us the 'One Armed Paper Hanger Award' tonight when they went off shift. Apparently, it was a poster with a cartoon of a guy doing many things at once and cussing out the ground. They complemented us on getting so much done last week. This is the way the mission should have run – with everyone enjoying themselves."

JC: "We worked a ten-day week, and the tenth day was supposed to be a day off. Because of our ugly schedule, we got so far behind that for the first four tenth days we gave the day back. We said 'Schedule us a work day, we won't take a day off [to catch up].' It was around the fifth day off that I said, 'No we are going to have a day off. We haven't had a [full] day off since we've been up here. We need to get our sense of humor back because we are getting inefficient.' And so we took the day off, with Ed doing solar physics anyway. Bill and I worked too [on the photography and an evening EREP] but we weren't working against a schedule. It was during that day off that Ed was doing a TV thing and he mistakenly misconfigured the communications systems. We went over one of our ground stations with our radios off. The press picked up on that immediately and called that mutiny, saying that we were a real crabby group. We were getting real testy, we had mutinied and said we weren't going to work that day, and we had turned off our radios. That got into the press and we have never lived that one down. But that's not what really happened that day."

Into the home stretch

A weekly review began, starting on Day 56 with a 7-day go-ahead being passed up to the crew through to the end of the mission. By Day 56, 62% of the original schedule had been completed, and in some areas Jerry, Ed and Bill were exceeding the pace of the SL-3 crew. In the medical field, the Skylab 4 astronauts were showing better responses on the cardiovascular system, despite the expected slight increased heart rates for all three men. In the first ten days of the flight, both Ed and Bill had lost about 4 to 5 lbs (1.8-2.2 kg) in body weight and were now holding at that point almost seven weeks later. By contrast, Jerry had actually gained 0.5 lb (225 grams). The accurate measurements of weight reflect the loss of bodily fluids, and photographs and leg measurements reflect the redistribution to the torso and upper extremities. The medics told the crew that the overall reduction in fluid volume was approximately two pints (one liter), which caused the production of red blood cells to diminish for a while. The influx of fluid to the upper torso caused by weightlessness signaled that the crew were edemic (excess body fluid). The body tries to get rid of the excess, resulting in loss of about one liter, and reduces thirst for awhile. Red cell production depends on the concentration of red cells per volume, so when fluid volume goes down, the concentration goes high and the body stops making red cells. Blood work during that time showed that production diminished until the proper balance was achieved and then it started up again.

Due to the reduction of gravitational forces on the vertebral column, body height increases by about 1.5 to 2 inches (up to 5 cm). This did not have a noticeable effect on the fit of the crew's clothing, although they did notice increased force from the pressure garment neck ring on to the shoulders. Two thirds of the mission was now completed, and as the crew became more relaxed and productive, the countdown to recovery had begun. The remaining milestones for the crew were to complete their program, closeout Skylab and head for splashdown. What would come next would have to wait. For now, all thoughts focused on a successful end of mission and a safe return home.

The 'Christmas Tree' made from food cans erected by Jerry and his crew and televised on December 24, 1973. Jerry and his crew were only the second crew to spend the festive season in space, after Apollo 8 five years before

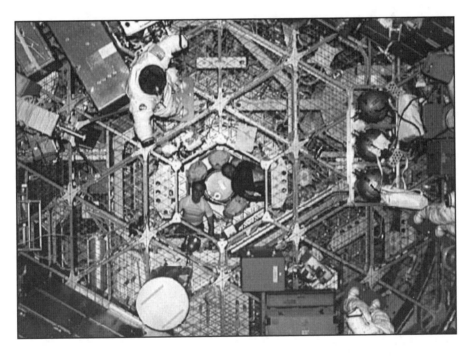

Ed and Jerry peer through the access from the Wardroom in the orbital workshop.

Jerry, Ed and Bill around the wardroom table.

Ed upside down with Jerry on right.

Chapter 9

Coming Home - The Third Month

By early 1974, the crew aboard Skylab had settled into their daily routine and were fast approaching new milestones in space endurance. Each day the program of experiments, housekeeping and maintenance continued. Ed Gibson was still searching to record his first solar flare from the ATM, while Jerry and Bill ticked off the targets scheduled for the EREP passes. All three were beginning to think about life after their mission, but there was still a month to go, followed by several weeks of mission debriefing and public touring. Skylab would consume their lives for some time after getting their feet back on *terra firma*. It was this 'firm ground' that was the focus of one of Skylab's major science fields and one that caused a few ripples through the intelligence community in early 1974.

One Photo Too Far

The Earth Resources Experiment Package (EREP) located in Skylab's Multiple Docking Adapter consisted of a suite of six different instruments that surveyed the Earth in visible light, by infrared and using microwaves. In addition, the astronauts used hand-held 35- and 70-mm cameras to obtain images of opportunity. The EREP package included a terrain camera which was based on a lunar topographical camera flown on Apollo 13 and 14, together with a multi-spectral scanner and cameras. Orbiting at an inclination of 50°, this offered an unprecedented opportunity for a US manned spacecraft to record the topography of most of the world's land masses.

The policy of open public awareness that NASA generated when it explained the purpose and benefits from the Skylab EREP program was in sharp contrast to the national security driven policy of the intelligence community with regard to data obtained from high flying aircraft and surveillance satellites. In 1965, photography taken during Gemini missions resulted in a directive from the Director of Central Intelligence that all subsequent Gemini images be reviewed prior to public release. This procedure was to remain in force during the whole of the Apollo and Skylab programs. In addition to reviewing the planned US targets for the Skylab photography experiment program, the intelligence agencies evaluated the use of such photography for its own purposes. From the mid-1960s, astronaut photography of the Earth became the subject of a joint CIA, DOD and NASA agreement. This agreement restricted the imaging of ground objects to a maximum ground resolution of 20 meters from low Earth orbit. When Skylab equipment was developed that improved upon the ground resolution beyond that agreed to in the joint agreement, an accommodation needed to be reached.

Jerry having fun in zero g

Jerry and Bill Pogue demonstrate their new-found abilities in balance and super human strength aboard Skylab.

In 1971, NASA requested a relaxation of the agreement for Skylab missions. It was suggested that a review of the proposed photographic targets for the first manned mission (Skylab 2), and presumably the second and third, would allow the intelligence communities the opportunity to inform NASA of areas the astronauts were not to photograph. It was acknowledged that imaging of targets of opportunity might also inadvertently capture some classified areas, but as all images were to be reviewed prior to public release any 'accidental' photograph could be vetted prior to formal distribution. NASA was informed that one area definitely not to be imaged was the USAF facility at Groom Lake. This facility, though often linked to conspiracy theories and science-fiction stories, was used for high security development of new aircraft and systems. Groom Lake, also known as Area 51, was the home of developmental and test flights of the U-2 spy plane, the Blackbird Mach 3 aircraft and the F-117 stealth fighter. This 'secret' base had been known of for years before the USAF finally acknowledged its existence in 1999.

This facility became the subject of an April 1974 memo within the CIA regarding a photo taken of the area from Skylab. During the post-flight reviews of Skylab 4 imagery by the intelligence community, it was found that the crew had, apparently inadvertently, taken a photo of the facility which, at that time, did not officially exist. The image was taken early in 1974, probably during a photo of opportunity shot. This oversight was not immediately known outside of the intelligence community, the US Department of State and NASA, as the photo and correspondence was classified for many years. It was not until 2006, after the internal memo had been declassified, that American space policy author Dwayne Day wrote an article on the incident and memo. The photo still remained classified, however. Day noted that the memo revealed that there was no agreement about the fate of the photo. The DOD elements insisted it should be withheld from public release, while NASA wanted to release it, without fanfare, and allow it to sit in the Sioux National Repository, Nevada, hoping it would not be exposed for what it was. The disclosures of the Skylab incident also generated a paper by James David of the space History Division of the National Air and Space Museum, who had reviewed both astronaut and intelligence community photography during the 1960-1974 era.

The opportunity to photograph areas of Earth for scientific purposes was one of the main reasons for flying Skylab in the first place. The national and international political, military, national security and privacy issues this potentially infringes upon have to be balanced against the environmental and ecological objectives publicly promoted. This delicate balancing act continues today for the current International Space Station program.

JC: "The Skylab 4 crew was never informed of nor chastised for imaging the 'forbidden fruit.' We have no idea which camera was used. I'm guessing that the Hasselblad and Nikon lenses weren't sharp enough, so that leaves the Earth Terrain Camera or one of the EREP cameras. If it was the latter, then it occurred during an EREP pass, and the crew could have somehow been forewarned since EREP passes were carefully planned out on the ground well ahead of time."

EXTRACTS FROM THE DIARY

Mission Day 57, January 11, 1974: "If this mission had gone as originally planned I guess we would be on the ground by now. We've passed our 56-day goal, and there are three more – 60 to beat the SL-3 crew, 71 to beat Al Bean's record, and 85 days to meet our mission plan. It doesn't really seem like 57 days. We've been so busy that the time really flies. I just finished the food inventory tonight, and we have more than enough food to finish the mission. The only tight item is urine sample bags. Someone goofed somewhere and we came up 23 bags short. I discovered that last week. Now we're saving our urine over 1½ days instead of one. Didn't sleep worth a hoot last night. Fell asleep about midnight after logging my comments. Then, about 45 minutes later, I woke up for no good reason and was wide awake until about 02:30. Dragged around all morning but felt better this afternoon after I had lunch. Had an EREP today right over Northern Baja, California and New Mexico up to the Great Lakes. Didn't get to look out though so I missed the best US pass of the day. Got to see Mt Rainer and Mt Adams again today. They're really beautiful. I wish I knew for sure where Mt. Shasta is. I think I've seen it, but I'm not sure. It's not marked on our charts. Talked to JoAnn and John and Josh tonight. Sounds like John's boat is coming along pretty good. He sounds very proud. That's good. He needs a good accomplishment like that for his self confidence. JoAnn sounded like she's getting tired of the mission. I don't blame her. Now that we've settled down to a routine the time will probably hang heavier on her than on us. Hope she can stay occupied and happy for the rest of the mission time. Twenty-eight days does sound like a long time – hope we can stay busy and happy too. It's amazing that we haven't had any arguments up here yet either."

Mission Day 58, January 12: "Sometime tomorrow we tie and then exceed the SL-3 mission time. That will make two records for us – longest EVA and longest mission. As I understand it we could lose that other CMG [Control Moment Gyro] tomorrow and still finish the mission with 85 days. There's enough TACS left and CMS propellant to hang on for about 28 days. Got to see some of Northern Europe today. It had really been cloudy the last three weeks. It's not too snowy, but it looks kind of dark. Cold all the time. Bill and I have been trying to locate the Arkansas property and haven't been successful yet. I want to get a picture of it with our 300mm lens. Sure hope we can get up to Arkansas this summer. I'd like to go as soon as school is out and spend a week or so. Would like to start thinking about a house up there – some sort of 'A' frame or the like. Got to get rid of the bus first I guess so we'll have some money to put into the house. To get rid of the bus it looks like we're going to need a paint job this spring, and the earlier the better to catch people when they are in the buying mood."

Mission Day 59, January 13: Found out the SL-3 crew went 60 days, so we don't tie them until tomorrow. [The ground] says they'll probably call us tomorrow at about 18:00 CDT to tell us when we make the mark. Too tired to think tonight. Didn't sleep well last night and had a long day today.

Mission Day 60, January 14: "Two EREP's today – weather messed up Bill's pass up the East Coast, but mine through Colorado was pretty clear. One of my targets we couldn't find. All the valleys and gorges look the same from up here. Hope the weather clears up for a while so we can get our licks in while we're over the US. It's kind of funny – Bill and I are wishing for good weather over the ground so we can see to photograph, and Ed is looking in the other direction praying for a flare. We got our call tonight about breaking the SL-3 record. As of 01:10, ours became the longest mission ever flown. Al Bean's personal record is 69.5 days so we'll soon have that one too. Dick Truly read us up a nice little résumé of how long it took to break each one of them. The medical people are a bit worried about us as of yesterday. We did blood draws yesterday, and Ed found that our hemoglobins were down from last time. We didn't think much of it, but the guys on the ground went bananas when they got our report. Seems like the drops were significant. We've got to do some more tests for them the day after tomorrow. I guess if it begins to fall any more they might terminate the mission. That would sure be ironic – we fight to keep the equipment running and then have to terminate for medical reasons. The funny part of it is that we all feel great – no colds, no nothing!! Talked to JoAnn today. She's ready to go back to Arkansas too. Guess we better start making plans when I get back. Need to get the white elephant [the bus] painted."

The US space endurance records were:

* May 1961 - 15 minutes (Alan Shepard) Mercury 3 suborbital - first American spaceflight. Record lasted for 9 months
* February 1962 - 4 hours 55 minutes (John Glenn) Mercury 6 - first American orbital mission. Record lasted for 8 months

* October 1962 - 9 hours 13 minutes (Wally Schirra) Mercury 8 - doubled US endurance record. Record lasted for 7 months
* May 1963 - 34 hours 19 minutes (Gordon Cooper) Mercury 9 - first American 24 hour flight. Record lasted for 2 years 1 month
* June 1965 - 4 days 1 hour 56 minutes (Jim McDivitt, Ed White) Gemini 4 - first extended duration American mission. Record lasted for 2 months
* August 1965 - 7 days 22 hours 55 minutes (Gordon Cooper, Pete Conrad) Gemini 5 - first week long mission. Record lasted for 4 months
* December 1965 - 13 days 18 hours 35 minutes (Frank Borman, Jim Lovell) Gemini 7 - first two week mission. Record lasted for 7 years 5 months
* June 1973 - 28 days 0 hours 49 minutes (Pete Conrad, Joe Kerwin, Paul Weitz) Skylab 2 - first one month mission. Record lasted for 3 months
* September 1973 - 59 days 11 hours 9 minutes (Al Bean, Owen Garriott, Jack Lousma) Skylab 3 - first two month mission. Record lasted for 4 months

Jerry, Ed and Bill surpassed Skylab 3 crew/mission record and Al Bean's career record of most time spent in space and become the new world record holders. This record lasted for 21 years for the Americans but the Soviets surpassed this record in March 1978, four years after Jerry and his crew came home.

Mission Day 61, January 15: "Flew T020 today [the foot controlled maneuvering unit]. What a *Rube Goldberg* contraption that has turned out to be, what with all the side straps and cross straps – *ad nauseum*. (The Rube Goldberg comic strip was very popular in Jerry's youth. He was an eccentric inventor. He would, for example, roll a ball down a chute which, after it went through four or five mechanisms, would result in a loaf of bread being sliced or a light turned on. As time went by, the term came to be used to describe any unnecessarily complicated system.) Bill and I started on it at 07:00 and finished at noon. There were a lot of last minute changes in the hardware, and in the checklist, and we wasted a lot of time sorting it all out. Once we got past that the day smoothed out. Later, we televised the view out the window. We got some shots of Korea, Japan and China, then picked up the Aleutians. Later we televised the West Coast from northern Santa Barbara to the top of Baja. It was fun. Bill said we sounded like a travelogue. I think we'll try to do that more often. Sounds like it was well received on the ground. Latest word of the hemoglobin scare is that it's not as serious as thought at first. We still have healthy valves. Just the sudden drop bothered them. I find myself thinking more and more about getting away from NASA and finding a low key job where I would have more free time available. The Corps is out, I think. The idea of teaching still appeals to me, but I wonder how long I would like to do something like that. I guess I don't even know myself well enough to know whether I'm the driver who just needs some time off or just a moderator who has been in over his head the last 8 years. I've enjoyed my work – never regretted it – but it has been a strain. I wish I could figure out whether I need a change or just a rest. I guess I'll have to stay around NASA for at least a year. May have to stay in the Corps for two years in order to retire as a Colonel."

JC: "After Skylab there was to be only one more Apollo mission (ASTP) and then a lengthy hiatus (4-6 years) before the Shuttle could fly. I didn't want to hang around for another 8 years for my second mission. I had pretty much ruled out staying in the Corps because I no longer fit the career pattern, so I would not be promotable, but the idea of teaching had been running in the back of my mind for a couple of years. In the end, I left the USMC in 1975 but remained with NASA until 1977, mainly because I hadn't made up my mind what I wanted to do when I 'grew up', and I was enjoying the work of defining the Shuttle cockpit." (See Chapter 10)

Mission Day 62, January 16: "Today was a busy day but not too taxing. Lots of busy work, but no pressure. Got through another LBNP run with no problem. Dr. Hordinsky filled us in a bit more on the hemoglobin situation. Seems our loss in percentage has exceeded SL-3 but not SL-2, so we're pretty much still in the spread. Glad to hear that the guys on the ground have settled down. Tomorrow I fly M509 suited. I've been dreading it for weeks. Will sure be glad when it's over. It's so much work for so little gain that I just hate to do it. (JC: "It was just an exercise in futility. We knew it could never amount to anything. Flying it suited was just icing on the cake. We already knew that it performed beautifully, and being suited wouldn't change that significantly,"). We're trying to weasel out of the suited T020 run but it looks like we won't be successful. Nuts.

Looking forward to the end of the mission. Note the protruding veins in the forehead and the puffy eyes, a result of the fluid shift in the body in weightlessness.

Thought some more today about teaching. Maybe I should stick to working with young people as an avocation rather than a vocation to see if I can get a better feel for it. I wonder if JoAnn and I (or just "I" if she's not interested) should try teaching Sunday School."

Jerry was reviewing the book *Man the Manipulator* in the little free time he had and made some personal observations in his diary entry for MD62: "As I read these books on modern psychology I keep thinking 'Why couldn't I have been exposed to this when I was younger? Maybe it would have freed me up in my relations with others and I'd be a different person now'." Jerry also took time to reflect on his relationships between himself, JoAnn and his children, adopting the approach explained in the book: "I see my main patterns are Dictator with the kids, shifting from Protector to Weakling with JoAnn, and Nice Guy with friends and associates. Overall there appears to be a hint of Judge. That's quite a combination. Up here [on Skylab], I seem to be the Judge and the Calculator. So far I've used up everything but Bully and Clinging Vine."

Jerry was also reading *Self Renewal* by John Gardner. The theme of this book is the realization that, if you get to a point when you feel your goals have been achieved, stagnation begins. Thus the essence of life, according to Gardner, is not achieving your goals but striving for them. Gardner also discusses those who are more method orientated than goal orientated, for whom stagnation results from too much time spent on perfecting one method, rather than being creative and forging on to the finish or finding alternate ways to complete the job. Jerry wrote that these observations certainly crystallized his own ideas: "I guess that's me alright. I frequently lose sight of the forest for the trees."

JC: "The books that I read were ones that I had been meaning to read for some time, but I just never had time to do it. I figured (unrealistically) that I would have plenty of time to get to them in my spare time during the mission. When I realized that 8 hours of sleep time were about one hour more than I needed, I reserved that hour to read and objectively mull over the things I read. Then I took the time to apply what I learned to myself. I'm glad that I did it, because the things that I learned about myself were valuable. I learned to tolerate both my own foibles and those of others."

EXTRACTS FROM THE DIARY

Mission Day 64, January 18: "Today's our day off. I goofed off last night and didn't make any entries in this erudite piece of literature. Had a fairly slow morning yesterday. Did an LBNP run on Bill, but after lunch things really tore loose, and it was the wildest Chinese fire drill I've ever seen. The M509 test plan had many more things to do than we had time or consumables to do it with, so it didn't take long for their neat plan to fall apart. We got everything set up fine and got started on time. I was suited, pressurized on a light umbilical, and the SOP [Secondary Oxygen Pack] was strapped across my thighs under the PCU [Pressure Control Unit]. I flew it out to the middle and did the calibration maneuver. Did one set wrong so repeated them. Then the N1 [Nitrogen 1] tank got low early so flew down to Bill to change it. From there things turned to worms. Bill couldn't get the tank connected to the recharge station. Meanwhile, as I waited, the battery began to run down. Shortly after, we finally got going again but had to stop and change the battery. Then a short time later the tank got low again, and after that the SOP ran out of oxygen and we had to reconnect the umbilical. Then the second battery pooped out. Bill was behind the whole time and had a wild look in his eyes and a frantic expression on his face as he zipped around me tending hoses, changing tanks and batteries. He looked so funny I nearly died laughing in my suit. When the SOP ran low on oxygen it let the suit depressurize. Then Bill got even, because he jammed the umbilical onto my oxygen port before I could throw the PCU lever to ABSOLUTE, so the suit repressurized to 3.9 psi instantaneously with a whoosh! Then I was in trouble trying to clear my ears. Meanwhile, old 'Hollywood' Gibson arrived with his TV camera to show the folks at home what a happy expression old Jerry had on his face. What he got was a beard and two eyeballs popping out! The whole exercise is recorded on film and I can hardly wait to see it when we get home. It ought to be a scream. After the debacle was over, we spent the rest of the day and evening putting everything away. The day after tomorrow we can unpack it all again for Bill's suited run. Today was a nice relaxed day. Had plenty of time to exercise, take a shower and look out the window. We did an EREP pass right after lunch. Talked to JoAnn and Jeff last night. I guess the little ones were in bed. I guess Jeff is still doing well in school. Hope he doesn't decide he's a hot shot and neglect his schoolwork. He really can name his college if he keeps his grades up. Sounds like he's doing well in track too. Good old John – still plugging along. Acting just like Jeff in school – never turning anything in."

Jerry found the M-509 AMU experiment a valuable exercise. The unit was quite easy to control, but the foot-controlled unit posed difficulties, especially since it was not gyro-stabilized. He recommended this option for the development of any future foot-controlled unit. The work conducted on the M-509 experiment led directly to the development of the Shuttle MMU that was tested and operationally demonstrated on three missions in 1984, and to the current SAFER rescue unit flown in support of ISS EVAs.

JC: "The M-509 AMU was easy to fly. It wasn't any problem at all. The thing that was a pain in the neck was the hand held nozzle. This was not part of the manned maneuvering unit but an extra thing to see if we could use a hand held maneuvering unit to control ourselves. It turned out to be a miserable flop. It took too much skill and training to get to the place where you could properly rotate yourself around your centre of gravity or translate yourself. It was the same with the foot-controlled unit. The thrusters were so far away from the centre of gravity that it did not work. But the manned maneuvering unit itself was a really good unit."

Mission Day 65, January 19: "Only 20 days to go. The last week will be a frantic scramble to deactivate, I'm sure. I've already told the guys on the ground to start working up a shopping list for us to start on early in our spare time so we can take the heat off later. I guess we've got about two weeks of normal routine left before we begin to seriously think about re-entry. I'm going to have to start spending some time in the CM just looking things over and thinking out entry. There's not much margin for error there, and I'd like for it to be as perfect as the rendezvous. Had a nice day today. Spent about 3 hours at the ATM panel watching the sun and spent a total of about an hour looking out of the window. Started out the morning feeling mean and cranky. My head was congested (sinuses) and I was getting headache so I took a Sudafed and snapped right out of it. We were scheduled for another EREP again today but got weathered out, so the day kind of loosened up. I felt pretty fine. I guess I'm getting a short-timers attitude while pedaling the bike. I began to think of what I could say when we get back to Ellington. Sure want to use a different line than all the rest have used there. There have to be some different ways to thank everyone for their help without using all the trite expressions. Everyone else is probably tired of it too! Think, Carr, Think!!"

JC: "About three weeks before the return, we began the process if getting ready to go back. We started doing the rehearsals. We would go down into the Command Module, get into our seats and go though our re-entry checklist. All we did was just touch the switches [without activating them] that we needed to and go though the procedures."

Mission Day 66, January 20: "Had a long hard day today – EREP this morning and suited M509 this afternoon. Bill flew the M509 today, and I was the observer. I ran my head off, and we had a more dignified run this time than last. Got quite a bit more done."

Mission Day 67, January 21: "Ed finally got a flare today. He was beginning to despair. The active side of the sun was about to turn away from us and after about mission day 70 there is little or no chance of our seeing a solar flare. Active Region 31 had been spluttering and starting all day, and this evening it was my turn to man the ATM panel while Ed exercised. Well Ed really didn't want to leave the panel, so I told him he could have my turn, but if he got a flare on my pass he owed me a bottle of scotch. He agreed and 15 minutes later he had a solar flare on his hands. Bill and I are really happy for him. He wanted a flare so badly that we've been giving him as much time as possible.

Our Earth resources observations are much more satisfying to us anyway. Had another EREP pass today. That makes about 30 of them we've completed out of 33 scheduled. Our batting average is pretty darned good. The only bad part is we aren't getting the particular sites we trained for because of bad weather. Hope the next two weeks are more fruitful."

Mission Day 68, January 22: "I got two small flares on the sun today! The first one came so fast I couldn't believe it was real, so I didn't get the kind of data I would have liked. The second one I was ready for and caught really early. Ed had gotten the big one so far and that's good. Our sick CMG (#2) has been acting up off and on all day, so I wouldn't be surprised to see it fold pretty soon (maybe even tonight). We've got enough gas for our thrusters to keep us up for a while and the CSM can hold the fort for a while after that, so we should be able to finish up the mission. Actually, after we beat Al Bean's record I'll be ready to come home any time soon. We should probably tie him either tomorrow or the day after. Once we make that one we'll have a record that won't be tied for many years. It's really hard to believe that we've put in 68 days up here. We're all getting a bit anxious to get home now. That'll probably make the next couple of weeks go slow. Ed did another hemoglobin test on us today, and the downward trend has leveled and risen a bit. That should ease a few minds. I really feel quite healthy. I'm really curious to see how one-g will affect me. You can tell from the medical conferences that [the doctors] are getting lathered up over recovery."

Jerry wrote in his diary that a new 'benign procedure' had been tried a few days earlier which involved putting a catheter in the arteries to measure cardiac output. Drs. Eduard Burchard (subject) and Chuck Berry (investigator) of the Life Sciences Directorate had tried it out. It was supposed to be perfectly safe and has been performed on thousands of patients with heart problems, but Burchard's heart went into a seizure and stopped for 20 seconds. It was claimed that this was a very rare reaction, but it still was proposed to be used on Jerry and his crew after recovery. It was soon "shot down," as Jerry observed: "Thank God! I sure don't want them punching holes in my arteries. Deke has been fighting for us since they first proposed it in October. Looks like he's got them where he wants them now." Jerry also started wondering about the planned reorganization of the directorates at JSC now that the Skylab program was almost over: "Can't imagine why they'd wait 'til Skylab is over. That would be too sensible."

Mission Day 69, January 23: "Today was Bill's birthday. He had forgotten it (so he says) but his family hadn't. It really bounded up Bill's spirits. He just hummed around here all day. I guess it will be some time tomorrow when we break Beano's record. Our CMG #2 is getting sicker and sicker. Seems to be accelerating towards failure, so I wouldn't be surprised to lose it at any time now – maybe tonight. I'm sure it will wake us up if it does fail because the computer will have to shift over to TACS control. That'll set off the ACS warning system and the jet noise will undoubtedly wake us. Sounds like we're inside a big insulated garbage can and someone is outside banging on the wall with a sledge hammer. We started doing things today that signify that mission end is near – microbiology sampling of our bodies this morning and this afternoon Bill took samples from all over the spacecraft. I've been gathering up the food for days 84 and 85 so I can pack it up when the menus show up. Had another LBNP run today with no symptoms or problems. My hands have stopped peeling too, so we must be pretty well adjusted to the [very dry] environment. The ground gave us some data on our exercise, and it appears, based on oxygen consumption, that we're in better shape now than on launch day or even pre-launch for that matter. I'm sure glad that artery puncture procedure got put away. I was going to fight it if it hadn't. The three of us agreed to refuse it. Now we don't have to."

Mission Day 70, January 24: "Today's the day we beat Beano's record, but so far no one has said anything. Bruce [McCandless] advised us today that we were completing our 1000th revolution [of Earth] but I don't equate that to Al's record. Flew the T020 this morning and got some good data. Now maybe once and for all we can get rid of that *Rube Goldberg* rig and get on with a new generation of maneuvering units. The only good idea in T020 is the foot controls, leaving the hands free. From there on things are bad. Hope someone can take the best ideas out of both maneuvering units and come up with a good piece of hardware for the future. Had another EREP today and then afterwards I started another science demo. I was making thin liquid films and it turned into a disaster. Couldn't get the fluids to leave the syringe and stick to the wire. It was if it had stage fright. The ground troops have also asked me to do my gyro bit again in such a way that it can be edited for national TV and for a Skylab summary film. They said my last one was the best of the Skylab programs but couldn't be shortened without ruining it. That pleases me that it was so well received. I really worked hard on it. Didn't get to look outside today for very long at all – I feel cheated. Here we are 70 days into the mission, and I still consider looking out the window to be the high point of the day. The gyros are still hanging in there. CMG #2 is still deteriorating, but since we can't do much to preserve it we're just going to press on, business as usual and see if we can finish the mission on time. I guess I really don't care now. My last goal was Beano's record, so now we just need to get down in good shape. Don't really care if it turns out to be early."

Mission Day 71, January 25: "We finally surpassed Al Bean's record last night at a little past midnight. Dick Truly informed us this morning that we are the new world champions. Then this morning none other than Al Bean called to congratulate us. It was nice of him to do it and I had a hunch he would. That's the real 'nice guy' thing to do. I think I would have probably done it too if I had been in his shoes. (Jerry and the SL-4 crew actually did so later, to Grechko and Romanenko for their 96-day world record in 1978 and to Thagard for his US record in 1995.) Our sick CMG is still plugging along in spite of indications that its bearings are going bad. We're still pressing on as usual hoping that it can wait a few more weeks to fail. We might as well get as much done as we can. We had another EREP today, and tomorrow, which is our day off, we're scheduled for a couple more. That's the only area we're behind in, as I understand it, so we'll be doing as many as possible in the next week or so. I really don't mind the EREP passes. They're fun. Bill and I started a TV program today which explains EREP and what we're up to. I'm the 'show and teller' and Bill's the cameraman. We sent down our first installments this afternoon. Hope it sells. Staying up late tonight. Bill and Ed went to bed early so I can quietly sit by the window and watch the world go by. Flew over Peking on the last day pass, then over Tokyo, and after that saw only water and clouds until Samoa."

Mission Day 72, January 26: "Today was another one of those 'days off.' Had another EREP pass and got assigned as VTS operator so had a good time seeing the sights. I really enjoyed myself. Luckily the weather was clear. Bill and I spent a lot of time today getting some TV science demos done, so we did not get as much window watching done as we would have liked. Ed spent the whole day in the MDA working the ATM panel – what ever turns you on! He got a good night's sleep last night (10½ hours) so I let him go for 7 day passes today. Bill and I just stayed away and did our thing. Finished up my TV short course on gyroscopes. That gyro is sure fun to play with. I'm going to bring it home as a souvenir. Talked to JoAnn tonight. Enjoyed hearing of her party last night celebrating our breaking Bean's record. She had Beth [Williams – widow of C.C. Williams] and the girls down for the night and Jennifer gave them rides on her horse, Sugarfoot. Sure glad Beth has gotten back with us again. She's a lot of fun. I hope we don't bring back too many sad memories for her."

Mission Day 73, January 27: "Had a long day today. Got behind tonight about dinner time and then just messed around for a while doing acrobatics and looking out the window. Next thing I knew it was bed time and I still had some things to do. What a dummy! Anyway now I'm all square, so I can get to bed. Julie [Gibson] told Ed tonight that we're due in Ellington two weeks from tonight. Man, that doesn't seem very long at all! At the rate things have picked up the last couple of days the next two weeks ought to fly by. Hope so. I'm ready to leave before this old buggy gasps her last. We have the EVA scheduled for MD-80 and after that we'll be closing things up in earnest. [CAPCOM] Bob Parker told us yesterday that there are only about five science days left and that includes 7 EREPs."

Mission Day 74, January 28: "The pace is picking up. They're trying to get the last of the squares filled on the mission plan. The ATM work is slowing down, and the EREP is frantically trying to get finished. The weather is killing us there. That's a real shame because we have plenty of film left and there's a lot of data to be gotten. Though the schedule is full we are managing to keep up okay. Bill has been having sinus problems lately, but seems to be getting ahead of it with Actifed. He had some pretty rough headaches a couple of days ago. I guess Ed is really seriously thinking of leaving NASA as soon as possible. If he could get a job offer in California I guess he would probably snap it up. Bill and I are both inclined to sit tight for a while. I don't think I really have any choice for a year or so. I think I have to put in one or two years as a colonel before I retire on a colonel's pay. I sure do need to look around though. I think I really would like to teach – college or high school, but I'd also like to earn a real good salary for a few more years to see all the gang through college. I'd better gather some data – some pros and cons for each course of action. I'm sure I can't do both. Retirement pay does give me more flexibility though. JoAnn and I need to give this a lot of thought. I'd probably be quite content doing a lot of different things. What we really need to get a handle on is where we want to be for everyone's sake. JoAnn's desires for her future need consideration too – like what kind of work or activities she wants to commit to. It's plain that the *house frau* bit is out. I think we all like Houston so well, maybe I better look closer by for starters. I'm convinced that we are started in the right direction in our relationship with the kids – particularly the older ones. We need to get over calling the younger three 'little ones' and start teaching them to actualize or have them learn with us."

Jerry was going to let Jennifer, Jeff and John read the "teenager bit" of *Man the Manipulator* to try to understand each other and the parent-teenager relationship even better. Jerry was also expressing the hope that Jennifer would be able to get some independence with her job and dating. Once he was back home he was committed to including Jennifer in more 'adult' activities, with JoAnn and himself. "We're inclined to lump her with 'the kids' and she's pretty much of an adult." In addition, Jeff and Jamee were beginning to "stretch their wings" and Jerry realized that they needed help in developing their sense of responsibility for themselves, rather than just reflecting those of their Mom and Dad. All this sounded good, but just how Jerry was to actuate it was not so clear.

"I've tried to think of what things up here mean to me, but seem to be dry unless I have someone as a catalyst to trade ideas with. Bill and I have had a couple of good discussions, but when I try to put something down it doesn't come. I've been a detail man too long and lack creativity in areas other than engineering. Need to loosen up more."

Mission Day 76, January 30: "Judging from the flight schedules we've been getting lately, it looks like the square fillers are frantically trying to get their last licks in before we close up shop. We're getting all sorts of screwy odd jobs like weighing before and after exercise, and all sorts of photos of this and that. Slowly but surely we are shooting up all the scientific film. Now if we can just get everything stowed in the CSM we've got it made. Tomorrow is another one of those 'days off'. We get a grand total of 4½ hours free. The rest is taken up with EREP, a TV press conference and other neat tasks. Can't say I'm looking forward to the press conference. The last one was traumatic. We've got to try to keep this one from getting too heavy. Hope we get an opportunity to inject a little humor into things. It's amazing to me that the three of us are still getting along so well. There has been hardly a harsh word between us. Everyone does his share, and we continue to get along pretty well. We're three pretty different guys, but we manage to tolerate each others foibles pretty well. Got a couple of opportunities to look out the window today."

Jerry managed to get a good pass over California from San Francisco to Imperial Valley which he particularly enjoyed. He managed to pick out Vandenberg AFB, Edwards AFB, Santa Barbara and Big Bear. Then later in the day he flew over Buenos Aires and was treated to a spectacular view of the Falkland current mixing with the South Equatorial current. "It was a beautiful swirl of blue, dark and light green, and a faint maroon. Best view we've seen yet." During that evening he viewed a run down the Western Pacific and managed to snap some more pictures for the Marines. He was only a few shots short of a collection of bases he had been stationed at during his years as a Marine pilot prior to joining NASA.

"I'm starting to collect the items I want to take home. We're scraping off the entire Snoopy decals around here and will share them with the other guys who have flown Skylab. (The famous cartoon character was the Manned Flight Awareness mascot. The Snoopy decals were attached for quick recognition to various lockers that contained personal items assigned to each of the crew, such as clothing and food. Jerry's decals were red, Ed's were white and Bill's were blue.) Guess I'll have to start a little list in the back of this book so I don't forget anything. I think Entry Day is going to come all too soon. That's one trip I need to be all packed for."

Mission Day 77, January 31: "Today was a lousy day off. We got a lot accomplished though. Did an around-the-world EREP this morning. Ran the radar altimeter and other instruments over the US. Weather was bad though at our US sites. Tomorrow is to be our last EREP. Spent most of my spare time updating our entry checklists, and the three of us spent 1¾ hours in the CM talking our way through the de-orbit and entry sequence. All three of us dreaded today's press conference. We took a long time answering questions so there would not be too many. They were good questions and we felt pretty good when they finished with us. None seemed antagonistic or argumentative or accusing. I'm sure the press will treat us okay when we get back, but Ed and Bill are still skeptical. Actually, Ed's really worried, and I don't think he should be. He has done a great job scientifically, and his colleagues should really be proud of his contributions to solar physics. The only place we didn't cut the mustard, as I see it, was from an operational standpoint, and we had a lot of help on the ground on that score. I also shudder every time I think about our foolish attempt to cover up Bill's illness – stupidity personified!"

JC: "About three days before it was time to re-enter, we began the process of adjusting our body clocks, because on the day of return we had to get up at about 2 or 2:30 in the morning (Houston Time). Probably a week before the re-entry, we spent all of our time repacking the Command Module, moving all the things into the CM we were to take home; all of the magnetic tapes, all of the urine specimens and fecal specimens that we would have, all that scientific data that we gathered. When we left, the CM was packed as tightly as when we flew up there."

Mission Day 79, February 2: "Flew our last EREP yesterday. Probably a good thing. S191 started acting up along with S190 so we barely got our data. I was the VTS operator, and my targets were all clear! I got all the US targets and two of the four Central American volcanoes. That was a good way to end the Earth resources surveys. I spent the rest of the day doing LBNPs and in the evening the last S201 look at Kohoutek. I'd only gotten about three hours of good sleep the night before so I really passed out at bedtime. Then damned if I didn't wake up at 04:30 this morning. Couldn't sleep anymore, so I got into a science fiction book [Al Bean's copy of Arthur C. Clarke's *Childhood End*].

The Apollo CSM docked to Skylab. Photo taken by Jerry on the last EVA.

Spent the morning in the CM checking it for return and stumbled onto a problem with Battery A. Looks like the battery is okay, but the circuit breaker bus won't carry a load unless you hold it in. It doesn't pop – it's opening internally. Good thing we didn't discover it weeks ago. They'd have made us come home then. We've still got four other batteries left, so I'm not too concerned. We may have to power down some unessential equipment to save power, but that's no sweat. I just want all the thrusters and engines to run properly. After 85 days away from the CSM you lose a lot of familiarity, and we will be hard pressed to cope with some complicated failures. Did my last LBNP run. We'll be glad when those are all done. What a colossal bore! EVA tomorrow. Am really looking forward to it! We're taking two cameras out – a Nikon and a DAC [Data Acquisition Camera, the movie camera] so we should get some good stills and movies. I get to be the 'wing walker' again. Great! That's a good way to end a mission."

EVA 4, February 3, 1974

Initially only two EVAs had been planned for the third mission. The final one was supposed to occur on MD 54, some weeks prior to their return to Earth. On

Skylab's unoccupied OWS during final fly around.

this EVA, the astronauts would have retrieved all the ATM film cassettes. By the time Ed and Jerry stepped out for their final EVA, it was actually mission day 80, only a couple of days before they were to come home, and there were 16 tasks to perform. Jerry and Ed used the back up 'clothesline' film transfer device to relocate equipment across their worksites. In between completing their ATM film retrieval tasks, Jerry demonstrated a hand-over-hand movement along a tether. Then they completed the EVA atmosphere photography they had started during their first EVA. Another Micrometeoroid Particle Collection experiment was deployed in the hope that it would be collected by Shuttle astronauts visiting Skylab in the early 1980s.

EXTRACTS FROM THE DIARY

Mission Day 80, February 3: "EVA Day. Last one of the mission and the most fatiguing. Came in bushed. It didn't look like it was going to be more than just routine – just a film recovery, some T025 comet photos and some S020 solar photos. Then they added in some movies for documentary purposes, some still photos, and removal of several samples from here and there. We stayed out about 5½ hours and did an awful lot of finger work and hand manipulations. The ends of my fingers are now really sore from pushing up on the ends of my gloves. [This EVA] really brought home the fact that we're doing something really unique and will probably never ever do it again. From now on it will be all memories. Some time from now we will look back on the mission and will have forgotten all the bad things. We'll only recall the good and interesting times and repress the rest. I'm already looking back and seeing more of the pleasant than the unpleasant."

The ATM during final CSM fly around.

Mission Day 81, February 4: "Four more days to go! The pace is really picking up now. I spent the whole day doing the experiment on flammability of materials. It's really pretty interesting. I didn't finish though. Only finished 30 of 37 samples, so I left the equipment set up in the hopes of finding a little time tomorrow to finish it up. We solved our CSM battery problem today, so now we don't have to sweat the power problems coming home. Must have had corrosion or contamination in the circuit breaker so we burnt it out by holding the breaker in tight while applying a 20 amp electrical load to the circuit. Worked beautifully. Finally got a chance to be at the window as we came down the California coast, and the weather was clear! Got pictures from Fresno to Guadalajara. Even tossed in Santa Ana for good measure [Jerry had been trying to photograph his home town for weeks]. Really enjoyed that pass. Go to sleep you dummy.

You've got a hard day tomorrow!"

For the three weeks prior to entry, the crew rehearsed procedures where they could during the few hours built into the daily schedule. They also began the process of moving the equipment back into the CM. The crew were supported in the reloading process by a team on the ground who knew where everything had to go and told the crew where to store each item. If things were stored in these locations, they could precisely work out the mass and centre of gravity of the CM to ensure a safe entry and landing. There was no margin for extra items.

Mission Day 82, February 5: "Well I was right. Today was a doozer! I started out fast because I knew if I got behind it could be brutal. It didn't help. I still ended up working until 10:30 tonight. I guess the biggest time burner was not finding the things where they are supposed to be. I took a load to the CSM and tried to pack a locker, but that was impossible. When you're trying to put six or eight loose items into a locker you need four hands. Everything keeps floating away. Bill was scheduled to help me this afternoon so I just had to do other things until he could get there to help. Then we got the job done right. Only got a short opportunity to look out the window. I feel cheated on days that I don't get to look outside. Sure am anxious to get finished with all this and rejoin the world. Speculating is fun but you need to be where it is too. Someday we'll be in a position where it will be a routine trip to shuttle up to a space station and spend a week or so then go home again. No special training needed or anything. I guess our kids will see that day but I don't think our generation will. It'll take too much money to develop."

Mission Day 83, 6 February: "I think we're over the hump. Today was rather relaxed compared to yesterday. I got the CSM pretty well tidied up and stowed, so we're ready for the deactivation sequences tomorrow. Going to bed 2 hours early and getting up at 05:00 tomorrow for a short day, and then back to bed in the afternoon and up again in the evening to start entry day. Sure hope we don't run into many snags. If we have no problems with the CSM the entry should be a piece of cake. Problems will hurt us because of the long periods of time that we've ignored CSM systems and operations. There are some things about this place that I'm going to miss, but for the most part I'm glad to be getting rid of it. The weightlessness and window gazing I'll miss. The rest won't be missed. I'll include EREP VTS in with window gazing. My sleeping pill is getting to me now."

This was the final entry in Jerry's diary with entry just over a day and a half away.

Jerry's Take Home List.

In his personal preference kit, Jerry collected a number of items he wanted to take back home with him in two bags and a pouch. These included:

1. Family photos (bag 2)
2. Snoopy decals
3. Back-up crew patches (pouch)
4. Christmas presents
5. C.C.'s labels (pouch) (These were flight jacket patches with C.C. Williams' name and rank on them. Jerry carried them for his wife, Beth).
6. M509 checklist (Bag 1)
7. Gyroscope
8. Swiss army knife
9. Scissors
10. Silverware (4 pieces)
11. Patches off my coats (Bag 1)
12. Decal form stowage book.
13. Tapes (18) (Bag 1 and 2)
14. Pens
15. Pencils
16. Skylab t-shirt

...and his diary......

Mission Day 84 – the shortest day

The night prior to entry, the crew performed some get-ahead tasks by powering up part of the Command Module. Members of the flight control team monitored their progress and then watched over the CM from the ground as the crew spent their final hours asleep on Skylab.

When Jerry and his crew had departed, there were no plans to revisit the OWS, though there was some contingency to ensure that, if a revisit were possible, the station could support at least a short stay. It was planned to leave the station in orbit, and there was a desire to keep it oriented properly to allow a possible visit by a future spacecraft, perhaps the Space Shuttle. The proposed 21-day Skylab 5 mission was abandoned when it was decided to extend the Skylab 4 mission from 59 to 84 days. A review of onboard systems revealed that it was impractical to support another crew for any longer than a few hours docked to the station.

Despite being the final crew to live aboard Skylab, Jerry and his crew still packed away everything where it belonged over several days prior to departure. They also took samples from around the OWS and stowed them in a return bag in the CM prior to sealing the hatches. One of the experiments they had to do on the last day was to go around the workshop and conduct a microbiology survey. They took little swabs and made scrubs from various surfaces, sealed them in vials and stowed them for the trip home.

The final day was basically spent putting everything away, shutting down all the systems and collecting clippings of bits of metal from various parts of the workshop. The crew intended to have small medallions made from the metal, so those who flew and worked on the station would have a commemorative medallion which contained metal from the actual workshop that had flown in space.

Mission day 85 – the last day on Skylab

When the crew received their wake up call for the last time, they were all ready to roll. CAPCOM Story Musgrave (Maroon flight control team) passed on the appreciation for all the work Jerry, Ed and Bill had done and wished them a safe return home. CAPCOM Bruce McCandless (Silver team) advised the crew that in order to complete their travel voucher, the mileage from lift off to splashdown was expected to be 30,561,000 miles. The only problem was that since it was all government transportation there would be no reimbursement for that. Jerry also observed that this included government-provided quarters and meals too!

As Jerry continued powering up the CM and performing a systems check of their spacecraft, Ed and Bill closed out the remaining items in the Orbital Workshop. This included passing the command capability of the ATM to the ground. As the ground confirmed the smoothness of the closeout procedures, CAPCOM McCandless voiced the Silver Team's appreciation for the three months work. "We've enjoyed very much working with you over the last three months. We wish you luck in your re-entry and future endeavors, and we'll be seeing you when you get back to Houston. Wish you a fair wind and a following sea."

Finally shutting down the vehicle was emotional for all three astronauts, as Jerry recalled. "We realized that it was he last time that we would do various tasks. We also put away a bag of items which we called the 'time capsule.' We used one of the white urine stowage bags and tied it down close to the hatch. We included a checklist and some photos and that was supposed to be the time capsule. The time capsule was a planned activity."

The three men then floated into the CM, shutting things down behind them. Once inside the CM they disconnected the umbilicals that were crossing the door sill, put them inside the workshop, closed the hatch and sealed it. There was no official ceremony, though Jerry did comment to the ground that they would leave the key under the front door mat so that the next crew could get back into the workshop.

A few hours later, with all three men inside the CM, and with the hatches closed and tunnel depressurization complete, they began taking off their helmets, gloves and pressure garments. A successful test of the Reaction Control System qualified them for use during departure from the Skylab. Jerry confirmed that he found a "good moon," which helped to define a horizon for navigation and orientation checks.

CAPCOM Bob Crippen, on duty at the console in Mission Control, updated the crew that it was time to leave. "The party's over; time to quit the fun and come home to go to work. Looks like a good day for you guys to come home. People out on the NEW ORLEANS [prime recovery ship] are sitting there waiting for you , and you've got fairly calm sea today." Jerry was particularly pleased to hear that report.

At 10:22:36 am on MD 85 (February 8, 1974), Jerry received the 'Go for undocking' call from the CAPCOM, with a request to "Say goodbye for us; she's been a good bird." Jerry replied, "She sure has." Just over five minutes later, Jerry radioed: "OK, we're undocked. We realized the docking probe extended a little bit too soon. It caught on the capture latches, [and we] had to release them and thrust our way off."

Jerry had finally got to fly the CSM again after three months attached to the OWS. With Bill taking photos and Ed helping him, he moved the CSM in a complete circuit around the station at about 50 to 100 feet (15-30 m) distance. When this was completed he fired the RCS system to back further away and initiated the re-entry sequence. Following the re-entry checklist he fired up the computer and commenced the procedures to begin the re-entry. All proceeded smoothly. During the fly-around the crew spent time taking film footage to document the OWS as they last saw it. Jerry operated the DAC, and Bill used still cameras to film the departing workshop. Jerry was amazed at the speed the camera was using the film: "Holy Moses, I've used half the film already. I'm going to cut it down to two frames per second. It's ridiculous." Ed asked if either Bill or Jerry could picture climbing around on the "outside of that thing." Jerry replied that he could and had pictures to prove it!

Ed added his comments over the open 'A' channel on the appearance of Skylab. "I'll tell you, this vehicle sure looks like it's been worked over. A lot of tender loving care has gone into the exterior of this thing." Ed reported that he could see two colors in the original parasol shade erected by the first crew. Originally covered by the sail erected by the second crew, more of the parasol had been exposed when the sail had pulled away a little.

> **EG:** "It's [the parasol] got two shades of brown; one very dark, and one just about a lighter tan. It's been a good home Crip."
> **CC:** (Crippen): "Yes, sounds like it. You guys occupied it long enough. Everything's looking good here. You are GO for the Sep [separation] maneuver."
> **EG:** "You can tell Al Bean and the guys they did a great job putting that sail up. It's very symmetrical."
> **CC:** "Very good. I'm sure they'll appreciate the words of their good work. Certainly helped out there to keep you comfortable."
> **EG:** "Yes. There's only one little spot not too well covered and unfortunately it was right by my sleep compartment."
> **CC:** "They arranged that especially for you Ed"
> **EG:** "I believe it. The only thing to look forward to is a bigger and better one [space station]. It's been a really useful machine. Hate to think we're the last guys to use it."
> **CC:** Well, it certainly did a great job. Served its purpose... along with some real fine guys running it for us up there. We appreciate all the good work."
> **EG:** "That's the whole NASA team that did that."

Bill Pogue agreed that it had been a good home as they flew around the station for one last time. He noted the OWS was dark against the black of space. Realizing that the return to 1-g was going to be "a real shocker," he decided to stow away some of his equipment and cameras before the de-orbit burn.

Jerry decided to remain station keeping and "watching the bird" for a while as they had about 26 minutes before the scheduled burn maneuver. Ed commented that it was not very comfortable looking out of his window [the centre hatch window] "No wonder I couldn't see [the OWS] during docking." Jerry replied that this was just as well as it would have frightened him, to which Ed quipped: "That's right. I was scared just watching the beads of sweat on your forehead watching you use the hand controller..."

It was left to Jerry to make the final observations as they viewed Skylab for the last time: "There it goes, fading into the sunset. We ought to have gotten a few really good [pictures] in there [during the fly around]. Okay, I guess we can start the maneuver now. The bird's kind of far away." The first of two burns separated the spacecraft from Skylab. The second would slow them enough in their orbital velocity to start the descent to Earth.

Leaving Skylab

Bill remarked that he was rather looking forward to that burn, while Jerry thought it would be a "real ring-ding-doozer." Ed suggested keeping strapped tightly against the seats because of the Gs imparted during the de-orbit burn. Looking around, he suggested that before the final commitment to entry they should move all the loose items to the 'floor' of the CM under their seats, "…because when we hit [the g-forces of entry], it's only going to really move down there rapidly." Things might break [or hit them], urine bags may split, and then they might turn apex down into the water (Stable Two) and have all the items fall to the front of the spacecraft, only to fall back when the flotation bags were activated.

As Jerry maneuvered the CM to the burn attitude to separate from the station, Ed commented that he could not see the OWS anymore as it disappeared out of the field of view from the five windows as they turned: "Sure has been a heck of an adventure. It's tough to top that one. You don't realize what you're doing if your chugging away at it day to day." Bill added, "Yes that's right. In some respects, it's almost incredible."

As Ed checked for stars through the sextant, he noticed the OWS pass into the view finder.

A 'kick in the pants' burn

It was a very good de-orbit burn. Ed observed, "I have a feeling we're not going to bound out of this spacecraft like gazelles."

Jerry had tested the gimbaling of the Service Propulsion System (SPS) at the back of the Service Module and was in no doubt that "she was wagging her tail." The horizon check was right on the money and that pleased him. Just over 3.5 hours after undocking from Skylab, they received the 'GO' for the de-orbit burn. Jerry replied, "so are we." Firing the RCS system jets resulted in a spectacular ice crystal show down the left side of the spacecraft.

The crew recalled the experience of the burn as a "kick in the pants," and after 84 days in zero-g it certainly got their attention. As they descended, Ed Gibson observed that they were getting so low it looked as though they were going to scrape the ground. The view reminded him of the one out of the window of a T-38 jet. CAPCOM Bob Crippen reminded him that they had been flying 'up there' pretty high.

> CC: "As a matter of interest, you're five minutes from perigee (the low point of the orbit), still going down. Perigee's running right around 90 miles I guess."
> JC: "Better suck up our landing gear, I guess."
> CC: "Right, [and] I can give you a little update on the weather. As I told you earlier, it's really looking good out there. Cloud's got a cirrus broken layer, and visibility 10 miles. Wind is out of the north-northeast at 10 knots. Got 1 foot waves on 3 foot swells. Temperature's about 59 degrees."
> JC: "Hey that's land lubbers weather there."
> CC: "Roger. Always nice in southern California."

With the re-entry burn completed, the CM had to be separated from the SM because they were now committed to re-entry. As Ed and Bill read off from the checklists, Jerry was pulling the circuit breakers and completing the separation checklist. He then finally punched the button that separated the CM from the SM. Now flying just the CM, the crew were heading for home. "When I grabbed the hand controller to begin to reorient the CM for re-entry with the heat shield forward, the spacecraft would not move." He had to switch to a back up system, which solved the problem. It wasn't until they had splashed down that they realized that Bill had earlier called out the RCS circuit breakers on the Service Module, A, B and C. What Jerry had done was to pull the *Command Module* circuit breakers A, B and C which were adjacent to them. "That was a real classic lesson in human engineering, because three sets [of circuit breakers] were labeled very much alike and were right next to each other. Luckily, we had a system that allowed us to go to Back Up and reorient the spacecraft. Then we went into entry mode, and everything worked out fine."

Being committed to the descent, there was very little time to resolve the problem – barely a minute or so. "When the controller did not work it seemed a lot longer." As long as he had entry attitude control, Jerry was not unduly worried, but Bill noted that Jerry was frantically working the hand controller and nothing was happening. He told Jerry that his heart really sank for a moment or two, as did Ed's. It was a while before it dawned on them what the problem had been. For now all they wanted to do was get down.

After CM/SM separation, the SM thrusters are automatically activated and thrust it away from the CM, placing the SM on a different descent trajectory. When Jerry found that his controller was not working, they were still pointed forward toward re-entry. "So we immediately shifted to the back up procedures, which

Splashdown of Skylab 4

Bill read off to me. We completed those, shifting to our secondary control system, and with a sigh of relief we flipped the CM right around, got her all set up and went on to re-entry."

As the spacecraft began to enter the upper reaches of the atmosphere, communications began to black out. As the two parts of the CSM separated, Ed commented that he heard it go, while Jerry radioed, "OK, here we go. Bye bye…"

Re-entry

The ride home was as dramatic as the ride up to orbit had been three months before. Jerry called upon the training that he had been conducting since first coming to the Astronaut Office seven years previously. The long hours of academic and theoretical study, and time in simulators, was put to the test as Skylab 4's CM dropped out of orbit.

JC: "You just oriented the Command Module and you went into the final entry mode on the computer. The computer oriented the spacecraft and kept you pointed in the right direction (and angle) and just guided us all the way in. Meanwhile, I had an entry maneuvering system called EMS with which I was watching everything the computer did. If it didn't follow the trajectory on the entry monitoring system then I was to take over manually and follow the EMS trajectory. There were not many Gs – probably 2.5 to 3 Gs – but we were all wearing girdle G-suits. That was one of the precautions they took because they realized we were probably not going to have much tolerance against the onset of Gs. When we started re-entry and began pulling the Gs, the girdle would inflate and squeeze on our belly and legs, pushing the blood into the upper part of our body to keep us from fainting. We all noted that our peripheral vision began to narrow a bit as we were coming in."

Jerry's onboard comments recorded on the CM tape recorder reveal his verbal impressions of the view out of the window during entry: "Horizon's coming into the window in good shape now. Can you see the glow? We're starting to warm up. I can see the horizon now…0.5 G… OK were starting to pull Gs… Oh that's gorgeous! This is something I'll never forget… OK the G load is 1.1… 2G…We're going just about at 3Gs…I've got some gold right up in front of us. Beautiful gold… Doing beautifully… We're just under 3Gs. Looks like we've peaked our Gs."

Ed Gibson commented on the effects of the build up of Gs as they came into the atmosphere: "Sure wish that elephant could climb off my chest. Man that's hard. I tried picking up my head and it was next to impossible."

Jerry later recalled the entry sequence: "Looking out the window you don't see anything until you begin to get into the atmosphere, and then you begin to see some of the ions and some of the blue flame that comes around from the heat shield. Actually it starts as sort of a rosy glow, and then it gets more rosy and red with a few blues in it. Quite memorable."

Recovery helicopters approach the bobbing CM after splashdown

Recovery Para rescue divers around the CM.

The pressure garments were left on Skylab, so the crew re-entered wearing soft brown flight suits. Just two years after the loss of the Soyuz 11 crew who were not wearing pressure garments when the cabin atmosphere was lost, NASA were confident enough to allow their crews to come home without pressure garments for Skylab entry.

JC: "It was a pretty smooth ride. I was surprised. You could hear the spacecraft control system fighting to keep our CG oriented properly. It did a good job of it, and we had this constant deceleration going down. We noticed a little peripheral vision beginning to squeeze in, but it never did come close to blackout. All three of us were able to carry out the duties we had to do during re-entry, of throwing switches and monitoring things."

After emerging from blackout, Jerry called Houston to re-establish communications:

JC: "Hello Houston, how do you read? We're doing fine."
CC: "Happy landing guys."

Monitoring the altimeter, Jerry was waiting for the point to manually deploy the parachute system. The two drogue parachutes were deployed first to stabilize the spacecraft, and Bill Pogue noted the distinctive jolt inside the Command Module. At 24,000 feet (7,300 m), Jerry initiated the deployment of the three 83.3-foot (25 m) diameter main parachutes, which unreefed at about 10,000 feet (3,000 m): "Mains are unreefed... open full... oh look at those beautiful chutes," Jerry exclaimed, looking through the CM window. Twenty five years later, the experience was still a memorable one for Jerry: "You feel a pop and a jerk and then you are just floating in and coming down very gently."

Jerry noted that everything looked good at 7,000 feet (2,100 m) as they heard the recovery helicopter calling the descending spacecraft.

Recovery 1: "Skylab... Skylab... This is Recovery... Over."
JC: "Hello Recovery, Skylab loud and clear."
Recovery 1: "New Orleans, Recovery. Stand by to mark splash... R1. MARK. Splash!"

Skylab 4 CM-118 was in the water just 176 miles (283 km) southwest of San Diego, California, so it would not be long before the crew would be back in America. They landed just three miles (4.8 km) from the recovery vessel, which according to Gibson was not bad for their first attempt. Jerry jokingly claimed that splashdown was so accurate that they had to move the carrier out of the way. It was February 8, 1974, at about 11:17 am EDT. Skylab 4 had completed 1,214 orbits of the Earth in 84 days 1 hour 16 minutes and had logged about 70.5 million miles (113.5 million km).

Ocean Recovery

The space mission may have been over but the new mission for all three astronauts, that of adjusting to life back on Earth after a spaceflight, was just beginning.

JC: "We hit the water and immediately flipped over nose down, so when I looked out of the window there was nothing but green water out there. We were hanging in the straps, and I said 'We've gone to Stable Two. We have a procedure.' (reminiscent of the Stable Two exercise we practiced at JSC). We whipped out that procedure, and I activated the air pumps that would fill the bags. They inflated and flipped us over on

Jerry steps out of the CM on the deck of the recovery ship, Bill Pogue is at far left.

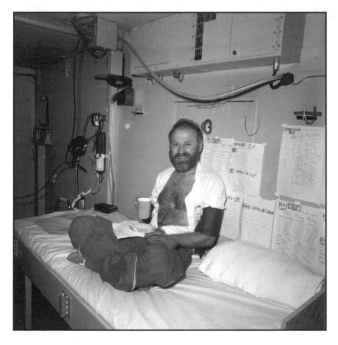

Jerry takes a coffee break during post-landing medical tests on board the recovery carrier.

our back (Stable One). Now we could see the helicopters come by and the frogmen were soon outside. Once in a while you could hear the frogman bump us as he inflated the flotation collar that went around the spacecraft. Then he got up on the collar and looked in the window at us. He was making faces at us while we were lying in there. We waved back at him, saying we were all OK."

Over in the Carr household, everyone was listening to the radio during entry and landing, with Rusty Schweickart translating for the family. There was a sense of sadness, if not anger, that, despite the record-breaking flight duration and the fact that it was the end of the Skylab program, the landing was not going to be covered live by the major TV networks.

JoAnn: "I had a household full of people. We didn't tell Jerry that he wasn't being covered, because I thought that was just the final insult, the final slap in the face. He didn't need to know that. Jeff became very annoyed and even wrote a letter to the networks. He wrote a nice letter to the presidents of all the networks, who answered him. They said it didn't mean that what his father had done was not important, it just wasn't newsworthy."

Jeff Carr recalled his response to the lack of TV coverage in a 2002 interview: "I had talked to my Dad. He had a strong sense of accomplishment at the end of the mission. I felt a strong sense of pride about my Dad coming through, and I wanted every one in the world to feel the same pride that I had. But I found out about a week before the re-entry that the networks [ABC, NBC, CBS] had decided not to cover the splashdown.

"I was very disenchanted, very disappointed. I couldn't understand that. It did not make any sense to me at all. I just couldn't fathom how, in a time when our country was so desperate and in need for positive reinforcement to focus on values that are good, when we were so

The Carr family celebrate the end of the mission - l to r are John, Joshua, Jennifer, Jamee,
Rev Dean Woodruff (the family pastor), JoAnn, Jeffrey and astronaut Rusty Schweickart.
The family dog, Rags, is taking it all in stride. Jessica is out of frame.

preoccupied with domestics and social issues, we couldn't celebrate the landing. I tried to express that the best way I could and sent a letter to each of the television presidents. I didn't know if I expected to make any difference. I think I was a little surprised, impressed to a degree, when I got responses from all of them. But it was almost as if they had compared them. Obviously they didn't, but the theme was the same. My argument was sound, they wrote. You have good reason to be very proud of your father's accomplishments, and what the crew have accomplished. But the decision not to cover the splashdown live was not based on these accomplishments but on the newsworthiness of the event. I thought, 'What's the damned difference?' That was my first lesson in public relations, my first lesson in the power of the media to set the public agenda. It was the first indication to me that choices were being made about what was important to people in this country, but not necessarily by people who had their priorities straight.

"So it didn't change anything and left me with a very foul taste in the mouth. I've come through my career understanding the nature of the media, and I have come to respect them to a degree. But the pain is not any easier. It is still a very deep motivation for me to find a way to eliminate the media from setting the agenda and breaking down the barriers between the space program and the public, so that they can set their own agenda and make up their own minds about the space program and how relevant it is. That experience still remains a very defining moment in my life. It had a very significant effect in shaping my career."

Strapped into their seats, Jerry and his colleagues would remain inside the CM during recovery to help with their adjustment back to 1G. Jerry recalls that, "We didn't have any problems with loose items in the CM at all." Then the primary recovery vessel, the carrier USS New Orleans, came alongside and the hoist was hooked up and raised the spacecraft out of the water and onto the hangar deck. Two NASA Flight Surgeons, Jerry Hordinsky and Eduard Burchard, climbed into the CM and checked over the astronauts. They put on blood pressure cuffs, and checked their heart rates, respiration rate and blood pressure, as Jerry

John uses a model helicopter, Command Module and a glass of champagne to demonstrate how they snatched his dad from the sea, though a crane on the side of the recovery ship actually recovered the crew in the Pacific.

remembers: "That probably took 30 minutes or so, and then they said 'Well, it looks like your blood pressure is good enough for you guys to get up and walk out of here.' So Ed went out first, and Bill went out second. I went out last because the naval tradition is that the captain leaves his vessel last. We got out of the spacecraft, and they had chairs on a big fork lift platform. So we sat down on the chairs and the fork lift took us away. We went to the medical area where they lowered us down on the deck and each one of us had an escort to hold on to as we walked.

"It felt pretty clumsy. We were used to being butterflies, but now we had appendages like our arms and legs and a big watermelon for a head. We learned very quickly on the carrier not to nod the head 'yes' or 'no,' because you could feel this watermelon moving around. It also caused a little bit of equilibrium shift, a little vertigo. So we felt pretty clumsy, but we felt good. We had exercised a lot during the mission, and our cardiovascular systems were in really good shape, so we were able to walk in 1G without fainting or falling down."

With the wind blowing in their faces again after three months cooped up inside the spacecraft, the sights, sounds, and smells of life on Earth became more apparent to the recently returned astronauts.

JC: "Oh it was great. I remembered, having been aboard ship, what the smells and sights were like, so for me it was very familiar to feel all the sensory perceptions of shipboard life. It was great to be back, and I remember sitting in the chair on that fork lift looking up and seeing all the sailors that were up on the flight deck looking down, waving, saying hello and all that sort of thing. It never seemed crowded or busy at all. There was nobody around us on the hangar deck. They just moved us over to the sick bay, and then each of us had our own doctor looking after us.

"Having done the first EKG on me, they allowed me a cup of coffee. We had landed at 9 or 10 in the morning and for most of the morning they did medical stuff on us, checking us out and doing the same protocol we had been doing while we were in orbit. Then they let us stroll around the deck a bit. We went over and looked at the spacecraft, and marveled at the scorches on the bottom of it. Then we went back and had dinner. We were pretty much separated from the rest of the people again because of the quarantine requirements. The first night I was sleeping in a state room, and I had a doctor sitting in the room with me.

The crew descend the steps of the Air Force aircraft that brought them to Ellington AFB near JSC.

His job was just to sit up all night watching me to see what I did. When we talked the next morning, he said 'You were really restless last night.' I said, 'I just felt like everything inside me was hurting, and I just could not get into a comfortable position.' He suggested that was probably because the muscles inside the rib cage hadn't been supporting any viscera for months, and now all of a sudden these muscles had to do their job again. It was really uncomfortable sleeping that first night, and for the next couple of nights, before it subsided."

The next morning, Jerry found out that it would take him some time to get used to being back on

Celebrations at MCC Houston mark the end of Skylab 4 and the program.

Skylab 4 crew report to congress.

JoAnn tweaks the 84 days of beard growth of husband Jerry during the welcome home ceremonies at Houston.

Earth when he almost dropped a urine specimen. He caught it in time before it smashed on the floor. Getting used to space took a while, and obviously getting used to life back on Earth would also require a period of adjustment.

He had talked with JoAnn and was excited about the recovery, asking if the family had seen it on TV. But of course they had not. Jennifer Carr recalled almost giving the game away when telling Jerry she was getting extra excited from the champagne she had sipped, but saw the look on her mother's face and suddenly stopped. At the time, the family was being filmed for the 'GO SHOW,' and the concern that Jerry could find out was etched on the faces of JoAnn and the children. It was also voiced on the program commentary. John asked why, when they had seen all the other astronaut Dads come home, they could not see their Dad.

JoAnn: "I knew he was going to find out ultimately, but I didn't want to ruin the moment for him by saying it wasn't important enough to cover. It was very strange, because I had watched all the other splashdowns on TV, but this time I wasn't really even quite sure when it had happened. Rusty was trying to let me know what was happening, but I was saying, 'Is he on the carrier? Where is he? What's happening?' I couldn't see it, and I couldn't really tell by what I was hearing, so Rusty was translating. Within a couple of hours we got pictures from Associated Press of them getting off the spacecraft, and when I finally realized they were on the deck, my head just dropped. I thought, 'I really need to be by myself. I don't want to be with all those people.' But there was no way to get away from them, so it was tough. It was a surprising relief to know his feet were on the deck, and it didn't even have to be *terra firma*, as long it was on the deck of the carrier. But he looked like a beat up old man. He was all stooped over and creaking along."

John Carr was really upset about not seeing his Dad land live on TV. Years later, he is still upset that his Dad's landing was not carried live. "I'm sure it would have upset him too. I will never forgive the media or whoever was responsible for that not being televised. They took a piece of history away from me and from anyone who had any interest in the program. I was disappointed back then. I still am today."

Back on Earth

The realization that his first space mission was over was quite a high point for Jerry. "We were glad we had done a good job. We'd had a very successful mission, had accomplished everything we had set out to accomplish plus some others from experiments we had set up ourselves. We felt really good about it. The next morning, when the New Orleans docked at San Diego, they had a big dock-side ceremony as we were getting ready to leave. There were about a thousand people on the dock with bands playing. We were each asked to stand up and say something. I don't really remember too much of what I said, but I remember what Bill said, and that was interesting. He said 'I'm in love with the whole human race now.' Because we had been up there and had seen the world from the perspective where you can't see the boundaries between the countries, unless it happens to be a river or something like that, we were all profoundly struck by the universality of the human being. Wherever you are, everybody has the same problems, the same needs, and when we returned we were very much filled with that feeling of universality of the human race. That's what Bill was talking about."

Skylab 4 -for the records

As Jerry, Ed and Bill became accustomed to their first hours back on the Earth, so their achievements were entering the record books. Despite the difficult start, they had met or surpassed all of the mission goals, a notable return for their efforts. An evaluation of the utilization of man hours on their third flight revealed the accomplishments of a rookie crew who soon became veterans of the program.

Skylab 4 man-hour utilization

Activity	Hours	%
Medical activities	366.7	6.1
Solar observations	519.0	8.5
Earth resources	274.5	4.5
Other experiments	403.0	6.7
Sleep/rest off duty	1846.5	30.5
Pre/post sleep and eating	1384.0	23.0
Housekeeping	298.9	4.9
Physical training/personal hygiene	384.5	6.4
Other (EVA etc)	571.4	9.4
TOTALS	6048.5	100.00

In the performance of experiments, the crew again recorded impressive totals over their three months in space:

Skylab 4 Experiment performance

Activity	Hours	%
Solar astronomy	519.0	33.2
Earth observations	274.5	17.6
Student	14.8	0.9
Astrophysics	133.8	8.5
Man systems	83.0	5.3
Material Science	15.4	1.0
Life Science	366.7	23.5
Comet Kohoutek	156.0	10.0
TOTALS	1563.2	100.0

Jerry, Ed and Bill had returned with 73,366 frames from their solar observations; 19,400 frames from Earth observations and 100,000 feet of magnetic tape data from Earth and solar observations. It would be many years before such a scientific return would be gathered on an American space mission again.

The records also revealed that Skylab 4 had set new world, mission and individual crew accumulated time in space records, along with a single mission endurance record of 84 days. They had observed and photographed Comet Kohoutek and set an EVA endurance record that would stand for 22 years. They had increased the previous length of US time in space by 50%, and would hold the world record for over four years and the US endurance record for over 21 years. It may have been a frustrating and difficult start, but it was a remarkable and rewarding ending.

Chapter 10

After Skylab

Now that the final mission to Skylab (and the longest spaceflight in history thus far) had ended, the mission hit the headlines again, this time celebrating its success and its application to the future. The Director of the Skylab Program, William C. Schneider, stated, "We have shown that there is no man or machine limitation to whatever we want to do in space. The only limit is ourselves, not our ability to do the work and not our technical knowledge." Schneider added that the end of the three manned missions to Skylab marked the beginning of a new scientific mission. The evaluation of the masses of material gathered by the nine astronauts over the previous nine months was just getting underway.

President Richard M. Nixon sent the astronauts a message that stated: "Your mission has brought to an end one of the most scientifically productive endeavors in the history of human exploration." Former Chief astronaut Alan Shepard, reflecting on the twelve years since he became the first American to fly in space, told the media, "Back in 1959 and 1960, when there was a great deal of cynicism about man in space, we would have found it difficult to believe we could make such strides in such a short time. I think we've accomplished much more than even the most optimistic of us would have predicted."

The Home Coming

Skylab 4 had splashed down on February 8 close to the Californian coast, so the voyage back to American soil was not that long. Following a good night's sleep aboard the *USS New Orleans*, the three astronauts were scheduled for 12 hours of medical examinations and experiments, with a further three hours scheduled for the morning of February 10. The press reports highlighted their activities for their first day back on Earth:

"So, you come back from three months floating around in a steel test tube in space and what do they have you do?

Pedal a bicycle upside down
Balance yourself on a one-inch rail
Spin around 30 times a minute in a rotating chair
Go through a six-and-one-half-hour physical examination
Eat the same old dehydrated food you've put up with for 84 days
Call your wife and tell her you won't be home for a few more days."

The results of these initial examinations showed that the Skylab 4 astronauts had come home in better condition than either of the previous two Skylab crews. Though shaky for a few days, they adapted to 1G (though still on-board ship) quicker than their colleagues had. Exercise in space was clearly important, and the doctors were soon indicating that, although a lot more work was required, there seemed to be nothing to prevent attempting a mission of up to 18 months. Of course, no American missions were planned or funded for this duration in the foreseeable future, but the Russians were planning such flights and monitored the American data from Skylab with great interest. The doctors also suggested that the crew could quite easily have extended their mission, medically speaking, for another 25 days or so, though it was clear that Skylab itself would probably not have been able to do so. Systems were beginning to fail and supplies were running low. The crew themselves agreed that they would have taken the extension had it been offered, though probably not if it had been in the last week or so of their mission. Once they had begun to prepare to come home, they would not have wanted to delay it any further.

The *New Orleans* docked at the North Island Naval Air Station San Diego at 08:30 am on February 9, but the crew remained in isolation aboard the helicopter carrier to undergo further medical examinations. A public ceremony was planned for that afternoon, but the astronauts remained on a stage on the carrier deck to accept the congratulations and thanks of military and civilian dignitaries. There were nearly 3,000 San Diegans cheering in warm sunshine as the three astronauts were introduced. One of those to welcome the astronauts home was Freda Carr, Jerry's mother from nearby Santa Ana, who had not been able to speak to her son since his return from space.

The Mayor of San Diego had declared the day "Skylab 3 Day" (confusingly referring to Jerry, Bill and

Ed as the third Skylab 'crew' instead of using the mission designation, which was of course Skylab 4) and told them that despite losing the inch in height that they had grown during the mission, "Your performance has added to your stature in a way that will never shrink."

Jerry said that the mission of Skylab 4 had laid to rest for all time the notion that man had no place in space. He was proud and satisfied with the job they had accomplished and called the final Skylab mission "The culmination of a dream. We feel we brought back a good deal of data and feel we've done something to benefit man back here on Earth." The celebrations were, according to Jerry, "A glorious day for a Californian boy to come home to."

Following the ceremony the three astronauts were flown by jet back to Ellington Field, Houston, for a reunion with their wives and further celebrations of their achievements, as well as continued medical examinations.

Back in Houston, JoAnn was celebrating too. When news came through that Jerry was finally back on the deck of the recovery ship, tears flooded into her eyes. "We've been living this thing minute by minute. They are home safe, now its time to have a party," she told reporters. With friends, family and colleagues of Jerry, including Al Bean, Jack Lousma, Stuart Roosa and Rusty Schweickart attending, the champagne was passed around. All at the Carr home were there to lend their support to the final stages of the mission, a family support role that Jerry himself had played during earlier lunar missions. JoAnn admitted she was still feeling a little shaky from the closing events of the long mission and that their immediate plans when Jerry finally returned home would be focused on the family.

Jerry, Ed and Bill finally arrived home at Ellington AFB near the space center on the evening of Sunday February 10, 1974. Deke Slayton boarded the C9 jet to formally welcome the three astronauts back to Houston. The three astronauts emerged to the cheers of a crowd of 500, the music of an Air Force Band and the embraces of their wives. Director of the Johnson Space Center, Christopher Kraft, told the onlookers: "You guys did a super job. These guys and myself may never see that record broken in space. It will be a long time before someone stays that long in space and does as great a job as they have done."

Jerry could go home, but had to remain apart (no closer than 25 feet, or 7 m) from the children for a week to reduce his chances of exposure to any germs while the post-flight medical tests continued. On the way home from Ellington, JoAnn recalled noticing that Jerry was very sensitive to any sounds, because there was little to hear up on Skylab. He was busy looking at everything as if he was on the planet for the first time. "It was like he was from Mars or something. When we got back to the house, Jerry would hear and take note of every tick of our big clock." Jerry continued eating NASA food for the next week, but for his first Earth-bound meal he enjoyed a fine Mexican dish, his favorite food. He was also looking forward to Jennifer's apple and cherry pies and JoAnn's lasagna once the requirement to eat Skylab food was over.

It would take a while to readjust to life back on Earth, and there had been no training for this part of spaceflight. For years, Jerry had not been part of the mundane 'nine-to-five, Monday to Friday, work and home routine' and the family were used to it. Before the flight, JoAnn was interviewed for the Today newspaper and commented that Jerry's absence during the mission over Thanksgiving and Christmas would not be a problem. "I'm an old Marine wife, and he's been gone before. He was gone for 13 months in Japan and holidays seemed to happen whether he was home or not. Christmas will come even if he is in orbit. The holidays are so chaotic and noisy around here with six kids, it won't be dull." But what neither Jerry nor the family could have prepared for was the adjustment to life after the spaceflight. Nothing would ever be the same again. Thirty years later, JoAnn reflected upon this.

JoAnn: "I think it was very hard for him. I don't think it was as hard for me, because my attitude was that whatever happened, it was to be about me now. I was going to have my time to do something, because I had been metamorphosing all this time into something other than Suzy Homemaker. Now I was ready to make my break. I think it was difficult for him and I firmly believe that there is something called post-flight depression, just like post-natal depression, that nobody ever recognized then. Jerry had focused on this one thing [the mission], and it was like having a 7-year pregnancy. He was focused on this one event and now it was over and it's never going to happen again. Those things that were fun, just after the flight, were the basic little rewards for all the trouble we went through, but that didn't last long."

JoAnn recalled that the kids were really upset because the dog could be at home but they couldn't. Jennifer recalled: "I had had to move across the street, but had to move back home early because I got sick. But I still had to stay away from him. I couldn't hug him or touch him or anything. I just remember thinking it wasn't fair that the dog could be there but we had to leave. I could never understand that."

Jennifer Carr did not immediately notice a change in her father once he got back home. "He was still Dad and it was life back to normal... like back in the Marines. His voice was very different because all the blood had gathered in his head, so he sounded as though he had a head cold. I helped him cut off his beard so he could shave it off and his face was very red. Mom could go out to Ellington and hug him but none of the kids could. I just always thought it was tough that the kids could never touch him. When he came back, all six of us had to move out because he could not be around us for two weeks. Our dog, Rags, could stay, but his kids could not."

Jeff Carr was never really sure if the quarantine issue was a NASA requirement or whether the wives cooked that up. "Something about needing to be quarantined in their homes with their wives for a week... Come on, I was old enough to evaluate that score. But on the other hand, I remember the first night he spent at home. I needed to go back and get some things while I was staying with my friends. I went over to our house, stood in the garden and looked though the kitchen window. I talked to my Dad though the insect screen. He was bearded and had the biggest smile on his face I had ever seen. He was home and my Mom had a bigger smile on her face. She was back there cooking dinner for him, so I had the chance to exchange just a few words before going back to my friend's house. It was a few more days before we wound up back together at the house. I remember the adjustments that Dad had to make. Reality set in. Gravity set in, in this case. He'd drop things, he'd turn corners too soon and too tight... We ate off paper plates for a while. My memories probably exaggerate, but I remember it being a funny time."

Jessica: "I remember that we were not allowed to touch Dad for the longest time after he was back home. We used to look at him through the window and smile and wave. His hair and beard were red. Weird. I felt a certain sadness about him."

Jamee Carr recalled debriefing her father 'over the years': "I know we asked questions about him getting used to gravity again. I remember him telling us funny stories about setting his razor up in the air and letting go and setting a glass in the air and dropping it, but I don't remember hearing a whole lot about it. It's hard to see your parents as anything other than parents. Whenever I see him, I think he's Dad; he's good at it, and I am proud of him and what he accomplished."

Debrief and a few weeks in a barrel

For the next six months following the end of the Skylab mission, Jerry was involved in post-flight activities and public affairs that took up a lot of his time. Known as 'a week in the barrel', these public tours and events could be more stressful than a spaceflight at times, with one event after another. While the information was still fresh in their minds, the crew spent hours being debriefed. This included producing a crew mission report and technical crew debriefings on their mission and experiences. The technical crew debriefing was an internal NASA document, a transcript of the crew discussing aspects of the mission and hardware with members of the training and procedures division, as well as representatives of the Astronaut Office.

The debriefing followed a prescribed format that covered pre-flight preparations and activities, training, simulators, powered flight, the rendezvous and docking with the space station, and activation of the orbital workshop and power down of the CSM. They also described a 'typical on-orbit day', and accounted for any unusual activities, before covering the deactivation of the workshop and power up of the CSM, separation, entry, landing and recovery. The crew also covered CM system operations, OWS system operations, the experiment program, EVA, flight equipment and documentation planning, visual sightings, interaction with Mission Control, human factors and anything that would affect the performance and execution of the mission. This debriefing, completed by every crew, provides a valuable reference for follow-on missions and a growing database of human factors engineering data from which to learn and develop future programs. From analyzing the mission air-to-ground recordings, transcript tapes from onboard tape recorders, Mission Control debriefings and experiment and PI feedback, a full summary of the missions, extending to many volumes, could be assembled. With Jerry's flight being the longest to date, this information amounted to a small library of data.

A week after their return, the crew gave their first official press conference at JSC in Houston. It was here that the details of the error in sequencing the hand controller circuit breakers was first mentioned. Jerry told reports: "We had that long moment of silence as I twiddled the hand controller and nothing happened... our hearts fell and our eyeballs popped, but we immediately moved to our next line of defense. Though there was an instant of stark realization that something was amiss, we did not feel we were in any great trouble."

Bill Pogue responded to questions on their condition after such a long spaceflight: "We worked pretty hard during the flight to maintain physical condition." The crew and doctors were 'delighted' with their medical test results following splashdown "We're running again and, aside from some very minor things, just maybe a sore calf and a little bit of lower back sensitivity, we're in great shape," Pogue assessed. Ed Gibson revealed that shortly after the flight they lost a little weight, but it soon returned. All three crewmembers learned a lot about how to exercise and stay in condition during the flight, which would have direct application to further long duration missions.

Following the press conference, a major Astronaut Office (CB) debriefing of Skylab was conducted on February 26 at JSC, with all the Skylab and ASTP astronauts in attendance. This closed meeting was an informal 'bull-session', with those who had flown on Skylab discussing their experiences and findings and their relevance to future CB operating procedures and application to future programs.

On March 20, President Richard M. Nixon completed a two-day visit to Texas by touring the Johnson Space Center, hailing the forthcoming joint mission with the Soviets in the summer of 1975, presenting the Skylab 4 astronauts with the NASA Distinguished Service Medal and inviting them to spend a weekend with their wives and families at Camp David in Maryland. President Nixon, who had authorized the Shuttle program in 1972, also mentioned that he would like to volunteer as one of the first non-career astronauts to fly on the Shuttle during the 1980s. But although his vocal support for the program was there, presidential commitment to expanding American exploration of space was not so evident.

Jerry, however, was more concerned with completing the every-expanding diary of appointments that seemed to fill the schedule over the coming weeks. It seemed that his hoped-for period of quiet planning for the future would have to wait a little longer until the circus of post-flight celebrations and activities settled down a little.

On April 23, 1974, Jerry's nomination for permanent promotion to the grade of Colonel USMC was among those approved by the US Senate. Three days later, the Skylab 4 crew returned to KSC for a reunion with their launch crew, and as guests of Center management at a luncheon with the NASA and contractor personnel inside the Vehicle Assembly Building (VAB). The affair was hosted by KSC Director Kurt Debus.

Gerald Carr Day

On March 29, 1974, Jerry's home town, Santa Ana, organized a 'Gerald Carr' day to celebrate the achievement of their hometown astronaut. Jerry and his family stayed at the Saddleback Inn and were accompanied by a NASA representative for three days. The event commenced with awarding Jerry the 'Keys to the City'. Breakfast at the Elks Club, sponsored by the Santa Ana Chamber of Commerce, was followed by visits to the Willard Intermediate School and Santa Ana High School, where Jerry presented mementoes of his flight and movies of his three-month trip into space. Telling his audience, "I must admit I was scared," when the Saturn lifted off, Jerry added, "The rocket's lift-off was incredibly noisy and bumpy… like riding on a train with square wheels." Weightlessness was "like floating in a swimming pool where you can't touch the bottom," and orbiting the Earth over a thousand times was similar to "sitting on top of a flagpole, but with a view you wouldn't believe." One of his former teachers at Willard told Jerry and the enthralled audience, "Jerry, I've had a lot of good students, but you were just out of this world."

During the day's events, Jerry was accompanied by JoAnn and all six children, his mother and grandmother. After an informal lunch at the Charles W. Bowers Museum, hosted by the City's Mayor Tom Patterson, the family rode in a small motorcade though the main portion of Santa Ana to the Plaza of Flags. There, under the direction of the Orange County Board of Supervisors, a ceremony was held. That evening, a reception was held at the Saddleback Inn, to which various guests including family, as well as city, county, state and federal employees, were invited. This was followed by a Recognition Dinner at the Elks Club, where approximately 500 guests saw Jerry present the film of his mission and honored him with presentations of plaques and proclamations. Jerry was also presented with a music box that played the tune 'Around the World in 80 Days'. Jerry's mother also was honored during the day's celebrations. Still a resident of Santa Ana, and employed as an auditor for Tower Manufacturing Company (a manufacturer of farm implements), she received a plaque from the City of Santa Ana which included the inscription, "The hand that rocked the cradle helped put a man in space." Jerry told the crowds that although he was born in Denver and had lived in many places as a Marine and as an astronaut, "Santa Ana is home because it is where I grew up as a boy." Jerry also told reporters that once the debriefing of the Skylab mission had been completed, he would probably join the Shuttle program.

Jerry also was the featured speaker for the Council Eagle Scout Recognition Banquet, held on May 1, 1974 at the Disneyland Hotel. He also received a Merit Award from the University of Southern California in April for "Worthy achievement which has reflected credit upon the University and its alumni." Jerry in return presented the University with a plaque of his mission, which included a flag and patch flown on the record-breaking mission and a photo of Southern California taken from the space station.

Several weeks after returning to his office, Jerry received a *per diem* check. As he read it he realized that, in typical government fashion, his trip on Skylab had been considered 'Temporary Additional Duty.' Consequently, he was eligible for reimbursement for extra expenses incurred during his TAD. There is normally an allowance for housing, but Skylab was considered government housing. As well, transportation costs and meal costs are reimbursable, but government transportation and meals had been provided. The last item, Incidental Expense, amounted to two dollars per day for incidental out-of-pocket expenses, so Jerry's TAD reimbursement check amounted to $168.00. He had a good laugh over it, turned it over and endorsed it to the John Glenn Senatorial campaign and sent it off. Years later, when visiting Glenn in his Senate office Jerry spied the check in a frame on the wall. John had never cashed it!

On May 24, Jerry accompanied Ed and Bill to the Seattle Day at Expo 74 in Spokane, Washington. During the event, Jerry reported that people around the country were beginning to think that the space program was a good investment in the future, and that the images of Earth taken during the mission were perhaps one of the most important returns from the flight.

Jerry's other awards for his Skylab achievements included the NASA Astronaut Gold Pin. The gold Astronaut lapel pin was traditionally awarded to those who had completed their first flight. Up until then the lapel pin they wore was silver. This was done at the "Pin Party," an evening dinner affair attended only by all of the astronauts and their wives. Here, the newly flown astronauts got their gold pins, and they handed out mementos to the back up crew and the support crew. The entertainment consisted of some sort of skit or presentation put on by the back up and support crews, teasing the prime crew for their mistakes and telling the audience how much better they could have flown the mission. There were lots of laughs all around.

He also received the Navy Distinguished Service Medal; the Navy Astronaut Wings; Chicago Gold Medal; Boy Scouts of America Distinguished Eagle Scout Award; Robert J Collier Trophy for 1973; City of New York Gold Medal; Marine Corps Aviation Association Exceptional Achievement Award; Dr Robert H Goddard Memorial Trophy for 1975; the FAI Gold Space Medal; the De La Vaulx Medal and the V.M. Komarov Diploma for 1974; the AIAA Haley Astronautics Award for 1974 and the American Astronautical Society 1975 Flight Achievement Award. Later, he would be presented an Honorary Doctorate of Science from Parks College of Saint Louis University, Cahokia, Illinois (in 1976) after giving the commencement address. In 1997, he was inducted into the Astronaut Hall of Fame.

Camp David

One of the more memorable post-flight events for the family was the official invitation to the Camp David complex for the weekend by President Nixon during the summer of 1974. JoAnn thought the Camp David trip was fun: "I was looking through the guest book there, and of course I didn't recognize the significance of it at the time, but there were all these men in the news signing on for a conference there. It was very eerie, because you never saw anybody. If you went to bowl you would come back later and everything was in perfect order. After a swim in the President's pool, you would leave the towels out and go to lunch. When you came back everything was put away, and there were new towels. You never saw anybody, but it was obvious somebody was around. You felt like you were being taped all the time. I was riding a bike up and down between the president's house and the log cabin we were staying in and suddenly armed Marines in combat gear came running out of the woods, crossed the street and ducked down behind the stone wall. Standing there with my bike looking at all this, I said, 'You guys are just practicing, right?' They wouldn't answer me, so I turned around and rode off the other way. We heard later that someone had tried to breach the fence."

Jessica: "Camp David. I really remember that. Two movies: Dustin Hoffman in *Papillon* and Bo Hopkins in *Culpepper Cattle Company*. A private screening room and all the popcorn you could eat. I liked hiking around the grounds and driving the golf carts. That was surreal even for a little kid. Wish I could go back."

Reflecting on Skylab

As Jerry and his crew 'performed' in several post-mission appearances, so the first results from his mission were being revealed, and the impact of Skylab as a whole was becoming apparent. Though created in the heyday of Apollo lunar missions, and often in their shadow, the early findings from the three Skylab missions were revealing a wealth of data and results that would keep the scientists busy for years. The investment and effort in that program was as important as that made in the Apollo lunar missions, though many would not realize this for years to come. Jerry has always been very proud of what Skylab, and his mission in particular, accomplished.

JC: "Most of the experiments went as we had expected them to go. That is, except the new ones they added at the last minute. We had to fumble our way through the procedures and work them out. That's why simulations are so important. There shouldn't be any surprises, and there weren't really many, other than inventory control problems. Most of the experiments worked well, although we had a few that failed. We did a little repair work here and there and got some of them to come back and gather some data, but we had one or two experiments that failed and just never got a bit of data, which is a disappointment. But we had fifty-three experiments, and to only lose one or two, I thought, was pretty good."

He found moving around in zero-g to be effortless, once the body had become accustomed to the new environment.

JC: "It's very much like the sensation you get when you're skin diving or just floating around in the water, except the resistance is much less. Regarding the handling of equipment we found something rather peculiar, but again not too surprising. We found it much easier to handle very large objects, which on the ground would weigh 100, 200 or 300 lbs. Very bulky objects were much easier to handle than the little bitty pieces, because the little pieces were inclined to float away so much more quickly than the large objects with the high inertia. We also found that if you ever dropped a small object or released it by accident, your first one-G instinct is to swing at it and snatch it before it falls away. So you instructively swing below it in order to catch it because of gravity, and what you end up doing, usually, is hitting it with the top of your hand and batting it away. So then you've only made the situation a lot worse. We found ourselves on many occasions having to unlock from the position where we were located and go chase a small item that we had batted away because we dropped it. Years later we tried to reassure the International Space Station people during the early stages of design that moving large objects would be very easy if they just take their time. They finally realized our point when they got into underwater procedures development."

Jerry believes that the most important contribution of his Skylab flight was the medical experiments. The physiological experiments were most interesting to the crew, watching how their own bodies adapted to the weightless environment.

JC: "We proved that human beings can live in a weightless environment for an extended period of time, and of course it has subsequently been proved that you can stay up for at least a year and a half. But I think we gathered the data that, medically at least, gave the Russians and others the understanding and the courage to say, 'Okay, we can stay up for longer periods of time.' I think that was a real breakthrough. Back in the Mercury days, the doctors weren't even sure if an astronaut could swallow, defecate or urinate in weightlessness. We have come a long way since then. Our experiments were very rigorous. They were very well prepared, and the data was very well taken. I think that solved a lot of problems."

Jerry also firmly believes that what he and his colleagues accomplished in Skylab, and the countermeasures they developed, were important steps towards longer flights on space stations and eventually to the planets.

JC: "If you're going to Mars, can you go there weightless? If so, are the countermeasures that we've developed going to be adequate for that? The problem is that although you can stay up for a year, when you get there you've got to be able to accomplish something. Then you've got to turn around and fly back, and that's another year. A space station is going to really be good for finding answers to those kinds of questions."

Jerry recognized that the solar physics community had gathered important data that answered a lot of questions in the world of solar physics. Equally, the Earth Resources and Earth Observation Programs created a wealth of data that scientists would be working with for years to come. "We learned quite a bit about food supply and crop estimation from NASAs's Earth resources program. Other areas of interest were in geological or mineral resources discovery." Oceanography and meteorology were also benefiting from the Skylab program, as was the study of the Earth's environment, especially in the fields of pollution and the way man utilizes the resources and environment in urban planning. Jerry ranked this number three in his list of importance from their flight, particularly citing the combination of Earth observations with EREP: "The two programs went beautifully together."

Perhaps the most important, and most overlooked, return from Skylab was learning to live and work up in space. Jerry became very passionate about adapting what he had learned on Skylab to new programs, both while still within NASA and after he left.

New Soviet space stations

In June 1974, the Soviets launched a new space station called Salyut 3. This was a military-class station and therefore the information concerning the activities of the only crew to board the station were restricted. The pair of Soyuz 14 cosmonauts spent just two weeks aboard the station. It was the first successful Soviet space station mission and the first crew to board a Soviet space station since Salyut 1 in 1971. At the end of the year, the civilian-orientated Salyut 4 was launched, and as Jerry continued his technical assignments in the Astronaut Office over the next three years, the Soviets began to gradually increase their experiences in long duration spaceflight. Between 1975 and 1977, operations aboard Salyut 4 and the subsequent military Salyut 5 provided a useful database of operational experience upon which to plan more ambitious manned missions. The next generation of Soviet space stations started with Salyut 6 in late 1977.

Space Shuttle Assignments

By the late summer of 1974, Jerry's commitments to Skylab had been completed, and he was reassigned to Astronaut Office support of the early development of Space Shuttle. There remained a small team of astronauts still working towards the joint docking mission with the Soviets (ASTP), many of whom had worked on the later Apollo lunar or Skylab missions, including supporting Jerry's flight. The assignments to the ASTP had been announced in January 1973 at a time when Jerry was committed to Skylab. ASTP completed the Apollo-era assignments by the summer of 1975. The next program was Space Shuttle, and most of the Astronaut Office were involved with that program, its support hardware (upper stages, early payloads, EVA etc), and the European science laboratory (Spacelab) that was being developed to fly in the payload bay of the Shuttle.

Jerry prepares to participate in Shuttle Orbiter egress simulations at JSC in 1977

JC: "As I began my Shuttle support activities (at NASA), I remember it was pretty quiet. I spent quite a few months just getting up to speed on what the Shuttle design was all about and what they were trying to do."

Reorganizing the CB

A major reorganization of JSC was planned for February 1974 and had been announced on December 7, 1973 while Jerry was in orbit. In May 1974, Jerry was named head of a support group assigned to the Shuttle cockpit design effort in the Astronaut Office. Though full time work on Shuttle development would not occur for three months after completion of his Skylab duties, in official NASA astronaut biographies between 1974 and 1977, Jerry was listed as being assigned to Shuttle development work. His assignments included Payload Support (mission phase), crew station hardware, Remote Manipulator System development work, and emergency egress evaluations.

In the early days of the Shuttle Branch of the Astronaut Office, Fred Haise (Jerry's old commander on the original Apollo 16/Apollo 19 crew) was Lead Astronaut and point of contact in working out early design proposals and the practicality of placing astronauts on a Shuttle whose design was still evolving. Jerry and Bill Pogue's tasks would involve human factors – the human interface with the controls and

A sequence of three photos showing
Jerry using a side hatch egress system
from a mock-up Shuttle Orbiter.

displays on the Space Shuttle. According to Jerry, the 'human factors' area was a whole new ball game after Skylab.

JC: "As a result of Skylab, I became a convert to human factors. Before that, in the early years up to probably 1970-1972, people looked at human factors engineers as weirdoes, the guys that always wanted to touch and feel things. We thought it was kind of silly to measure the distance from your wrists to your elbow and how you reached for things. But in Skylab, we began to realize the importance of the relativity of the human to the piece of machinery he or she is working with. It became important to understand what the weightless environment does to your skeleton and how it affects the way you do your work. So that precipitated my interest in working the Shuttle cockpit design and doing it rather extensively. That was really my swansong as a NASA Astronaut."

There was very little training equipment nor mock-ups of the Shuttle to work on in 1974, as there was no real detailed Shuttle design as yet.

JC: "As I remember, we were helping to develop the requirements to be put into the request for proposal for human factors design. That is, how the cockpit should be laid out and what sort of functions should be done from within that cockpit. We were doing a lot of things on foam core mock-ups, where we just took pieces of foam core, sliced it up and taped it together to resemble the way we felt things ought to be.

"We came back from Skylab with this concept of the neutral body posture, which really hadn't been understood very much before then. The fact is that when a human is weightless, the muscle structure pulls him or her into a sort of simian crouch, which means your actual floor to ceiling height is probably 4 to 6 inches (10-15 cm) less than it would be when standing vertically on Earth. We made the point that what you need to do is to make all of your work surfaces compatible with this neutral body posture. So those were the factors that we took into the design of the Shuttle cockpit. Another factor which was particularly relevant to our own experience was the layout of circuit breakers and switches, so that the kind of mistake I made on re-entry would not be repeated. We tried to think very carefully about what it would take to design a cockpit in which switches with similar names were not close to each other and where the shapes of switches and controls would be indicative of the function, much like the design of a modern airplane cockpit. In most planes nowadays, the landing gear handle has a wheel on it, and it's round like an aircraft tire. The purpose of this is to make sure you realize when you are grabbing for it that it is indeed the landing gear that you are operating. In the case of the flap handles, you don't use just a little toggle switch. A lot of planes now have a handle that operates the flap that feels like a flap. Those kinds of things we tried to put into plan in the design of the RFP (Request for Proposal) for Space Shuttle. That's pretty much what I did from late 1974 until I was ready to leave in 1977."

Jerry also worked on the Shuttle ejection seats, developed to support the Shuttle Approach and Landing Test (ALT) flights and the first Orbital Test Flights of the vehicle in space. Two ejection seats were fitted into OV-101 Enterprise (for the ALT program) and OV-102 (Columbia) for the Orbital Flight Test (OFT) program. Designed to provide emergency atmospheric escape for the commander and pilot, they were active during the first four missions in 1981-1982. They were deactivated during STS-5 when the first four-person crew (and first Mission Specialists) flew in November 1982, and were removed from Columbia before she flew again on STS-9 (Spacelab 1) in November 1983.

JC: "I worked on the ejection seat program for a year or so at White Sands in New Mexico, watching prototype seats being tested. We studied a lot of the research work done by USAF Colonel John Stapp, who is famous for his participation as a subject on high-G ejection studies on rocket propelled sleds. We observed some rocket sled tests done on a prototype that was considered a candidate for the ejection seat system for Shuttle."

There were also evaluations of the ejection system used during Gemini and other high-performance ejection systems of modern jet aircraft. Jerry worked with the contractors charged with developing the Shuttle ejection seat system as part of this assignment.

JC: "One other side job that I got during the period was when the Shuttle simulator was to be built. I was named Chairman of the Acquisition-board Source Evaluation-board (SEB). That was a monumental job, trying to get the requests for proposals put out in the proper form and then get it modified. When the

proposals came in, the task was to evaluate them and select the company who would build the 40-million-dollar simulator that is used today. That was administratively scary, because you had to be so careful to ensure that the decisions made were totally unbiased. You had to be very, very careful that everybody was treated fairly and equally and that no mistakes were made in the selection process that would leave NASA open to legal action. So we had administrators, lawyers and procurement specialists breathing down our necks."

This acquisition of the Shuttle simulator took up a lot of Jerry's time during his final year at NASA. It was frustratingly painstaking work from which he got little enjoyment. In addition, Jerry worked on evaluating different methods of getting out of a Shuttle in the event of a crash or emergency landing away from normal post-landing support equipment at primary landing sites. The crew compartment of the Shuttle is quite high off the ground with or without the landing gear deployed. This meant that some way had to be devised for lowering a crewmember (able or injured) out of the cockpit from the side entry hatch or the emergency overhead windows. Jerry participated in these early evaluations as part of the cockpit design work in 1974, and participated in exit simulations both in 1G mock-ups and in a water tank during his final months at JSC.

Two photos showing Jerry completing an emergency egress from overhead windows of the Orbiter, descending down the Pilot side of the mock-up.

Jerry also was involved in evaluating the optimum number of crewmembers on a Shuttle crew.

JC: "We always had the plan to carry at least four and up to seven people. I don't think that changed anywhere down the line. We always had the idea that there would be a regular airline cockpit-type seat layout, where you have a two-man crew up front and a flight engineer sitting behind and between to help keep track of things."

The development of the 'third seat flight engineer' was something that Bill Lenoir and Joe Allen evaluated during the first flight of Mission Specialists onboard STS-5 in 1982. The role they developed on that mission has been followed throughout the Shuttle program since then. The commander, pilot and mission specialist/flight engineer operate normally as the 'orbiter crew', while the other mission specialists and payload

Jerry reviews footage of the egress tests along with his colleague, one of the trainers who took part in the exercise. The heel attachments on their shoes are for an ejection seat, so these tests were conducted on equipment for one of the early Shuttle test flights with ejection seats.

specialists normally operate as the 'science or payload' crew. The fourth seat on the flight deck of Shuttle is also occupied by a mission specialist, with a role to play during the ascent and entry phases in support of the 'three-person' flight deck crew.

Jerry was also assigned to early development work on the Shuttle robot arm, developed by the Canadian firm Spar Aerospace. The arm was designated the Remote Manipulator System (RMS). This 'Canada Arm' has proved to be one of the most versatile and useful tools on the Shuttle program over the years and has been used for the deployment and retrieval of payloads, the construction of the International Space Station and in support of EVA operations, particularly during ISS construction and Hubble servicing missions.

JC: "I did a lot of work in the development and tests of the RMS. We used a big wooden Shuttle payload bay mock-up, which still exists in Building 9A at JSC, and we had some hydraulic simulators of the RMS. We practiced different methods of how to go about aligning the systems, and how you could see to align a payload so that it could either be retracted or inserted into the trunnion fittings, which were called Payload Release Latch Assemblies (PRLA).

"From the human factors standpoint, we were looking at how the operator at the back of the mid-deck could look out and see what he was doing and how the control ought to work. One of our conclusions was that the translation handle, the one the crewmember uses to manipulate the RMS in translation, should be a cube. Then the person would never have to look at the control itself. You could feel the cardinal directions of control while you are looking outside or at video, in order to align your payload and put it where you want it. The operator does not need the distraction of having to look at his control to make sure he is not cross-controlling."

Jerry believes that some sort of RMS system would have been of use on Skylab to assist in some of the EVA work, and at one point such a system was contemplated but not developed.

JC: "It would have helped because we did have some repairs in areas where there were no foot or body restraints. We had to innovate. For example, on the side of the vehicle facing the Earth during the EREP passes, we had the microwave antenna. One of the drivers on the microwave antenna failed, so Ed and Bill went out on that first EVA and pinned the antenna in such a way that it could only be rotated in one direction. There were no foot holds or hand holds in that area, so Bill had to stand attached to temporary foot restraints and hold Ed by the legs while he did the work on the antenna. That was the only way they could do the job.

"The RMS work I did involved looking at all kinds of different schemes that would help us figure out how to see what we were doing. Putting a huge payload into the payload bay with the rear window obscured by the payload itself was pretty much like trying to put an elephant in your shirt pocket. We ended up needing all those remote cameras – elbow cameras, the wrist camera and the bulkhead cameras in the payload bay – in order to see what we were doing. We did a lot of analysis and evaluation as to the different modes of RMS operation. There was a rate mode and a direct mode. As well, we evaluated what we would have to do if we experienced a joint lock-up and how to *avoid* such a lock up."

Retiring from the Marines

On September 1, 1975, Jerry retired from the USMC after 22 years service. When he reported his decision to leave the service, the Corps invited him to Washington DC to have the retirement ceremony at an evening parade in his honor at the Marine Barracks. Jerry was the reviewing officer for that parade, which was a proud moment for him. "That was a fine way to leave the Corps," he recalled years later. John Glenn was there as well, as Jerry's special guest. Both Jerry and Glenn talked about the ceremony and how much they both enjoyed the event. John made the remark that his wife, Annie, would not let him carry any fountain pens when he went to the Marine Barracks to review the parades, in case he was tempted to sign up again. Twenty years later, Glenn must have carried a pen to NASA, as he signed up for a second spaceflight at the age of 77. At that time, there was also speculation about flying a former Skylab astronaut in his 70s. One of the more compelling arguments for both ideas was the interest in comparing medical data with that taken when the astronauts flew in their 30s or 40s:

JC: "I always felt that I would like to have flown a mission to the International Space Station for two reasons. First, because of the medical data comparison opportunity, and second, because Bill, myself and several others spent nearly 13 years working with Boeing on the development of the Station. I would have liked to have experienced the results of those labors. Not that Pat (Jerry's second wife) would have let me. She was not about to let a perfectly capable studio assistant slip the harness and go gallivanting off into space!"

The only requirement imposed upon Jerry by the USMC after his retirement was to maintain up-to-date location records. They made it clear that he was "on retainer," which meant that he could be recalled at any time for a national emergency. "I think I'm safe now; they don't need a 75-year-old in Falujah."

Leaving NASA

By 1977, Jerry was looking to new challenges away from the Astronaut Corps. "The main reason I left NASA was because it looked like we were facing six to eight more years before the next flight. I just didn't want to wait that long for another flight. I was also very interested in the idea of beginning to ply my trade as a professional engineer in the commercial world."

On June 22, 1977, NASA issued a press release revealing Jerry's decision to leave NASA on June 25. He would be joining the Houston consultant engineering firm, Bovay Engineers, Inc., where he would be the corporate manager for business development. At the time, he was heading up the Design Support Group (Space Shuttle) within the CB at JSC Houston. He had been an active astronaut for 11 years 2 months. With his departure, there remained only 27 active astronauts on the rolls at JSC, and a recruitment drive was being conducted that would see the first 35 Shuttle-era astronauts enter the program in January 1978. In the press release, Jerry explained his decision in leaving the astronaut program:

"Leaving a career in aviation and spaceflight is a difficult thing to do, but I feel that it is time for me to take a new direction. I wouldn't trade the past 22 years as a Marine aviator and astronaut for any other experience, and I look back over those years with great satisfaction."

Had Jerry stayed in the astronaut program, it is reasonable to surmise that he would have received the command seat on an early Shuttle flight. Crew assignments for the first orbital flight tests were being evaluated at the time and those crews would be formed from the active pilot astronauts, mainly from Group

5 (commanders) and Group 7 (pilots). Jerry could have been assigned to a command seat on STS-3 through 6, which at the time were designated as OFT Flights. These flights finally flew, as he predicted, in 1982-1983, five to six years after he had left. He may also have been assigned to an early Spacelab flight, which would have seen him go through a six- to eight-year training program, and ten years subsequent to Skylab 4.

Looking back at that period, Jamee Carr felt that her Dad was never really happy in the last few years at NASA and in his first positions in aerospace. She assumed this was probably due to the difficulties of his flight and the knowledge that he would not make a second spaceflight for a long time. "He never seemed to settle. I think he was searching. It was also a difficult time at home, and he seemed to be searching from the time he came home in 1974 to when he set up his own company some years later." Jamee speculated that the difficulties on Skylab may have precluded Jerry's assignment to an early Shuttle mission.

John Carr was not surprised to see his Dad leave NASA before another spaceflight. He thought that flying the Shuttle after commanding a small Apollo spacecraft and Skylab would not appeal to his Dad. "I think he was pretty much finished with NASA. There were a lot of new things starting, like the Shuttle, but that's like flying a commercial airliner. Why would you want to drop back down to flying a ten-seat aircraft after flying a military jet? He had to move on to bigger and better things. Back in those days, NASA was not a normal day job."

Jeff Carr thought it would have been great if his Dad had stayed and flown the Shuttle, but he was skeptical about the design right from the early days. He realized that it could end up killing somebody. Always the one member of the family that was tuned in to the space program, Jeff followed that interest into a media career that would take him to work at NASA, where he would become the JSC Director of Public Affairs years later. Jeff remembered watching John Young and Bob Crippen fly the first Shuttle mission, STS-1 in April 1981. At that time, Jeff was in college at the University of Texas. "I remember thinking it was amazing to touchdown on a runway. It was just a few years ago that my old man got dumped in the Pacific Ocean. Now they were landing right down the runway. I remember being stunned, sitting there with my mouth open. This was no longer the space program my father was part of, this was a whole new ball game! By that time, my Dad had made his decision to go on to something else. I knew he couldn't wait for years for another flight, and I understood that, but I remember feeling excitement about the space program again for the first time in 7 years."

Jennifer Carr always thought her Dad would fly again. She actually thought he would go up quite quickly after his first mission, because his colleagues were doing so in the late 1960s and early 1970s. But of course the program was changing and flights were being cut, leaving only the Shuttle which was many years in the future.

JoAnn never thought Jerry would stay on to fly the Shuttle. "He was talking about teaching, looking around at options. But Shuttle, from that vantage point, seemed like a very distant possibility. It wasn't even scheduled at that time. He could have stayed, but I think the Skylab flight exhausted him. I think that the major post-flight depression that he went through, which nobody, including me, recognized, went a long way towards getting him out of the program. It probably didn't help our marriage either."

A couple of years after leaving NASA, Jerry was interviewed about his decision to leave. "After a period of soul searching involving my obligation to stay on, I feel NASA got their money's worth from the mission in which I flew. I was at a point where I had to decide what to do with the rest of my life. To fly in space again would take more years of training and preparation. I had to measure that against my physical condition – could I go through that again for another flight?" Jerry felt that he was at a crossroads in life and had to weigh up the opportunity to step into a new area, in effect a third career. "I had accomplished a major goal with NASA and I wondered 'How do I want to spend the rest of my years?' There were some things I hadn't done. I hadn't tried business, for example. I'd had a successful first career in the Marines, and a second one with NASA. I'm glad I made the decision to go with business. If I had stayed with NASA, I would have begun to feel stagnant. 'How could an astronaut feel stagnant?' is the question a lot of people might ask, but after ten or fifteen years even an astronaut can feel in a rut. I felt I should shift gears, not for a bundle of money, but for a challenge. It was time for another challenge. I also felt I should make room for some of the younger guys."

Fond Memories of a home in space

In July 1979, the Skylab space station re-entered Earth's atmosphere and burned up to destruction over the Indian Ocean and Australia. Since 1974 when Jerry had left the station, it had remained unmanned but controlled, awaiting a possible visit by the Space Shuttle. That hope faded with increased solar activity that increased atmospheric drag on the station, coupled with several delays to the maiden launch of the Shuttle.

Interviewed at the time, Jerry said he hadn't thought much about the impending demise of Skylab until he read a newspaper headline: "The Last Days of Skylab". That triggered a few nostalgic memories. "It was a great home and I hate to see it come in." He was also sorry that no one would revisit the station to retrieve the 'time capsule' his crew left shortly before leaving the workshop.

JC: "I especially remembered one night when I couldn't sleep and I roamed around the station. We were crossing the West Coast at the Oregon/California state line. I looked out the window, and there was full darkness over California where I grew up. The San Francisco bay area was very clear and looked like a big black bean with lights sprinkled around it. I could see the Golden Gate and Bay bridges, which looked like strings of lights. Monterey Bay was clear, and Los Angeles looked like a black velvet bowl filled with pearls. Then I saw a flashing green light south of Los Angeles and I realized I was seeing the airport beacon at San Diego Lindbergh Field. The entire view was spectacular, unforgettable."

Over the next two decades, Jerry would follow a new career, in aerospace and eventually in art, but he would forever be associated with the space program as an astronaut of the pioneering era of space exploration. Though not a moon walker, he was one of a small group of nine men who lived and worked aboard the Skylab space station. These men often attend reunions, meetings and commemorations of that pioneering program. One of the largest was the 25th anniversary of the program in 1998. The next year, Jerry participated in the NASA JSC Oral History Project, recording, sometimes emotionally, his personal story in the space program and his flight on Skylab. Many passages from that oral history are included in this book.

Each year he continues to undergo a physical examination at NASA JSC, contributing to the growing wealth of biomedical data on space explorers both during their years at NASA and after retirement. This program, called the Astronaut Longitudinal Study, provides a broad database of medical information both before, during and after spaceflight. Hopefully the data will provide a better understanding of the long term effects of spaceflight on the human organism, which is as important as the data gathered during their selection, training and missions.

Currently, Jerry participates periodically in the 'Astronaut Encounter' program at the Kennedy Space Center Visitor Center in Cape Canaveral and does a number of presentations in local schools, churches and service clubs in his community every year. For forty years, he has supported the Boy Scouts by writing letters of congratulations to newly designated Eagle Scouts all around the country.

Though now retired from the astronaut program for more years than he was active, Jerry is always remembered as a Skylab astronaut, and always will be. Former astronauts don't really retire, they just don't fly anymore. They tackle new challenges and strive for new goals, but none can be as high as that attained by sitting on a rocket and being blasted off the planet. Those three long months on Skylab changed not only Jerry's life for ever, but also that of his family. There was the life 'Before Skylab', and there was the life 'After Skylab'. The next stage of that was also 'After NASA'.

Chapter 11

After NASA

When Jerry decided to retire from NASA, he felt it was time to do something completely different than aerospace engineering, although he still wanted to remain in engineering. At a social function, he met a prominent Houston engineer named Harry Bovay, who was founder and President of Bovay Engineers. The company was involved in airport design, water and waste treatment plant development and several other engineering fields, and it had offices in six U.S. cities. Jerry and Bovay took an instant liking to each other and, shortly after the event, Mr. Bovay invited Jerry to join the company to assist in their business development efforts. Essentially, they hoped to capitalize on Jerry's reputation as an astronaut to get a foot in the door at the offices of those who were planning the large projects in which Bovay was interested. It would be Jerry's job to understand the project well enough to determine the company's capabilities for completing that work and then formulate a plan with the officers of the company to go after the project.

A Professional Engineer

For the next four and a half years, Jerry was employed by Bovay as Vice President, and later, Senior Vice President for Business Development. Jerry used his engineering experience and judgment to determine new areas for the company to focus its attention on. He sought out new engineering opportunities, determined if they were compatible with the capabilities of the firm, and then ensured that engineering proposals were prepared correctly. Jerry had a staff of four senior engineers plus clerical support, and his team was tasked with coordination of national, federal government and state government accounts, and with providing liaison among the six offices of the firm. He had additional responsibility for staying abreast of new fields of energy generation, particularly in the developing areas of solar thermal, solar voltaic, wind, ocean thermal, and geothermal energy. Near the end of his tenure with Bovay, Jerry was transferred from development of business to management of the entire Houston office.

Jerry served on the Board of Directors of the SUNSAT Energy Council, which was fostering the concepts of solar power satellites for the mid 21st century. Jerry applied to be a registered professional engineer in the state of Texas during 1978. He had learned a lot about the business of engineering from both sides of a proposal. At NASA he had been the Chairman of the Source Evaluation-board for the 40 million dollar Shuttle Simulator contract. Here he had dealt with people who wanted the work and evaluated their suitability for the award of the contract. At Bovay he was on the other side of the equation; one of those vying to be awarded the contract. Later, Jerry became responsible for the management of all Houston area operations, including marketing, production, quality assurance and a staff of 175 engineers, technicians and support staff. "It was very interesting. Learning to manage an engineering department to get a quality product out on time was a really stimulating challenge for me."

Moonlighting and Consultancy

A colleague asked Jerry if he had any spare time at Bovay that would allow him to complete a little side work in aerospace engineering, for Applied Research Incorporated in Los Angeles. He discussed this with Harry Bovay and received Harry's approval to occasionally help out with the aerospace program. Jerry did his moonlighting in Los Angeles as a consultant on human factors engineering. It involved a fairly small amount of time, but served the purpose of keeping him up to speed in aerospace engineering. In his work with Applied Research (and also with former astronaut Dave Scott's company in support of US Air Force space programs), Jerry applied some of the lessons learned on Skylab to some of the design activities.

Soon, Jerry realized he was getting homesick for aerospace and eventually decided to get back into it again full time. He resigned from Bovay in 1981 to work full time for Applied Research.

In February 1982, Jerry became a senior consultant at Applied Research Inc. He once again became involved with payload integration, crew training and the development of procedures. Jerry worked with the Air Force on the proposed military Shuttle missions. His activities were mainly crew station design, crew integration and crew training protocols; not for Shuttle itself, but for some of the payloads being developed to fly on the Shuttle on military missions. Jerry remained with Applied Research between 1981 and 1983.

A New Life

Following his space flight and during his final years at NASA, things were not the same at home for Jerry or JoAnn. At the time of the flight, JoAnn thought that they were in for the long haul towards manned flights to Mars. But that was clearly not to be. She also realized that Jerry was not too keen on staying at NASA to fly the Shuttle. He was soon looking around for options, with the Shuttle seeming to remain a distant possibility. Teaching school was one of the options they discussed.

Skylab 4 had not exhausted him, but JoAnn still believes that the post-flight depression was a leading contributing factor in Jerry's desire to leave NASA. JoAnn does not think that the Skylab mission itself changed her, but the experience of working with NASA certainly had: "I was greatly influenced by what was happening in the world around me, with the growing awareness that my children were soon to be gone, and I wanted my own identity. I did not want to have somebody else, or what they did or did not do, become my only way of relating to the rest of the world. I wanted my own thing. I think I was one of the first astronaut wives to go as far as I did, but I definitely think, at least in Jerry's class of astronauts, that there were a lot of wives striking out on their own a little bit. I think the previous groups really didn't do that. We were on the crest there, and so a lot of people did move out and do things on their own."

JoAnn decided to go back to school – law school. "Why law? I asked myself that very question," JoAnn recalled years later. "It seemed a great idea at the time." It was not something she really planned, it just worked out that way. "I got chosen for jury service, which was the first time I had ever been in a court room. I was very taken with that whole scene and several people said to me that I ought to go to law school. I thought, 'Could I do that'?" As Jerry was figuring out what to do with his life after spaceflight, JoAnn was also looking at new challenges and horizons. The marriage continued to deteriorate without them realizing. "I didn't catch it and neither did he. Things gradually went from bad to worse and we ended up getting divorced and went on to lead two very different lives. I was really aware that he deserved better than I was giving him at that point in time, but we just grew apart after the flight. I think that was hard on the kids. I don't think it was too bad for me, because I had a lot of things I wanted to do."

When Jerry and JoAnn finally decided to go their separate ways, the hardest thing was to tell the children. Jennifer was still living at home but was attending college nearby and was not aware of anything wrong. "It was very surprising to me when we sat down together and were told they were getting divorced. Josh was most upset at hearing it, I remember." John recalls seeing the split coming. "I've always been very observant, and over time I knew something was happening. I did sense that, and so did my brothers and sisters, I think. It was not a total surprise."

John surmised that both his parents had simply worked themselves apart, both dedicated to their sides of the association with the military and then NASA: "What people looked at via the media, generally, was Dad. The focus was on him and what he had accomplished in his career. But they never said anything about what had been accomplished at home – what his family had to go through. The astronaut wives may not have flown in space, but they overcame quite a bit to keep things going. It takes two parents to raise kids, and in the Marines and with NASA, Dad was not in a nine-to-five, five-day-a-week job. I don't think Mom ever got the attention and the gratitude that Dad did. I think that with her frustration talking to my father, and my father's frustration in understanding where she was coming from, they simply couldn't see eye to eye anymore."

Jeff, too, saw the signs: "I think it was a little more disruptive and maybe a little more unsettling for the younger kids than it was for us older ones. I saw it coming. I know that there was an irreconcilable frustration, an anxiety that I know my mother could not escape, and that my father could not help with. I had a similar experience as an adult, so I was able to relate to it later. A child experiences divorce and has a secret wish that their parents would get together again. But as an adult, I like to rationalize. I can see that my father is now very happy, and that my mother is focusing on the key values in her life. She made a decision that allows her to do that. It made me recognize that although our parents divorced, their children still had two parents that worked together to care for them, love them and care about their security and well being. I think with my younger brothers and sisters the divorce was a little more unsettling, and I think in a big way they were all looking forward to getting their own time with Dad. When I was a kid we went to ball games, went fishing and camping. Dad and I did things together before NASA intruded deeply into his time. I was lucky enough to have my Dad through those early years. The other kids, I think, were patiently awaiting their turn, so there was a great sense of frustration that it wouldn't happen."

Jerry and JoAnn were divorced in 1978. Following their divorce, JoAnn completed law school, passed the Texas Bar exam and obtained a job with a small downtown Houston law firm. However, she was not happy there, and after three or four months returned to the Clear Lake area and a local firm. This worked out quite well for her, and three years later she was able to set up her own firm, practicing law until the late 1980s. She subsequently became "totally burned out because I was way too idealistic about what I thought I could do." After closing her business, JoAnn moved to California and rented a house on the beach. Eventually she realized that she needed a job to afford to live, so she decided to take the California Bar exam. She kept the house in Clear Lake, renting it out mostly to her children. She told everyone that the only way she could "get rid of my kids was to give them the house and leave." She then worked at the Resolution Trust Corporation (RTC), established by the US government in 1990 to address the savings and loan crisis that the country was facing at that time. That work was finished after 6 or 7 years, so JoAnn moved back to her home in the Houston area and, for a short time, held a job with a non-profit corporation created by Boeing to provide educational materials for various museums across the 50 states. That job was very satisfying for her, but funding problems forced a premature termination of the program. So, JoAnn retired to a life of hiking, touring and traveling, but still with very close ties to her children and grand children.

The divorce from JoAnn was a difficult time for Jerry. He lived in an apartment about a mile away and tried to devote individual time to each of the children. He also buried himself in his work at Bovay and continued his affiliation with the Webster Presbyterian Church. Ironically, his destiny was to be determined by an old friendship with Hershel Musick from Santa Ana. Herschel had an aunt called "Pat," an artist who would have a profound influence on Jerry's life beginning in the late 1970s.

Jerry had promised Herschel that he would look up "Aunt Pat" (Pat Musick), who had moved from New York to Clear Lake to teach at the University of Houston. The recent widow of Jack Musick, former head coach of football at Cornell University, Pat was a professional artist, psychologist and mother of three grown daughters. After the loss of her husband, Pat decided to resume going to church on a regular basis, which had not been possible during his illness. She attended the Presbyterian Church in Webster, a suburb of Houston, near to NASA's Johnson Space Center. After attending for a few Sundays, someone pointed out to Pat that the person sitting in the front row was 'the astronaut', which she thought interesting. Nothing happened for several months, until one of the church deacons asked Pat to come over for Sunday supper to meet 'the astronaut'. Pat was not too enthusiastic about that and tried to get out of it, but when she was told that 'the astronaut' knew her family (the Musicks) from Santa Ana, California, she thought that might prove interesting, so she agreed to go along.

Jerry's divorce was not finalized at that time, and he was not up to meeting people, so he, too, tried to pass on the invite. Jerry was quite low at this point and was certainly not interested in getting involved with someone else. However, when the name Musick was mentioned, Jerry replied that he had known a Herschel Musick in California, who was one of his best friends. Then he remembered that he had promised to look up Herschel's Aunt Pat.

In the living room before dinner, while the hosts and two other couples sat smugly observing, Jerry and Pat started talking, comparing places they had known in Santa Ana and talking about Pat's nephew Herschel. It surprised both of them how often, during those years, they had been in the area at the same time without encountering each other. As the discussions continued about their favorite beaches in the area and other common places they had visited or knew, they gradually moved closer together on the couch. After about an hour they were sitting side by side, much to the pleasure of their hosts and guests. They were obviously warming to each other, much to the detriment of the evening's pork roast, which became over-cooked.

One part of the conversation would have a profound impact on their future lives. "I think," said Pat, "that the space program is mankind's most creative effort of the twentieth century." "Creative!!" replied Jerry. "It was nothing but a 'nuts and bolts' affair... strictly engineering." Pat, who taught the psychology of creativity, would not let that go. "No, you are wrong. It contains all of the elements of creative problem solving: identifying a problem, looking at all sorts of solutions (divergent thinking), and then, when settling on a plan, working it through to produce the final product (convergent thinking)." Jerry, who had never thought of himself as an artist, was introduced to the notion that all humans are creative and use creative skills throughout their lives.

From that day on, Jerry and Pat met during the 'coffee hour' social event after the church service. As Jerry's divorce was proceeding, he did not want to be seen officially 'dating' someone else until after the process was completed. Unaware of this, Pat realized that "something was happening here," as they both made a beeline for each other at any social gathering they were attending. However, the fact that Jerry never arranged either to collect Pat or take her home during this time period frustrated her.

In the summer of 1978, Pat had arranged to go on a Mediterranean cruise around the Greek islands with family friends, so she decided "to cross Jerry Carr off my list and explore other options." During the cruise, Pat did a lot of talking to herself. On the final night following the dinner at the Captain's table, she decided she wanted to get some fresh air and took her drink out onto the deck.

PC: "I was leaning on the rail looking out into the black when I noticed a little orange light off in the distance. I wondered what that was, as it didn't look like ship lights. Then it got a little bigger and bigger and pretty soon I realized it was the moon coming up over the Adriatic. It was just absolutely, unbelievably beautiful. As the moon reached about the ten o'clock position and turned to white, a shooting star passed right to left in front of it and then disappeared into the black ocean. As I looked at it I said to myself 'Watch out Jerry Carr, this is a sign!' and threw my drink into the Adriatic. I returned home and went to Church the following Sunday, I marched right on up to Jerry at coffee hour and said, 'I think its time for us to cook dinner together'. He said 'I think you're right.' Something had happened."

JC: "What had happened was that my divorce was finalized, so I felt like I was free to date. That was the beginning of the relationship."

Pat Musick was born Patricia Tapscott in Los Angeles, California, in 1926. She spent most of her childhood in Southern California. Her father was a traveling salesman dealing in life insurance, which meant moving around a lot. "We ended up doing a stint in Las Vegas, Nevada, when I was in 5th grade, in Albuquerque, New Mexico in 7th grade, and back to settle in Santa Ana, California for junior and senior high school. Santa Ana is where I called home." Of course this was Jerry's home town, but Pat was six years older and she was always leaving a school when he was starting. "But we did go to the same high school, the same university, and we sang in the same church choir. And my favorite nephew (Herschel) was Jerry's best friend in high school."

Living close to the glamour of Hollywood, Pat had a secret yearning for childhood fame. Though her parents recognized an ability to draw at an early age, Pat was never encouraged to develop her obvious talent – if anything, quite the reverse.

Encouragement and inspiration came from her mother's side of the family. Her mother's twin sister, Aunt Louise, spent hours drawing with her when she was very young. During a 1938 trip to Iowa with her father when she was 12, Pat visited a great uncle on her mother's side of the family, Ash Davis, who was an artist. Seeing his paintings of larger than life religious figures was inspiring and fired Pat's childhood determination to make art her main interest. Pat's world was opened to another type of 'art' by an Iowa cousin who introduced her to poetry.

An earlier influence on the path Pat would follow had come at the tender age of four to six. While her mother was very ill following the birth of Pat's brother, a Navaho Indian woman from the Arizona reservation came to help with childcare. The woman worked under a program created by the local Presbyterian church which helped Navajos improve their English skills. Pat's earliest memories of art and drawing are of the endless hours she spent watching this lady weaving beaded belts, headbands, anklets and wristlets. Pat was intrigued by the intricate patterns, colors and shapes formed from the woman's imagination.

In later years, this memory would show in Pat's work, through her interpretation of other ancient civilizations from Central America, Native North America and ancient European cultures. "I have a deep affinity and sense of relationship to the Native Cultures, both as people and as artists. "

Pat majored in art at Santa Ana High School during 1941-44. From there, she attended the University of Southern California (Jerry's Alma Mater) on a Fine Arts scholarship. In 1946 she married Jack Musick and then followed him in his career and the beginning of their family. They had three daughters, Cathleen, Melinda and Laura. Pat attended many hours of art classes at the universities where he coached. While at Dartmouth, she studied with Paul Sample for six years, and he encouraged her become a professional. Also at Dartmouth, she was influenced by Frank Stella, Robert Rauschenberg and Hans Hoffman. While at Cornell she studied with Alan d'Archangelo and attended lectures by 'Earth Artists' Robert Smithson and Michael Heizer. She received her MA in 1972 and her PhD in 1974 from Cornell.

Pat then became a teacher at Northeastern universities between 1974 and 1976, including a part-time position at Syracuse University where she taught art therapy. Pat had begun to merge her psychology studies with her art studies, developing a creative problem solving aspect to her work: how an artist would be motivated to create and think about his or her art. In 1977-78, she did a post-doctoral study in psychology at the University of Texas Medical Branch in nearby Galveston.

She was also very interested in the space program, developed after President Kennedy's challenge to "put a man on the moon by the end of the decade [of the sixties]." She painted two series of works on what man would see and look like as he journeyed into the cosmos. One day in 1974, she heard that the Skylab would make a pass over the Northeastern U.S. That night she took a blanket to the meadow on the top of the hill where she lived and sat and watched as the tiny light traveled so very fast from west to east, before disappearing from view. As she looked up she said aloud, "I wonder who those guys are up there and what they are like." Five years later, she married one of them.

In 1976, Jack was diagnosed with a fatal brain tumor and Pat moved him to Houston for medical care. She took up a post at the University of Houston at Clear Lake City, where she was assigned the task of designing a graduate program in Art Therapy.

Following Jack's death she found herself free to resume her own creative art with passion. Coincidentally, she was sensing a developing passion for a former astronaut.

From 1978 onwards, Jerry and Pat spent an increasing amount of time together, although they still lived apart. On September 14, 1979, they married at the Webster Presbyterian Church where they had met. A few of Jerry's former astronaut friends attended the ceremony, including Stu and Joan Roosa, Charlie and Dottie Duke, Ed and Julie Gibson and Jack and Gratia Lousma. Artist Harvey Bott and his wife Margaret were also there. All of Pat and Jerry's nine children also attended.

Jerry had been living in the apartment in Clear Lake while Pat's house was in Kemah, both near to the space center. They decided to buy a house southwest of downtown Houston, in the Kirby/Richmond area. Pat and Jerry decided to hold a couple of dinners for friends on the two nights following their wedding, since their honeymoon would not occur until December on a visit to Cancun in Mexico. One night was a dinner with all their friends from Church, which totaled about 50. The next night was a gathering of Pat's professional friends and colleagues from the art and university world, and Jerry's professional friends and colleagues from the engineering and aerospace industries, as well as the nine kids.

After several years living in the house they decided to move to a new property, which had room to expand Pat's studio, but remained small enough to take care of during their busy schedule. They found a property in the Heights area west of Houston. They would remain there for a couple more years before moving to Arkansas.

A Telescope for Texas?

In 1983, while still with Applied Research, Jerry assumed a new role as Project Manager for the University of Texas McDonald Observatory's 300-inch Telescope Project. Jerry was responsible for planning and controlling the project to ensure its successful and timely completion in 1985. The telescope was to have been sited in the Davis Mountains in Jeff Davis County in west Texas.

JC: "This was another overlapping project I worked on when the work at Applied Research began to thin out. I was approached by the Director of the McDonald Observatory at the University of Texas to see if I was interested in assisting in the process of developing a 300-inch telescope, which would have been the world's largest optical telescope."

The telescope was to feature a primary mirror that was to be only 4" inches thick and flexible in design. Computer controlled actuators attached to the back of the mirror would push and pull on the mirror as gravity, temperature, wind and ground movements affected its focus. Normally, conventional telescopes would have to be about one-sixth as thick as they are wide, so a 300-inch telescope would have to be almost 4 feet thick. At the time, the largest telescope in the world was a 236-inch facility in the Soviet Union; a conventional mirror of over 50 tons resting in a support structure weighing over 700 tons. The time required to adjust the mirror due to temperature changes resulted in distortion, and bad weather was a constant problem. This made the telescope difficult to use effectively. The conventional design technology, then decades old, was also used in the US Mount Palomar 200-inch telescope in California. Clearly, the new telescope would have to call upon new technology to improve the size of the mirror while lowering the overall mass. The plan was to use digital computers, laser beams and more state-of-the-art sensors and materials.

Jerry was very enthusiastic about the project: "It will be a significant feather in the cap of the state of Texas and the University to do it. It will bring a lot more scientific luster to the reputation of the university, as well as more research contracts, and more science opportunities to the students and faculty. It's an intangible, long-term investment. Everybody has the gut feeling it's the right thing to do," Jerry stated at the time.

Its projected cost was to be about $32 million (1983) US dollars, about one tenth of the cost of the Mount Palomar telescope. The most significant cost savings would be the thin mirror, a short focal length and the aiming system. Contributions from the major cities in Texas had raised about $3 million, but a Senate Finance Commission decided to cut $5 million from the state budget. This money had been planned as seed money for the telescope to help encourage private contributions, so the decision came as a big disappointment.

Working on the telescope project proved quite interesting for Jerry. Drawing upon his work at Bovay, he put his newly acquired engineering experience to good use. Unfortunately, the funding proved to be the downfall of the project, and after a couple of years there just wasn't enough money available to complete the job. After "running up against a brick wall," the project was eventually abandoned in 1985. "That's when I withdrew, and Pat and I set up CAMUS Incorporated, which was our own company, a family owned corporation."

CAMUS was named using the 'C-A' from Carr and 'M-U-S' from Musick. "When we put it all together and saw that it said *CAMUS,* we said, "That's a great French existentialist philosopher, and we're a kind of philosophical company. It's also the name if a good Napoleon brandy, so this has to be a good name!"

A House in the Forest

In 1971, Jerry and Bill Pogue occupied a two-man office within the Astronaut Office at what was then called the Manned Spacecraft Center. Both men had come from families who had owned "a shack in the woods" in the 1930s. Jerry's family had owned a little cabin in the Rockies during their years in Denver, Colorado, which they used at weekends. The two men had discussed their desire to continue this trend for their own families while still at NASA. Both astronauts therefore looked around for undeveloped land on which to build their weekend retreats. This would enable them and their families to escape the hustle and bustle of the growing Houston metropolis and the fast-paced technology of the space program. However, the Rocky Mountains were too far away from their homes in south Texas, and the wooded area north of Houston, called 'The Big Thicket', was still too flat for their plans. Jerry and Bill wanted mountains, rivers and backwoods country to pursue their interests in hunting, fishing and camping. It was also too hot and humid in Houston during the summer, so they looked further afield. They eventually came across the Ozarks in northern Arkansas. The town of Fayetteville was large enough for their needs and the site they found about 50 miles east was certainly remote enough.

The site they chose was 283 acres of land that a local rancher named Bill Stewart wanted to sell. Bill Pogue went to look it over during a family vacation to the area and returned to Houston to tell Jerry that it was a wooded area, very hilly, with a river going right through it. There were also three natural water springs. It was everything they had wished for. Jerry took his family to the site and they, too, were impressed.

Despite the good price, they could not afford all 283 acres, and instead offered to take all the wooded high ground and some land adjacent to what was called Kings River. The astronauts suggested that Stewart should sell the ranch house and the flat meadowlands as a separate and more valuable package. He agreed, and the two astronauts gained 193 acres of beautiful, forested uplands.

Even though Skylab training became more intense and took up most of their time during the mid 1970s, both families enjoyed the land during their latter years at NASA. As part of the divorce settlement from JoAnn, Jerry had retained his part of the land.

Bill and Jerry had not intended originally to build permanent structures on the land, but in 1979 Pogue and his wife, Jean, decided to build a home on the Arkansas property. After Jerry and Pat married, they looked to move out of the Houston area. Each December, they vacationed at a time-share property that Pat had brought in Cancun, Mexico. While they were there in 1981, they discussed where they would like to spend the rest of their lives. They were very interested in a plot of land with a 99-year lease in Playacar, near Playa del Carmen, between Cancun and Tulum on the Yucatan coast. This area was relatively undeveloped, and every lot, it seemed, had a small Mayan ruin on it. They almost took out the check book to put down a deposit when Jerry said, "Let's go back to the villa, think about it for one more night and come back tomorrow to do it." At the villa in Cancun, they drew a line down the middle of a sheet of paper and listed the pros and cons about Playacar. They also listed all the things they wanted to have in a home and those items that were important to them.

They both wanted to live close to nature, preferably near water, although it could be a river or a lake. Local hospitals were important, and they needed to be near a university for Pat's art and for intellectual stimulation, as well as near an airport for Jerry's aerospace work. They also wanted to be centrally located in order to reach their children, who were spread across the United States by this time. When the list was finished, Jerry remarked that the land he shared with Bill Pogue in Arkansas fitted more of the requirements than the location on the Yucatan peninsula. They decided to delay putting down the deposit and to talk with Bill and Jean Pogue, so they arranged a visit to Arkansas. While they were there, they looked over the land and discussed their idea with the Pogues. At this time, Pat had never been to Arkansas.

PC: "My idea of Arkansas was Henry Fonda and the Jobe family in the *Grapes of Wrath* movie. I thought Jerry was taking me off to the end of the world."

The Pogues had already recognized the potential of the area and were at the time living in a trailer next to the place where they were building their own home on the Arkansas property. As they reviewed their pros and cons list, Jerry and Pat realized that all the boxes in the 'pros' column were ticked by the land in Arkansas. They looked around and found a spot that was on Bill's south part of the property. After negotiations with the Pogues, they agreed to trade some sections of land in order to acquire that ideal building site. Then they found Tom Rimkus, who became their contractor. He had been recommended by Bill Pogue and lived down in the valley adjacent to their property. Tom, in turn, recommended Albert Skiles, a local architect, who specialized in environmentally friendly and solar passive homes. Over the next twenty years, Tom became an invaluable asset to Jerry's aerospace career, to Pat's art work and to the construction of the family home. It was to become a very special place for both family and friends alike.

Situated on a 200-foot cliff overlooking the valley (but not too close to the edge), the main complex was located along an old logging road. The main home was planned as a 1000-square foot, single-story building of open plan design, with an exercise lap pool located inside a covered extension of the building. A studio was to be built for Pat on one side of the main living area, above which would be a guest bunk area for family visitors, particularly grandchildren. On the other side, separated by a terraced garden and patio, an office and a guest room was planned for Jerry above his workshop and garage. They developed a 4-year building program. During the first year, the foundations and retaining walls were laid for the three main buildings. In the second year the studio was built. This was designed to include a small kitchen, bathroom and the bunk room, as well as a studio space and room for both Jerry's and Pat's offices in the same building. This allowed them to commute from Houston and live in during holidays for a year while the remainder was being developed. During the third year (1984), the main house was built, followed by Jerry's office/workshop block.

Pat and Jerry moved into their new home around Christmas 1985 and over the following year, all the terracing and the decking between the house and Jerry's office was built.

That following year was both a happy and a sad time. In January 1986, the Space Shuttle *Challenger* had been lost and the seven astronauts aboard killed, grounding the program for what became 32 months. When the accident occurred, Jerry was in meetings in the main administration building at the Johnson Space Center. The meeting had been temporarily adjourned to watch the launch on closed circuit TV, and Jerry realized instantly that the explosion of the solid boosters was an anomaly and that a disaster had occurred. He immediately sought out his son, Jeff, who at that time was a Public Affairs employee at NASA. He had been personally training Christa McAuliffe for the video portions of her educational mission and felt particularly close to her. Jerry found Jeff in tears and shared his grief.

Further family grief was to follow on April 25, 1986 when, just two weeks after her 75th birthday,

Jerry's mother Freda passed away in Hoag Memorial Hospital, Newport Beach, California. She had been suffering from colon cancer for about 18 months. Luckily, Jerry and Pat were in California for a visit, so Freda's sons and their wives were with her at the end. All the family had good memories of Grandma Carr. Jennifer used to spend nights out at her grandma's, and after the family moved to Texas she would fly out to California to spend time with her mom's sister and also with Grandma Carr. Jennifer often recalls breakfast at Grandma's, where fresh figs were brought from her back yard trees and served with cream. Jamee Carr lived with her grandma for a couple of years and has treasured memories of her time there. Jerry's mother was always very proud of her son's achievements and kept a scrap book of his NASA achievements, adding it to mementoes from his school days and scouting years. She wanted to keep for all the children and future grandchildren the early records of Jerry's life so that they could learn how he had progressed from Denver to California, to the Marines and NASA and into space. Freda, a dedicated crossword puzzle worker, enjoyed many hours working puzzles and looking at the view when she visited the Arkansas property.

Ten years later, progress on their Arkansas home saw the addition of a formal dining room by connecting the main home to the studio from an old patio area. This afforded spectacular vistas of the valley during meal times, plus the vantage point of seating on the newer outer decking area. Here, many reflective hours were spent looking at the view or using the permanent barbecue installed in the outer deck. There was a hole in the deck where a large hickory tree had been growing. It suddenly died and Pat and Jerry decided to fill the hole with a barbecue pit.

Nearby, John and Jerry had built a strategically sited wooden bench close to a protruding cliff edge that was a perfect place for quiet reflection and a personal view of the nature surrounding the visitor. A few hundred yards away from the house, a warehouse was constructed to handle Pat's growing art inventory, with a larger work and dispatch area as well as a small studio exhibition area. A barn, part of the original ranch located on the floor of the valley, was converted into a guest house and used initially by friends and family. Some of the happiest summer times spent by the Carrs with their extended family and friends were picnicking at the nearby riverside. Independence Day was often spent at this picnic site with friends, and a group reading of the Declaration of Independence was the highlight of the event.

The Barn was also promoted as a rentable guest property for vacationers, providing access to the river for fishing and rafting. This added a source of income to the property. The Carr property was located adjacent to a 15,000 acre wildlife management area. Access to the Carr's required a 6 mile drive east on a dirt road from the main Huntsville to Eureka Springs highway.

The property's south facing hill meant that snow disappeared early, but ice was a challenge. Three wood stoves and a 4000-watt generator kept the power going when main lines iced up and collapsed. This occurred at least once each winter, and once in a while power would be out for up to five days. Hurricane lamps and candles provided most of the light when the power went off. It was also advisable to carry a chainsaw in the vehicle just in case the single track road was blocked. Abundant dead trees provided ample firewood, and evergreens were a source for Christmas trees.

Pat described her nature retreat and source of inspiration in a 1997 exhibit catalog:

"The place where I live inspires my work. When I turn off the highway onto a six mile dirt road tunneled by oak, hickory, and sycamore, I enter a continuum of time. I emerge at the edge of a cliff lipping a river valley. In all of my vision, the world of nature reigns over mankind. The sandstone and limestone rocks that fill my view are 350 million years old. The woods that surround my home echo with the footsteps of ancient cultures. This is a place where prayers were offered to the trees before they were cut. And so for me, oaks stripped bare and waxed clean form knobbly-knee human bones. Cedar or oak shafts open gates to inner spaces. Limbs of pine form vessels containing ancient spirits. But the work is also about the future."

CAMUS in Aerospace

CAMUS was established in 1983 as a Texas corporation and was re-established as an Arkansas corporation in 1991. The corporation had two products. Pat's art was emerging as mostly sculpture rather than painting, so one line of business for CAMUS was in art production, marketing and emplacement. The other facet was the aerospace work to which Jerry wanted to return. Initially, Pat was established as the President of CAMUS so that it would be designated a minority owned corporation. Jerry began looking around for suitable consultancy roles, drawing on experiences from his years with the USMC, NASA and the aerospace business.

He contacted his fellow Skylab crew members with a plan to get into the fledging Space Station program. The astronauts realized they had a wealth of experience on long term living in space. In fact at that time (1984), the Skylab astronauts were the only American astronauts with long duration flight experience. Ed Gibson was not available as he was pursuing a different line of work, but Bill Pogue was. Together, they decided not to seek a role as employees to an aerospace company, but to remain independent as subcontractors, seeking contracts to utilize their experience in developing human interfaces for a future space station.

Through CAMUS Incorporated, Jerry and Bill offered their experience and service (along with other former astronauts and engineers who came in when they secured a contract with Boeing) to assist with the design of the Space Station, then called Freedom. Discussions were held with Boeing, Lockheed, Rockwell, Vought and others.

JC: "Boeing, I think, was really the only company that took us seriously. Most of the other companies were willing to use us for business development and to be spokespersons, but they really didn't see us as being 'in the trench' people who wanted to work with engineers."

Most companies only wanted to place Jerry and Bill's names on their list of 'known' consultants, who helped on the design of their space hardware, but without any real plan to actually call upon their experience in a hands-on manner. Boeing, however, did recognize the potential of having CAMUS help them to return to the aerospace business again. Boeing had had engineers working on quality control during the Apollo program and on the Lunar Roving Vehicle, but most of the workforce who had worked on the lunar effort had either retired or gone on to other pursuits. The company subsequently decided to compete for major space station contracts, but the younger engineers on staff had little or no experience in space operations. Some of the senior managers were their only experienced people.

JC: "Boeing tried to get us to sign on as full time employees, but Bill and I had decided this was not what we wanted to do. Our counter proposal was that we wanted to be a technical support subcontractor. Our task would be to work with young Boeing engineers 'in the trenches' on the design of Space Station. We did not want to be employees; we would take care of our own taxes, benefits and all that. Boeing could sign us on to do work, but we did not want to have the responsibility to deliver documents, analysis or studies. CAMUS would be consultant and mentor, and Boeing would generate the final reports."

One of the earliest projects focused on human productivity studies, developing what became known as the Manned Systems Integration Standard (MSIS) for Space Station. The prime contactor for this was Lockheed, with Boeing as the subcontractor. CAMUS was subcontracted to Boeing, working on the final review committee. This was during the Space Station Phase A activity, which was basically a fact finding stage of station development prior to defined contactor assignments.

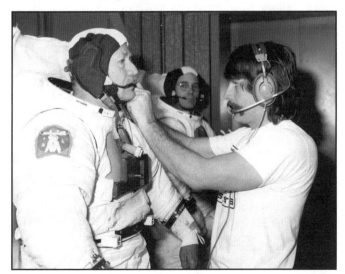

Boeing won a Phase B contract with the assistance of CAMUS. "We reviewed and digested everything that they did and gave them advice on what to do and what not to do. We helped them to know where the pitfalls were, and at the same time we gave lectures to all their engineers regarding weightlessness, zero-g body posture and how zero-g man/machine interface designs ought to be done. Then we helped with their Phase C/D proposal which they eventually won. CAMUS was carried in their proposal as one of the team member companies. All of a

Jerry, representing CAMUS under contract to Boeing, wears a Shuttle EMU pressure suit and is about to participate in underwater simulations of space station Freedom EVA activities in March 1986. His suited colleague in the background is Boeing designer Brand Griffin.

sudden, little CAMUS was being listed among the big support corporations on the Boeing Team."

From 1984 and for the next 15 years, CAMUS worked primarily as a subcontractor to Boeing, completing final work on verification analysis for Space Station. They also worked with Boeing on other side studies, involved with a return to the moon and human exploration of Mars. Though most of its work was with Boeing, CAMUS did associate with other firms on cooperative studies. This included Vought Aerospace on the evolution of the Orbital Maneuvering Vehicle control station, and with Lockheed on Space Station Maintenance Workstation issues, both in 1984. Between 1985 and 1988, Jerry served as a consultant with ALENIA SPAZIO in Torino, Italy, providing human factors and operations support for the Columbus pressured module design for Space Station Freedom. He also worked with TRW in 1987 on integration support for Space Station, and CAMUS teamed with Essex Corporation on studies of Advanced EVA Systems for Geosynchronous, lunar base and Mars exploration.

As the contracts expanded, so did CAMUS. The aerospace side of the company was divided into four branches covering Man Systems (Human factors and systems development and analysis); Launch Systems; Operations (Systems operations and analysis; payload operations and utilization; training, mock-ups and simulations; and EVA); and Advanced Programs.

In addition to Bill Pogue and Jerry, several other key personnel were subcontracted to CAMUS at one time or another to support the different contracts the company held. Those with spaceflight experience included:

* Skylab 4 science pilot Ed Gibson.
* Byron K. Lichtenberg, (Omega Aerospace Inc) who had flown as a payload specialist on STS-9 in 1983 and aboard STS-45 in 1992.
* Jack Lousma, who had been the pilot of the Skylab 3 crew in 1973 and commander of STS-3 in 1982.
* Harrison 'Jack' Schmitt, the Lunar Module pilot of Apollo 17 and one of the twelve men who had walked on the moon.
* Don Peterson, who had been a mission specialist aboard STS-6 in 1984 and
* Vladimir A. Solovyev, the former Russian cosmonaut who had been a flight engineer aboard both the Salyut 7 (1984 and 1986) and Mir (1986) space stations.

Jerry was never able to secure any work for Solovyev, and though he was technically associated with CAMUS for several years, nothing ever came from the association.

There were also several former NASA engineers who worked for CAMUS over the years:

* Milton L. Windler, former Apollo and Skylab mission flight director, director of the Shuttle RMS program; also Consultant to the Director of Alenia's (Italy) Operations and Multi Purpose Logistics Module (MPLM)
* Gordon Ferguson, former aerospace training specialist for Apollo, Skylab, Shuttle, TDRSS and Centaur. In Mission Control during Apollo, he served as the feared and respected Simulation Supervisor for flight crew and flight controller simulations.
* Glen H. Cress III, former Senior Engineer in the Space Station Program Office, Payloads Flight Controller in Mission Control, and Simulation Controller and Training Coordinator in the Apollo Program.
* Edward Pavelka, former assistant director of Mission Operations Directorate, NASA JSC, who worked in Mission Control as a flight dynamics officer in the Gemini and Apollo programs. He was head of all trajectory flight controllers for Gemini, Apollo, ASTP and Skylab missions. During the early Shuttle years he was responsible for all flight planning and flight data file procedures during the Apollo years.
* Ken Young, former flight dynamics officer during Apollo missions.
* Nellie Needham Carr, former lead payload flight officer for Hubble and Gamma Ray Observatory (Compton Observatory) deployments, Intelsat repair and re-boost mission, SpaceHab and EURECA retrieval missions. Nellie Carr was then Jerry's daughter-in-law, Jeff Carr's first wife.

Two persons never directly associated with NASA worked with CAMUS as well. Nicolas L. Shields left the Essex Company to take on a role as an independent subcontractor for CAMUS. As a USMC

Astronauts from the Gemini to Skylab era. L to r Gene Cernan (Gemini 9, Apollo 10 Apollo 17), Jerry Carr (Skylab 4), Jack Swigert (Apollo 13), Jim Lovell (Gemini 7, Gemini 12, Apollo 8, Apollo 13) and Walt Cunningham (Apollo 7)

Hospital Corpsman (USN) who served in Vietnam, he returned to the United States to gain a Master's degree in psychology and became a human factors specialist. He is now one of the foremost human engineering specialists in the United States and is employed as a Senior Fellow at Boeing. Tom Rimkus, who was the Carr's home contractor and later Pat's studio assistant, was educated as a computer engineer. For the corporation he provided administrative support and kept the computers running efficiently.

CAMUS and Space Station

"We worked alongside Boeing engineers who were doing space station work. We reviewed their work and advised them about weightlessness; what the effects of weightlessness are on the body and how it affects one's ability to address the piece of machinery one is trying to operate. We talked to them about the importance of scheduling and how to make system interfaces user-friendly so that one doesn't end up getting people wrapped around the axle when something goes wrong. I think a lot of that probably paid off. We've seen the space program morph from Space Station to Space Station Freedom to Space Station Alpha to International Space Station. We've been through the whole thing."

Jerry later recalled that this was sometimes a painful and frustrating process as space station evolved into what is now called ISS. In the hope of having some influence on the NASA side of space station design, Jerry even tried to secure work with Eagle Engineering and Hernandez Engineering, both based in Houston supporting JSC operations, but was unable to find the right kind of contracts to go after.

An associated line of work that gave Jerry a lot of satisfaction was his advisory role with AERITALIA which subsequently became ALENIA SPAZIO, in Torino (Turin), Italy. Jerry spent about four and a half years working part time in Torino with Alenia. As with the Boeing engineers, he conducted a lot of initial training of their personnel in the human factors group. He introduced them to zero-g posture and all the associated things that have to be considered in a weightless environment. The ESA Columbus module was Alenia's prime piece of equipment in those days. The design ideas were carried on into the Multiple Purpose Logistics Module (MPLM) that has flown successfully on the Shuttle to deliver and return supplies to and from the ISS.

Observers of CAMUS and the development of ISS habitability would surmise that Jerry and his colleagues had a significant input into the eventual systems and procedures flown on the station. In response, Jerry replies, "Both yes and no."

JC: "A lot of the advice that we gave Boeing on how the space station ought to be designed resulted in approximately 80% of our recommendations to make space station a better vehicle being thrown out. This was done not so much because of Boeing's unwillingness to consider, but because of the budget punches that kept hitting us. Each time a budget restriction trickled down from Congress to NASA to Boeing, something had to give. Consequently the top management at NASA and Boeing would cut another feature from crew comfort. As a result, the ISS today is nowhere near as crew friendly as it was on Skylab. On Skylab we had private quarters, a shower and a toilet that worked the first time."

Drawing on their Skylab experiences, Jerry and Bill Pogue tried to get a personal crew compartment for each person for privacy. There was a plan to include an uplink and downlink video system, and a personal communications control panel in each crew quarters. This would enable each crewperson to have a telephone/video conference with their family or with a scientist to work a science issue. CAMUS worked on an improved toilet system, the trash bags and the designs of the wardroom and bathroom. When the suggestion was made to include a shower with a 40 gallon capacity for three or four people to take a shower every day, Jerry and his colleagues noted that it was not feasible to carry that much water. They explained that the Skylab shower made do with only 5 liters of water per person once a week or so and was quite adequate. They won the water argument, but the shower was tossed out completely. Jerry explained, "The baby was tossed out with the bath water. They [Boeing and NASA] would listen to us, but when they got into a budget crisis those niceties were thrown out."

Jerry was frustrated with the lack of progress. Every year, budget change pressures forced a redesign, to the point where it was becoming almost ludicrous. In October, Congress would tell NASA what their new budget was (usually reduced significantly from the previous budget). NASA then went back to Boeing and CAMUS, who would spend the fall and winter redesigning items to fit the new budget constraints. By May the following year, when they fitted these changes into the new constraints, Congress would then criticize the space agency for changing the design again and again. With Congress trying to micromanage the space program, ISS was going nowhere fast. It was at this time that the Russians were brought into the program and Freedom became Alpha, then ISS. With ESA, Canada and Japan committed, there was even a real concern that NASA might pull out of the program and leave the partners holding the debris. The political pressure that fed the space station redesign program took its toll on what companies like CAMUS were trying to do – to build upon past experience to make ISS better than Skylab.

JC: "We had a lot of trouble getting the crew office to buy into some of the things we were recommending. They just felt they knew better than we did. We convinced them on some things, but lost a lot of battles with them in terms of things we thought ought to be done to maintain crew morale and productivity. Their experience was based on a ten day mission mentality, and they couldn't see a long-run picture.

"One of the major problems we encountered was that we could not get the crew office to pay much attention to Space Station until about 1996. Up to then, it appeared that those who were assigned to the space station program seemed to feel like they were out of the flight crew flow in the Astronaut Office, and they were busy trying to get back into the flow for assignment to a Shuttle flight. Turnover in the Astronaut Office in the Space Station division was phenomenal, and you couldn't get any continuity out of there at all. Consequently, Boeing and the other companies had to depend on us to provide input. More than once, to our frustration, our inputs were then overridden by the crew office and the NASA program office. Suddenly, we had a space station design that we at CAMUS were not very happy with. But we learned to live with it."

Other studies

CAMUS participated in a series of paper studies for future space activities and utilization in addition to the Space Station work. The Lunar Surface Mining Study looked at mining certain types of rock to extract water and oxygen. There was also work on a 'Space Transfer Concept Study' for vehicles designed to transfer logistics between Earth orbit and Mars orbit in support of a Martian exploration program. Another study looked at the design of an advanced Lunar Rover, which NASA thought at the time would be a test bed on the moon for later applications at Mars. It included remote manipulator systems incorporated into an EVA suit-type workstation attached to the pressurized rover. This would allow an astronaut to drive right up to an object and then put his or her arms into the 'suit' configuration to conduct or direct intricate work using mechanical hands.

CAMUS also assisted in a 1989 report that proposed a possible format for extended surface exploration at Mars. It suggested that two 45-day surface expeditions were far more beneficial to operations than a 20-day or 40-day surface stay. For the benefit of the report, a 2003 window to Mars was assumed, during which an 8-person crew would be dispatched on one vehicle, but with two landers. The 18-month mission included six months in Martian orbit. Two teams of four astronauts would complete a 45-day surface excursion each. An expanded EVA program envisaged 600 man hours per site, including surface sampling and analysis, setting up a wide variety of surface experiments, and possibly drilling and mining activities.

Earth and Sky Related

During those years of CAMUS work, Jerry did a great deal of traveling. In 1983 he was invited, along with 49 other astronauts and cosmonauts, to appear at the United Nations Headquarters to celebrate Earth Day. Six were chosen at the last minute to speak about the most important impression they had of their flight. Astronauts and Cosmonauts from Saudi Arabia, France, Germany, Spain and Russia each spoke in their native languages. Mary Cleave, representing the United States, spoke in English. Mary talked about the fragility of the Earth and its atmosphere, the necessity to be good stewards and an urgency to care for it. The space explorers, gathered on the dais in the General Assembly room, did not have translations of the other languages, but in the audience Pat, her daughter Cathy and granddaughters Caryn and Wendy had the advantage of simultaneous translation. When the program was over Jerry came down into the audience and

A meeting of the Association of Space Explorers (ASE) members. Jerry is standing in the second row from the back, third from right.

asked Pat, who had tears in her eyes, "What did they say?" "You're not going to believe this," she replied, "but they all said the SAME thing that Mary said." This really impressed Jerry because the statements were completely ad lib. No one had been warned that they were going to be asked to speak nor briefed on what to say.

This reflected one very important result of spaceflight, shared by almost all astronauts and cosmonauts, and one which President Clinton referred to as "the overview effect." Looking at the Earth from great distances gave them an understanding of its tenuous place in the vastness of space. Years after the early flights, as impressions of the flyers were gathered, this effect became prevalent and, as Jerry has said, "We didn't realize it at the time but we had become instant environmentalists."

During the nineteen eighties, Jerry had become a member of ASE, the Association of Space Explorers, which was an organization composed of astronauts and cosmonauts. The only requirement for membership was that one had orbited the Earth. Each year, one or more host astronauts, working with his/her government, arranged for a week of symposia, meetings, and sight seeing. Earth's environment became one of their primary issues, along with concerns for space rescue and cooperation. Through the ASE, Jerry and Pat visited the Netherlands, Germany and Austria. They were in Berlin on the first anniversary of the fall of the Berlin Wall. The group toured the Brandenburg Gate, where the Russians were noticeably silent as they viewed the great gate they had embedded into the Berlin Wall. Then, a visit to the ruins of Dresden and the reminder of American fire bombings of World War II produced the same reflective silence in the US astronauts. Those were shared moments… a communal realization of war's destructiveness and their longing for a peaceful world.

Pat's family had begun a custom of reading the Declaration of Independence aloud as part of their Fourth of July celebrations, and Jerry and Pat decided to incorporate that into their family life as a tradition. The ASE Congress in Berlin happened to span Independence Day, so they printed up copies of the Declaration and brought them with them to Germany. On the day, they ordered champagne and invited the astronauts to gather in the ballroom to celebrate. All stood in a circle and took turns reading a paragraph. The cosmonauts, who had been invited to attend if they wished, stood in a circle surrounding the astronauts with their three interpreters. There was a strange emotional feeling in the room as the men of different cultures, bound by shared experience listened intently to "revolution." When the reading was completed, champagne was poured and all toasted the USA, to Vladimir Dzhanibekov's loud exclamation, "Happy Birthday America!"

Jerry and Pat became close friends with several cosmonauts and their wives and vowed to learn Russian so they could communicate. Following the ASE congress in Washington, DC, the Zudovs (Vacheslav and Nina) came back to Arkansas with the Carrs, and the Kubasovs (Valeri and Ludmilla) with the Pogues, for a week's visit. It was a memorable time. Pat and Nina cooked together with the Russian-English dictionary between them, and the eight of them shared campfire cookouts at the Kings River, complete with vodka.

Jerry found interesting relationships between his views of Earth from space and his later travels. One

East meets West. Former astronauts with former cosmonauts. Right to left: Skylab 4 pilot Bill Pogue and his wife Jean, Russian cosmonaut Vyacheslav Zudov (Soyuz 23) and his wife Nina, cosmonaut Valery Kubasov (Soyuz 6, Soyuz 19, Soyuz 36) and his wife Ludmila, Jerry, and Pat.

of the first things he had recognized from above was the shape of Italy. This boot-shaped peninsula later became Jerry and Pat's favorite foreign country. They have made thirteen visits there in the last twenty-eight years. They love the people, the land, the Earth-centered food, the sense of family, and above all the art. Tuscany is their favorite spot, and they return each time to a villa in the small village of Pergine Valdarno, south of Florence, where they can become villagers and not tourists.

In the late 1970s and during the 1980s, Jerry made three trips to Japan on lecture tours. The first time he was accompanied by his daughter, Jennifer, and the second two by Pat. Having been stationed in Japan in his Marine years, he retained a fascination for the Japanese culture, so was pleased to be able to expose Jennifer to it. Later, when he and Pat traveled there, he enjoyed watching her develop a passion for sushi. On one memorable occasion they were invited to visit the president of Soka Gakai, a major Bhuddist sect in Japan with millions of members all over the world. Dr. Daisaku Ikeda and Jerry discussed at length the philosophical changes that occur in a person who has the opportunity to view the Earth from afar. Dr. Ikeda questioned Jerry regarding the environment and what could be seen from Earth orbit. Meanwhile Pat, the artist, was admiring a huge shoji screen that served as an artistic partition in the living room.

The president noted Pat's admiring glances and told her that it had been painted for him by a prominent Japanese painter who was designated a 'National Treasure.' Pat assured him that it was, indeed, a beautiful piece of work. A month or two after returning to the US, Jerry and Pat responded to a knock at their front door and met two Japanese gentlemen with a package. They identified themselves as representatives of Dr. Ikeda and presented a beautiful painting of the moon rising over a rice field. They stated that Dr. Ikeda had commissioned the same painter to do the painting for Pat and Jerry.

During the 1980's, when Jerry consulted with Alenia Spazio, he was able to take Jessica with him on one of his trips. This was probably the first time they had been able to spend a significant amount of time alone together. Pat accompanied him on several other trips and eagerly took advantage of the opportunity to study the great renaissance art in Torino and particularly in Florence. While Jerry worked, Pat took several side trips by train to Florence.

Jessica: "He did take me to Italy once. We actually had a great time! But I remember thinking on the flight over 'I don't even know this guy! How am I going to get through the next week?' One night he had a bit too much to drink, and he was a lot of fun. When he let his guard down he was a different guy. The trip brought us closer, but it would take many, many years and another trip to Italy to bring us together."

The Yucatan is another location with a relationship to space. The Carrs spend time there every December and over almost three decades have studied the Mayan culture in depth. They are intrigued by the many renderings of the 'Upside Down God' at Tulum, whose headgear resembles an astronaut's and whose body appears to be falling head down through space. Some Latin cultures have ruin patterns that could only be observed from above (the Plains of Nazca). Was there an ancient relationship between humans and the cosmos? They discovered that the Yucatan is also the site of the collision of a meteorite that signaled the end of the dinosaur period in our planet's history. They toured the 'observatory', the Caracol at Chichin Itza, where it is surmised that agriculture and religious events were planned based upon study of the heavens. They fell in love with the soft-spoken, peaceful Maya of today, their simple food and the beautiful land surrounded by turquoise seas.

Jerry was invited as a guest to the May 2001 Autographica show in Northampton, England, and Pat accompanied him. They planned to spend a few extra days afterwards exploring the English countryside and

Skylab's 25th reunion (1998) - l to right Jerry, Bill and Ed with their mission emblem.

The nine Skylab flight crew members at the 25th reunion in 1998. l to r rear (Skylab 4) Carr Gibson and Pogue; l to r middle (Skylab 3) Bean, Garriott and Lousma; l to r front (Skylab 2) Conrad, Kerwin and Weitz

Skylab 4 crew and wives at the 25th reunion. L to right Ed Gibson, Julie Gibson, Bill Pogue, Jean Pogue, Jerry and Pat.

viewing ancient stone circles, to add further inspiration to Pat's work. The visit included a tour of Stonehenge in Wiltshire. They found another interesting relationship between Earth and space in southern England, in the Cornwall area, the land of 'ancient circles.' Pat and Jerry loved their visit there to study the great stone circles that abound in the countryside. Far beyond Stonehenge, the most famous, are hundreds of megaliths placed in rough circles off of narrow dirt roads. Many of these seem to have a relationship to the locations of the sun and the moon, and indications of an apparent study of the cosmos. They hired a helicopter to fly them over Stonehenge and the great white chalk horse carvings in the earth, another sight that could only be appreciated by a 'view from above'. When the helicopter pilot learned that Jerry was an astronaut and helicopter qualified, he relinquished the controls, and Jerry had a wonderful time polishing up his piloting skills.

A New Frontier

Jerry finally retired from aerospace work in 1998 and took a new job – as studio assistant to his wife Pat. He has found a new interest that is quite fulfilling, supporting the woman he loves in her passion for sculptural art. He draws on his own engineering and technical skills learned in the Marines and in the aerospace business, and has renewed skills he first learned in the Boy Scouts decades before. "I'm enjoying welding, woodworking and running heavy equipment while I assist my wife in the design of her sculptures. I do a lot of the engineering design to help ensure that we have good solid bases and that they stand up straight the way she envisioned them," Jerry recalled in his official NASA 2000 oral history. This work in reaching new horizons is the latest chapter in the life of this amazing man.

Of space, time and the universe. Guests at the Autographica 2001 event in Northampton, England. L to right: Skylab 4 science Pilot Ed Gibson, actor Kenny Baker (Star Wars R2D2) actor Colin Baker (the sixth Doctor Who from the BBC TV series), Skylab 4 commander Jerry Carr and his wife, artist Pat Musick

Chapter 12

New Horizons

In reaching for new horizons in the 1990s, Jerry drew upon, and added to, the experiences in his life that had molded him. Jerry has long recognized that there have been a number of people who were important mentors throughout his career. Perhaps the first of these was the scout master who taught him independence, self confidence and how to take care of himself. Then there was the Marine Lt. Colonel in the NROTC program who became a role model and convinced him to be a Marine. At NROTC, Jerry learned the value of integrity, dedication, hard work and accomplishing what you set out to accomplish; in other words, to set goals in life and work hard to achieve them. When he was serving in VMF-122, Dale Ward, Jerry's fighter squadron commander, taught the future astronaut leadership skills: "He taught me that if you're going to have good leadership, you need a leader and you need people who want to be led. The leader has to very clearly let people know what he wants from them, and he may need to show how he wants them to do it. Then he has to leave them alone to do it and be willing to share the responsibility if there are mistakes."

When Jerry was at NASA, Pete Conrad was a leading mentor for him. Jerry greatly admired Conrad's free spirit. He led by setting clear goals and used his unique enthusiasm and humor to help people enjoy their work while achieving the goal. When working with Pete, Jerry shared in the enthusiasm to get things done, to make sure Pete Conrad was not disappointed. Another influential person during Jerry's NASA years was Wernher von Braun who, before his death in 1977, became a good friend. Jerry proudly recalls one conversation they had after he had flown Skylab. He was showing von Braun the Shuttle mock-up in Building 9A at JSC. Both men were inside the mock-up payload bay when von Braun proclaimed that the Shuttle was a "wonderful machine," one of which he had dreamed for years. Then he added: "But if I had been given the choice to fly any mission that NASA has flown, I would have chosen your mission [Skylab 4]." Jerry realized that this was an expression of Dr. von Braun's sincere conviction that human destiny lies in space. He was a vigorous proponent of space habitation.

In his official NASA oral history in October 2000, Jerry became emotional about mentioning those who helped him achieve his goals, especially his wife, Pat Musick. Jerry admits that when he met Pat he was a "technocrat;" an engineer, a technician, an astronaut, and a pilot. Through meeting Pat and getting to know her family and friends, a whole new passion suddenly opened up which he had never realized was inside him — an appreciation and love for art.

It took time, but Jerry brought experience from his former careers to that of Pat's artistic work, and the two blended perfectly. Jerry helped Pat to organize her career, to document and keep records and to understand the value of a back up plan. Up to then Pat had worked on one plan, and if that went wrong then the whole project "went down the tubes," something that frustrated and upset her. With Jerry's encouragement, Pat began to realize that it is a good idea to have an alternative option if things do not go right. Jerry helped bring that realization to the forefront in her early planning, so that if a process did not work she was able to adapt and overcome the setback, complete the work and still be happy with the outcome.

Over the next decade, Jerry and Pat worked together, harmonizing each other's strengths and overcoming any perceived weaknesses, to develop a program to teach the younger generations and to disseminate her sculpture to a wider audience, using talents and experiences gleaned from two very different backgrounds. They also found time to start a Christmas tradition of making special little loaves of fruitcake for family and special friends. The recipe they used was a cherished one that Pat's mother discovered and had hand-written for her. At this printing they are making twenty cakes each year. The number will undoubtedly grow as the family continues to grow.

The Fabrication Artist Inside an Astronaut

Pat and Jerry constructed the warehouse close to their Arkansas home to support the fabrication, storage, packaging and shipping of Pat's art. They included a small gallery in order to display some of the finished works. The teamwork involved is very impressive. Pat develops an idea for a new piece and explains it to Jerry, who first asks her, jokingly, "How much does it weigh?" They then decide which materials to use in achieving the idea and if it is at all possible to construct. This is how he worked at NASA; he was assigned a task, and he had to figure out how to achieve the desired goal. For Pat's art, 'figuring it

The main house in Arkansas, home for Jerry
and Pat for almost 20 years.

The porch where many a happy hour was
spent entertaining and relaxing.

The view from the cliff across the Kings River Valley.

out' was often quite challenging and included pieces that could weigh 300 lbs or more with the mixture of natural and manmade components.

Some of Pat's ideas were initially created as small sketches which her daughter, Laura, an accomplished graphics designer, transferred to a computer-generated design on a CD. The CD would then be delivered to fabricators in the Fayetteville area who could custom cut steel pieces for her using a sophisticated computerized water cutting process. Once their metallic elements were returned, Jerry and Pat set about the physical construction, assembling the metal framework, welding, stretching canvases and preparing wood and other materials.

Pat, as an Adjunct Professor of Art at the University of Arkansas, had several opportunities over the years to teach art students as part of an independent studies program. She and Jerry helped the students to develop new skills and also introduced them to the problems involved in operating a studio, including archiving and marketing their art products. These are two areas that are not covered in most college curricula. The graduated artist is suddenly confronted with what to do with the art he or she produces and soon learns that "Build it and they will come" doesn't always work out. The sale of art requires an aggressive marketing program and very thick skin.

Jerry also became an integral part of the marketing process, using his private plane, a Mooney Ranger, to ferry museum curators and other art professionals to and from the nearby airport in Huntsville, Arkansas. Spending weekends in the guest accommodations gave them plenty of time to view Pat's efforts in the gallery, and taking in the spectacular scenery from their bedroom window, the dining room or the patio did no harm either.

One piece symbolizes this blend of artist-turned-sculptor and astronaut-turned-artist. It was entitled, "The Source," a tribute to the space age. Its three white pillars are suggestive of the Shuttle External Tank and twin Solid Rocket Boosters. Frazzled ropes extending down from the pillars depict the launch blast on the pad. A plow atop the structure evokes the impression of a space station, something like Skylab. Therefore it symbolizes the agricultural source of our

The Barn - where guests could relax and enjoy the scenery, the nature and the solitude

The Sculpture Garden.

The nerve centre of CAMUS - the office where Jerry and Pat coordinated their various professional activities

planet's nurturance as well as the source of an expanding knowledge of outer space. It represents the hope both Pat and Jerry have for the possibility of cooperative space exploration for all cultures across the world. This sculpture is now in the NASA permanent collection at the Kennedy Space Center.

In reviewing the work of Pat Musick from the 1990s, you cannot ignore the influence Jerry has had on both her personal life and professional accomplishments. He has helped her to develop her art to achieve new levels of scale and inspiration. Equally, the influence Pat has had on Jerry has allowed him to utilize his engineering vision to find a new interest beyond that of flying and space exploration. He uses his talents and experience to support his wife's ever-larger projects. There have been cases of former astronauts who have left NASA or the space industry and who have struggled to find rewarding goals or challenges in life. Jerry has clearly found those new challenges in the love for his family and art.

The Ozarks Woodland Sculpture Garden

On their property in the Kings River Valley in Arkansas, Jerry and Pat had a small, 11.4 acre meadow next to the river, surrounded by forests of oak, sycamore and cedar. Inspired by David Smith's sloping sculpture field at Bolton's Landing, New York, Pat dreamed about a sculpture garden on that meadow that would be visible from the bluff near their home. To inaugurate the garden, a sculpture was installed called 'Iron Eagle', a work created by nine-year-old Michael Taylor Wilson, one of Jerry's grandsons. The dream was to get a local educational institution involved and to use the sculpture garden as a focus for Arkansas sculpture talent and student involvement.

Jerry and Pat approached the University of the Ozarks in Clarksville, Arkansas, located a two hour drive from the proposed garden site, as a potential partner in their plan. The Administration of the liberal arts college was both enthusiastic and supportive of the idea. From here, the program was designed to develop the garden from an idea to a reality.

Their plan for the garden was loosely modeled after a project in which Pat and Jerry participated at the University of Miami in Oxford, Ohio. The program they planned featured an honorarium of $4,500.00 given every other year in the autumn to a chosen master Arkansas sculptor. His or her design would be constructed on the site during the summer semester the following year. A three-week summer class in sculpture was offered by the University. Students worked on campus with their sculpture professor for two weeks, wrestling with the practicalities of designing sculpture for public places. Then they built maquettes (models) of their own designs. During the third week, the master sculptor, the professor and students were guests of the Carrs. Together each day, they worked long hours constructing the final sculpture. Each evening, dinner was followed by a colloquium, where guest artists from the local area described their work and philosophy to the students. At one of the colloquia, the students presented their maquettes for critique by the professionals... a superb educational experience.

Jerry and Pat funded the first year's program in 2000 to initiate the concept. Photo documentation of that program enabled them to secure subsequent community funding for future years with great success. Large and small companies and individuals pledged support for one or more years, enabling medium and long range planning. Once a year, a five-person committee, consisting of a broad representation from the State of Arkansas, met in Little Rock to review slides from the Arkansas State Artist Registry. From this review, a master sculptor for the next year's project was selected and then invited to participate. If he or she accepted, the alternative choice would become the invited sculptor for the following year.

To promote the program, Jerry and Pat contacted regional media services and invited them to attend an afternoon work session, then stay for dinner and the evening colloquium. Several newspapers, magazine and TV companies featured reports on the work at the Garden. In addition, students from the University of the Ozarks Communications Class spent two days producing a 12-minute video of the year's events. This video was to be used in the community solicitation effort the following year. Jerry and Pat also created a dedicated web site featuring the growing number of sculptures at the site, and pursued additional grants to support the future work. Consequently, an application to the White House Millennium Trails Council resulted in the garden receiving an historical designation as a Community Millennium Trail site for efforts to "Honor the Past - Imagine the Future". The certificate was signed by First Lady Hillary Clinton. For Pat, there was an additional honor. She was designated an "Arkansas Ageless Hero" for her efforts in developing the Garden by the Arkansas Blue Cross-Blue Shield.

Over the years the sculpture garden program has featured:

"Iron Eagle" – Michael Taylor Wilson – 2000
"Sting of the Scorpion" – Bryan Massey – 2000
"Great Circle Stanzas" – Pat Musick – 2000
"Visionaries" – Michael Warrick – 2001
"Kings River Totem" – Maurice B. Caldwell – 2001
"Keeper of the Campus" – Bryan Massey – 2002
"Broken Promises" – Greer Ferris – 2003
"The Portal" – Hank Kaminsky – 2004
"Final Place: - Pat Musick – 2004

In 2007, when the Carrs sold the property, the new owners elected not to continue the garden. All but Greer Ferris and Hank Kaminsky removed their works to other locations.

Stone Songs on the Trail of Tears

While flying back from England in 2001, Pat read the book "Arch" by English sculptor Andy Goldsworthy, in which he describes setting up his sculpture 'Arch' in relation to the sheep folds along an old sheep drover's trial in northern Scotland. Every day he would seek permission from the local farmer to arrange his sculpture in relation to the sheep fold and photograph it. These photos were then incorporated into a book that contained diary entries by the artist as well. Pat was intrigued with the idea of taking a sculpture along a trail and realized that she *had* to create a similar project. The minute she decided on a 'trail', she knew it would have to be the Trail of Tears, which had long fascinated her. She told Jerry, "We have to do this," and returned home to research the historic and tragic 'Trail of Tears' journey made by most of the Cherokee nation in 1838 across northern Arkansas. The Native Americans had been displaced from their homes in Georgia and the Carolinas by orders from President Jackson, and were sent by forced march west to the Oklahoma Territory. In tracing this event, Pat and Jerry were about to embark on their own 'journey', which would consume them for the next five years.

Author Dave Shayler with Pat and Jerry at their Arkansas home in 2002. In front are faithful companions Rocky (left) and Ladybug (right)

Author Dave Shayler and Jerry erecting Mesa, one of Pat Musick's sculptures, in 2002.

Pat Music's sculpture Mesa.

The first thing Pat had to do was to learn more about the Trail, and she located Arkansas historian Bill Woodiel, himself part Cherokee, who had studied sites in Northern Arkansas for the past ten years. During their field excursions with Bill over the next four months, the Carrs were able to identify 23 sites along the Old Military Road (St. Louis to Santa Fe) where the remains of early American travel were still evident, such as ruts in the path, road beds, road cuts, and imprints of wagon wheels in sandstone. They researched original Louisiana Purchase survey maps from 1830-1850 and compared them with contemporary county maps. Once a correspondence between the Old Military Road and the contemporary road was identified, Jerry, Pat and Bill journeyed to the site and started looking for evidence of the Trail. While hiking around the site area, they often found local land owners who would supplement their discovery with oral history stories which had been passed down from generation to generation about the 'old road' on their land.

The impressive sculpture used for the project featured five units, made up of two posts each with a bar between them and a stone hung from each bar. The materials used were oak, steel and native Ozark sandstone. The total design represented a procession of pairs of people carrying a stone on a yoke between their shoulders. They were symbolic of the struggle of those who made the journey… young and old, male and female. Pat named the sculpture 'Stone Songs', and this was featured in the title of the book that subsequently accompanied the project.

In March 2002, they set out to re-trace the original Trail. The weather was an important factor for Pat, as she wanted the photographs to recapture as much of the perilous conditions of 1838 as possible. On the days they were in the field in rain, hail, snow and mud, they tried to average photo shoots at two sites. At each of the 23 sites, they painstakingly erected the Yokes in a configuration that was dictated by the lay of the land (including water and ice), in a general arrangement that suggested a procession or journey heading westwards.

During his Skylab mission, Jerry's crew took over 2000 photos using Hasselblad and Nikon cameras. To document the journey of Stone Songs, he used a similar Hasselblad. Before shooting, he carefully arranged the sculpture with, as Donald Harrington wrote in the introduction of the book, "the same sureness that he had piloted the spacecraft." Following the photography, the sculpture was taken down and packed into the truck ready for the journey to the next site. They then returned the land to its former condition as they did not want to disturb the ecological balance of the forest.

The Carrs received financial help with the project through grants from the Arkansas Arts Council, the National Endowment of the Arts and a local Arkansas entity called the Happy Hollow Foundation. This allowed them to complete the second phase of the project. Each of the 23 photos was enlarged into a 24"x 24" mural size image, and these were developed into a traveling museum exhibit that included Pat's poems and diary entries written at each site. Over the next four years, the exhibit traveled to nine museums in the mid-West and Southern United States.

During these years, Pat compiled all of the material into a book manuscript. Donald Harington, the noted art historian (whose ancestors were on the Trail) wrote the introduction. The book also contains an essay by Bill Woodiel and Jack Baker, President of the National Trail of Tears Association. In 2006, the book, entitled *Stone Songs on the Trail of Tears, the Journey of an Installation,* was released by the University of Arkansas Press, and is now in its second printing.

In September 2003, the journey of the sculpture ended when Tyson Foods Inc. purchased the piece for its permanent art collection. It has been installed in front of the corporate headquarters in Springdale, Arkansas, close to the original Trail of Tears. Located next to the visitor's center, it honors the Cherokee Indians who long ago made that painful journey. It is available for public viewing.

But the story has not yet ended. Jerry, Pat and Bill are still doing book signings and lectures around the country, where they talk about the fact that the project has forever changed their lives. What started out as an art installation soon changed into a deep, personal identification with the Cherokee people and the suffering of all displaced people anywhere; "It was, after all, ethnic cleansing," they point out and a history lesson they hope they can share with many people.

Hanging up his wings

Jerry had continued flying after leaving NASA, using his Mooney Ranger since 1984 to support his business trips across the country. When he decided to wind down the aerospace side of CAMUS, he made a decision to end his flying career and hang up his pilot wings. On August 19, 1998, he made a 1.1 hour flight in the Mooney Ranger for the last time, ending a flying career that had begun on April 14, 1955 with his first training flight with an instructor pilot in a SNJ-4 aircraft. After Jerry soloed on May 18, 1955, he continued his USMC flying career as an operational Marine aviator until April 2, 1966 aboard a T-1A aircraft, his final flight in a naval aircraft before reporting to NASA and astronaut training. As a Marine, he had logged over 2290 hours flight time.

Colonel Gerald (Jerry) P. Carr, USMC (Retired) and former NASA Astronaut. Commander Skylab 4 (November 16, 1973 - February 8, 1974) with a model of his home for 84 days.

Jerry's first NASA flight in a Lockheed T-33A occurred on May 12, 1966 and over the next 11 years, he would add another 1230 hours in the T-33 and T-38 aircraft and 164.2 hours flying the Bell 47G helicopter, the latter in preparation for his

A Carr Family portrait circa 1986. Last row, L to R: Fred Carr (Ron's son), Josh (holding Ron's grandson), Jerry, Scott Farris (Ron's son-in-law) Pat, Christine Farris (Ron's daughter), John, Ron (Jerry's brother), Jeff. Front Row, L to R: Jessica, Jamee, Lois Stone (Jerry's aunt), Lois Carr (Ron's wife, holding twin grand daughters Rebecca Farris and Cassandra Farris), Jennifer (holding Jessica Farris, Ron's grand daughter)

A Musick-Carr Family portrait circa 1987. Third row, L to R: David Chase (Cathy's Husband), Tim King (Mindy's husband, holding daughter Hillary, Greg Wright (Laurie's husband, holding twin daughter Alisha). Second row, L to R: Cathy Musick Chase, Mindy Musick King, Jerry, Pat, Laurie Musick Wright (holding twin daughter Jennifer). Front row, L to R: Grand daughters Wendy Chase, Marley King, Caryn Chase, and Jamie Cicchetti (Laurie's oldest daughter).

assignments in the Apollo lunar landing program. His final NASA flight occurred on June 22, 1977 in a T-38 flight with Bill Pogue.

In his Mooney Ranger between July 1984 and August 1998, Jerry logged an additional 2567.8 hours flight time. Jerry's pilot log book shows a grand total of over 6568 hrs flying time in aircraft and over 2017 hours flying in space. This amounts to a career total of over 8585 hours flying time (or 358 days off the ground!).

Family Affairs

Over the past three decades, Jerry and Pat have watched their family continue to grow around them. In 2007, it had reached thirty-four members: nine children (six with spouses), twelve grandchildren (four with spouses) and four great grandchildren. Each of them has had a special life, and many have worldwide interests, which gives Jerry and Pat personal reason to hope that their dream of world peace will someday be realized.

Jennifer Ann Carr, Jerry's eldest daughter, lives in Texas. Her brothers, sisters and their families are a great joy in her life. She is an active member and Elder in her church, as well as a board member of More Light Presbyterians; a network of people seeking the full participation of lesbian, gay, bisexual and transgender people of faith in the life, ministry and witness of the Presbyterian Church. She loves to cook and bake, not only for family gatherings, but for her church family as well. She travels with her mom, and hiking ranks high on her list of passions. Scuba diving is another passion; there is nothing like being underwater in the absolute quiet and beauty of the ocean. She currently enjoys working as the bookkeeper for her younger sister's business.

Jamee Carr Wilson, Jerry's second-born daughter, recently earned her M.A. in Clinical Psychology. She is currently working towards the completion of her LPC internship, working with adolescent victims of trauma and providing general counseling. Jamee deeply believes in the value of every human being and their right to a happy and fulfilled life. She loves to travel and experience other cultures, creating masterpieces in the kitchen and watching her son grow into a fine young man.

Gregory Wilson, the husband and tireless supporter of Jamee and father to Michael, has been working in the inventory management field for over 20 years, the last 12 of those in the marine industry. Greg also loves to travel, plays golf and is the highly skilled, "go to guy" for home improvements.

Michael Taylor Wilson, the first of Jerry's two grandsons, was born in Laguna Beach, California. As Company XO in JROTC, Michael excels in the military arts and leadership and has contracted to join the Navy (sorry, Grandpa) upon graduation from high school. Michael enjoys international travel and has served as a student ambassador in Italy, France and England. He is an accomplished drummer and plays in a Christian rock band.

Jeffrey Ernest Carr, Jerry's oldest son and Jamee's twin, is most likely to do whatever he wants to, in spite of the advice of others. It's worked out okay, though. He is a senior executive at United Space Alliance (USA) in charge of Marketing, Communications and Public Relations. He was a Space Shuttle mission commentator in Mission Control for more than 30 Shuttle missions in the 1980s, served as News Chief including a tour of duty at NASA Headquarters in Washington, D.C., and became Director of Public Affairs at NASA/JSC before joining USA in 1996. He has dedicated his entire career at NASA and in the space industry to convincing others that space exploration is as essential to our species as food and water. He is motivated in his work by an urgent need to check and balance the inordinate and dangerous power wielded by the news media in setting the public agenda on space, combined with a deep fear of how badly others would do the job if he didn't. His chief priorities in life, however, are family, hunting and working on his ranch in the Texas Hill Country…in that order. He hopes to be retired from the space industry along with the Space Shuttle.

Mengo Carr Carr, Jeff's wife and stepmother to Betty and James, has been a working professional in the space program herself, since 1995 at NASA/JSC, then with United Space Alliance, and currently with The Boeing Company. Her maiden name is also Carr. She might just be the most like Jerry in that she is easily the most squared-away, obsessively organized one of the bunch, making the family wonder about a possible genetic connection. Her greatest passion is looking out for the people she cares about most. She plans and hosts many of the Houston Carr family weekend get-togethers, herself, when she's not planning Jeff's and her retirement at their Hill Country ranch. Failure is not an option for this one.

Betty Ann Carr, the first of Jerry's three granddaughters, was born in Texas on Martin Luther King Day in 1994. Some in the family think that her birth may have caused the San Francisco earthquake that occurred at that same instant. An accomplished young swimmer and soccer player, she also excels in the Language Arts – most recently with an interest in journalism. Her greatest passion is for people and animals, and she aspires (currently) to become a child therapist. Her enormous personal energy and Christian faith will serve her well in that pursuit. She is very proud of her grandfather but would just as soon look after the pets and children of astronauts while they're off exploring space.

James Robert Carr, Jerry's second grandson, is a budding student athlete, also born in Texas in 1997, who probably stands the best chance of following in his grandfather's footsteps. He recently won first place in the District Science Fair, but it may have been more the competition than the Science that motivated him. He plays league football, basketball and is a perennial baseball All-Star. His earliest impression of the space program was, unfortunately, from the Columbia Space Shuttle accident. His competitive nature and zeal for new experiences will certainly propel him beyond his fears, though, if there is something he sees in space. He also does amazing comic impressions and is an overly-accomplished class clown.

John Christian Carr, Jerry's second son, spent over 17 years in the restaurant management industry in Houston, Atlanta, and Dallas. For the past 10 years he has been a business owner in the landscape management industry. The devoted single father of two little girls, Cameron and Avery, he has a passion for the outdoors and working with others. He is known for his calmness, sense of humor, patience, compassion, and his ability to adapt to any situation.

Cameron Lea Carr, Jerry's second granddaughter, was born March 8, 2000 in Houston. She excels in school as a leader and a helper and is always there to please. She is most likely to follow her father's interests because she loves the outdoors and is fearless when it comes to wildlife and exploration. Cameron has been known to bring live animals, including snakes, to show her Father.

Avery Irene Carr, Jerry's third granddaughter, was born November 27, 2002 in Houston. She excels in school and is a one-woman wrecking machine! Avery loves to sing and dance and is most likely to become the next Hannah Montana pop star. She is unbelievably talented for a five year old and is serving notice that her father will have his hands full for years.

Jessica Louise Carr, Jerry's third daughter, lives a quiet life in Santa Ana, California with her partner Joy Ridout and four dogs: Riley, Jake, Peanut and Jak. She owns and operates a successful company in the high tech industry despite the lack of a college degree. Most recently, Jess and Joy have completed a 6-year renovation of a 1925 Spanish Bungalow. Having no prior home improvement experience, they took on the task of restoring this little gem into a classic, vintage home. Those six years of hard work turned into a passion for the restoration process that they will share with anyone who asks! Jess and Joy love to travel, drink wine, cook and hang out with their dogs.

Joshua Lee Carr, Jerry's third son and Jessica's twin, lives in California with life partner Ray Limon. Together they run a successful theatrical production company producing musical theatre all over southern California and Arizona. His strongest memory of Dad is his sandals crunching in the gravel on the way to an exciting day at the neighborhood pool! Being so much younger than the other siblings when the flight was going, he didn't pay too much attention to the details. He just knew that Dad was "at work" and would be home in a few months. He was, however, unhappy because John got the lead in the Go Show! Best lesson – take a stand for that which you believe in, even though the cost may be great. God reveals the answer through new challenges and presents opportunities for new growth.

Cathleen Louise Musick, Pat's eldest daughter, lives on Long Island in New York State. She is honored to be the mother-in-law of two phenomenal young men who are currently serving their country as United States Marines, as Jerry so nobly did. Cathy studied biology and marine science in college and graduate school, later continuing her studies in education to become a certified teacher and administrator. For the past eighteen years she has worked as an enrichment specialist in an elementary school on the East End of Long Island. She is a teacher, lead teacher, and peer coach, joyfully collaborating with a talented staff to provide an array of enrichment programs for all students. She is always amazed and inspired by the students' creativity and vision. Cathy's life is filled with love and the joy of family, with admiration for Jerry as the family great-grand father to her three new grandsons. Cathy serves her church, the First Congregational Church of Riverhead, as Deacon.

Caryn Patricia Rakov, Pat's second granddaughter and Cathy's firstborn, proudly inherits her middle name from her creative and talented grandmother. At 30 years of age, Caryn lives in Okinawa, Japan, where she is stationed with her husband, Bradley Rakov, who is a United States Marine Corps KC-130 command pilot. Caryn and Brad have lived in Okinawa for nearly six years, and in that time Caryn has learned the Japanese language, taught English to Okinawan children and adults, earned a Masters degree in Elementary Education, and taught Fifth Grade at a DoDDS elementary school on-base. Brad has flown thousands of hours in many countries in the Pacific region, including countless missions of in-flight refueling for both helicopters and jets, as well as participating in relief efforts, such as the Indonesian/Thailand tsunami and Philippines' mudslide disasters. Caryn and Brad also enjoyed unforgettable vacations to such countries as Australia, New Zealand, Thailand, mainland Japan, Singapore, and China. Their most recent achievement however, is their first child, Brayden Alden Rakov, born August 9, 2007 and growing every day! His parents are already looking forward to his learning all about the achievements and adventures of his Grand-Poppa Jerry!

Wendy Frances Porter, Pat's fourth granddaughter and daughter of Cathy Musick, was born in Long Island, New York in July 1981. One of Wendy's favorite hobbies is interior design, but she is currently in love with being a wonderful wife and mother. She met Jesse Glenn Porter while visiting her sister, Caryn, in Okinawa, Japan. Jesse, originally from Gilbert, Arizona, was stationed there as a young active duty Marine. Jesse and Wendy married on January 17, 2004 and were sealed for time and eternity in the San Diego Temple of The Church of Jesus Christ of Latter Day Saints on June 4, 2005. Jesse is currently stationed in Riverton, Utah, where the Porters are enjoying the ability to spend time as a family and taking advantage of the great outdoors. Jesse and Wendy have a 3-year-old son named, Chase Jesse Porter. Chase is a happy, energetic boy who loves to play, explore and be with his parents. Jack Glenn Porter was born on

October 30, 2007 and is Pat and Jerry's newest great-grandson. The Porter family is awaiting Jesse's orders to another duty station to see what's next in the journey of life for them.

Melinda Musick King, Pat's second daughter, lives with her husband, Tim King, at an elevation of 9,000 feet, high in the Rocky Mountains of Colorado. Mindy teaches 4th grade in Bailey, Colorado, and Tim is a senior manager for the Colorado State Parks Department. They enjoy traveling the United States and to countries around the world, and snowshoeing during the winter. Mindy and Tim frequently return to the Northeast to visit family in upstate NY and Vermont. Mindy is a breeder and handler of Parson Russell Terriers and has enjoyed showing her dogs across the nation. She has exhibited at Westminster and the Eukanuba National Championships, as well as at her breed National Specialty shows. Her dogs are consistently ranked in the AKC Top 20.

Marley King Wynne, Pat's eldest granddaughter and the senior granddaughter in all three families, takes full responsibility for naming him "Poppa Jerry". She currently works in the computer technology field with the Denver Public Schools, but hopes to start a family with her new husband, Thomas, in the not too distant future. In what spare time she has, she enjoys singing with the Colorado Symphony Orchestra Chorus and hopes to pursue an advanced degree in music with an emphasis in choral conducting.

Thomas Wynne, Marley's husband, is a Senior Software Developer specializing in Enterprise Geo-Spatial Data and Software Development. He received his Bachelor of Science in Computer Science and Engineering at the University of Arkansas in Fayetteville, Arkansas. As a member of the University of Arkansas programming team, he helped the U of A advance to the world finals for the first time in school history in both 1998 and 1999. Thomas now lives with his lovely wife Marley and their Welsh Springer Spaniel, "Streak," in Colorado, but hopes to move back to the Northwest Arkansas area in the near future.

Hillary Musick King, Mindy and Tim's younger daughter, is an International Studies major at the University of the Pacific in California, where she is on the Dean's List. She spent her junior year studying abroad in Senegal, Africa and St. Petersburg, Russia. Hillary totally immersed herself in both cultures, living with host families and taking classes in the native languages. While in Africa, Hillary interned in a women's health clinic and participated in World Aids Day activities. She plans to attend graduate school and then work in the International arena after her spring 2008 graduation.

Laurie Musick Wright, Pat's third daughter, was born in Denver, Colorado and raised in Hanover, New Hampshire and Ithaca, New York. Laurie graduated from Cornell University in 1975, majoring in communications in the College of Agriculture and Life Sciences. A graphic designer, she was employed by Cornell University, Exxon Corporation, and an advertising agency in Vail, Colorado. Laurie founded her own studio, LMW Design (lmwdesign.com), in 1986. She is the mother of three daughters, Jamie Ziehm (1977) and twins Jennifer & Alisha Musick Wright (1986). She volunteers for her church and serves on the President's Council for Cornell University. A Reiki Level One graduate, she is currently studying the hands-on healing practice of Bioengergetics and is a student of A Course in Miracles.

Greg Wright, Laurie's husband, attended Syracuse University. Because of a love of nature and the outdoors, Greg earned a bachelor of science degree from the College of Environmental Sciences and Forestry at Syracuse University in 1973. He has worked for over 32 years with the US Forest Service on four National Forests. Greg and Laurie married on June 12, 1982 in Leadville, Colorado.

Jamie Ziehm, Laurie's oldest daughter, lives with her husband Eric and their son Cole just over the Vermont border in Hoosick Falls, NY. Jamie and Eric met at Cornell University and now own and operate Higher Ground Farm, a full-service equine training and boarding center. Eric is also a partner in his family's dairy farm in Buskirk, NY. Their first son, Cole, is the second addition to Jerry and Pat's list of great grand-children.

Jennifer Musick Wright, one of Laurie's twin daughters, was born in Vail, Colorado in 1986. She is currently a junior at Wagner College in Staten Island, NY, majoring in both International Affairs and French. After a life-changing experience visiting Zimbabwe to volunteer at a private home for children orphaned by the aids pandemic, she traveled to Kenya and volunteered for three weeks at a public orphanage. While in Kenya, Jenn experienced corruption and deplorable conditions first-hand and was inspired to find a better way to offer orphaned children a healthy life. She created HEAL the Children Foundation and has organized a network of volunteers to design a program that will nurture the lives of orphaned children and in the process "Raise the World!"

Alisha Musick Wright, Jennifer's fraternal twin, is attending Beloit College and majoring in

Religious Studies. Alisha sings with the folk music group, Village Harmony, and has toured the northeast US as well as Bosnia and Herzegovina. Alisha will spend a semester abroad in Bosnia in the spring of 2008. She is studying world religions and the beauty of individual faith. She also carries the environmental science gene in the family and is passionate about sustainable living, herbal remedies, and being outdoors. One day, she hopes to combine her love of menstrual activism, herbal medicine and female sexuality in a career in midwifery.

A New Life

For some years, Jerry and Pat had been planning to leave their Arkansas home for a smaller, more manageable property, with room for Pat to continue her art work but without the work that the upkeep of 108 acres of land entails. In 2000 Pat was elected to the Alumni Board of Trustees at the School of Human Ecology at Cornell University for a four-year term. The thrice yearly meetings meant a return to the Cornell campus she loves and the renewal of many Northeastern friendships. She became aware of a strong desire to return to the area she had left in 1976, and Jerry agreed that he too would like to explore a new horizon. They began looking for a place that would be small, manageable and with room to work, but which would provide them with more time to travel. It was not until 2004 that they found what they were looking for in Vermont. A new retirement complex in the village of Manchester Center was in the planning stages. They viewed it and signed on! While the complex was being constructed over the next two years, the Carrs put their Arkansas home up for sale. In September 2006, they finally moved from their beloved Kings River home in Arkansas to Vermont.

In 2005, as part of their downsizing effort, Jerry and Pat donated their professional libraries to the University of Southern California and Cornell University respectively. Jerry was then confronted with how to dispose of much of his space memorabilia. With the help of his daughter Jennifer, he gathered a special collection of memorabilia for each of his children. They framed the items and presented them to the family.

Portrait of Jerry sitting at the cliff in Arkansas.

Portrait of Pat.

The move to Vermont was a complex one. The household goods came in the first load. Then, over a period of a year, it took three large truckloads to bring the studio and shop equipment and 36 crates of art. Jerry and Pat purchased a unit in a new light industrial building in nearby Sunderland, and they equipped it with a metal shop, storage area, work area and small gallery space. Their apartment is much smaller and easier to maintain than the 'ranch' at Huntsville. It is more compact for the two of them, even with their two dogs, Ladybug and Rocky who took over the laundry room as soon as they moved in. In the Spring of 2007 however, while in Huntsville preparing their Arkansas home for the final sale, their much loved dog, Ladybug, died and was laid to rest in the woods where Pat and Jerry feel sure her spirit will continue to chase rabbits. Pat and Jerry, and particularly Rocky, missed her very much. Rocky died shortly after the return to Vermont, but was later laid to sleep in Arkansas beside his "bug" where he has, no doubt, joined her in the bunny chase.

At this writing, the Carrs are happily ensconced in their two-bedroom condo in a complex that furnishes a library, work out rooms, lecture hall, pool, maid service and a wood shop for Jerry and Pat to work in. They live close to Laurie and her family and granddaughter Jamie and hers. They are enjoying shared time, especially with great grandson, Cole, whom they have nicknamed "Smiley".

They spend their days working on Pat's art (although at a reduced level), traveling and getting acquainted with their new community. They are both doing some writing and lecturing, which they enjoy. They attend the local Congregational Church and participate in many of the available cultural activities. They continue their Christmas fruit cake tradition.

Final Words

From Colleagues

Over the years, Jerry has gained respect and admiration from friends and colleagues from his careers in the Marines, at NASA, in the aerospace business and in partnership with his wife Pat Musick and her art career.

Often overlooked is the fact that Jerry commanded the first all-rookie American spaceflight crew since Neil Armstrong and Dave Scott flew Gemini 8 in March 1966. The next all-rookie crew was that of STS-2 in November 1981. Apart from the one-man Mercury missions, these have been the only all-rookie American spaceflight crews launched, and Skylab 4's 84 days in space remains the longest duration for such a crew.

Fellow astronaut Jack Lousma, who flew as pilot on the second manned Skylab mission and then commanded the third flight of the space Shuttle in 1982, recalled his long association with Jerry:

"Jerry and I are good friends and were the two Marine pilots selected by NASA to join the astronaut corps in 1966. We worked together on several projects in early-stage development, including Skylab and the Lunar Rover, before we flew on missions of our own. After NASA, I also worked occasionally for CAMUS, the entrepreneurial space consulting firm Jerry successfully pioneered. Jerry's demeanor is pleasant and easy-going, but it belies the determination and persistence that drives him to excellence and inspires those who know him. He combines the compassion and strength of character that draw people to him. NASA's confidence in Jerry's personal leadership and professional skills was demonstrated when he was assigned a Commander position without the typical prior spaceflight experience in a subordinate role."

That achievement of commanding a crew on his first spaceflight was reflected upon by Skylab 3 Commander Al Bean, with whom Jerry had worked on Apollo 12 prior to the Skylab assignment:

"For a flight crew to have a successful mission, they need to have the ideas and hardware and the enthusiasm of the back up and support crews. One of the ways that Pete Conrad, Dick Gordon and I felt very lucky was that we had Jerry Carr on our Apollo 12 support crew, not only during the training program, but also serving as CAPCOM during critical phases of the mission. As a result, Jerry was held in such high esteem by both the management and his colleagues in the Astronaut Office that his skills and leadership resulted in him being assigned the command of the third Skylab mission on his first flight into space. Skylab 4 surpassed all records for American manned spaceflight in terms of endurance and remained the milestone mission for the next twenty years. Amongst his peers Jerry achieved an excellent reputation and was admired for his integrity, for his method of approach, and for his determination to get a job done and done well."

Scientist astronaut William 'Bill' Thornton MD, who later flew on Shuttle mission 8 (1983) and 51-B/Spacelab 3 (1985) pointed to Jerry's contribution to the long term exploration of space, along with that of Ed Gibson and Bill Pogue.

"The SL-4 crew under the command of Jerry Carr produced what is arguably the largest volume of essential data on effects of weightlessness, by *any* space mission to date, including the first:

* documentation of fluid shifts, gains and losses, in weightlessness
* maintenance of body mass with adequate nutrition
* maintenance of body strength and mass with proper exercise and nutrition
* documentation of increase of stature in weightlessness
* documentation of weightless posture
* documentation of anthropometric changes in weightlessness.

Much of this was done in addition to previously planned and scheduled work and has resulted in numerous basic operational changes, such that all present and future travelers will owe this commander and his crew a debt of gratitude."

Portrait of Pat and Jerry sitting in front of their wall sculpture creation entitled, Earthmark.

The mission of Skylab 4 produced a series of results and an understanding of the complexities of living and working in space by a crew who, despite long training regimes, had not experienced actual spaceflight. Their approach to this and the determination to achieve the objectives and overcome many personal and professional challenges and hurdles was inspired by the character of the Skylab 4 commander, Jerry Carr.

Since 1974, the endurance record of human spaceflights has increased gradually up to 14 months, with expeditions to space stations of between 4-6 months becoming the standard tour of duty. That regularity in spaceflight operations stems in no small part from the effort and devotion of Jerry Carr and his crew to meet and overcome more challenges than they had originally trained for. In the annals of space exploration, as we return to the moon and reach for Mars, the milestone mission that allowed this to happen, and from which much was learned and is still being applied, was Skylab 4.

From Jerry

When asked what word he would use to sum up his life, Jerry quickly responds, "Fortunate." Jerry has always felt that opportunity is constantly flowing by, and whether or not one is able to seize it depends on luck and if one is prepared to act.

He has been fortunate to have a mom who gave him unqualified support in spite of trepidations about flight in jet fighters and spacecraft. Mr. Martini, his scoutmaster, provided him with the basic tools of leadership and self confidence, and his church founded him in spiritual meaning. He was lucky to have come into contact with Lt.Col John Finn who persuaded him to become a Marine and with Lt.Col. Dale Ward who had great confidence in him and who, by his example, completed his education in leadership.

Jerry was, indeed, fortunate to have been selected to be a NASA Astronaut. He has said, "At the time I made my application there were literally hundreds of pilots equally as qualified to be astronauts. I don't know for sure what it was about me that caught NASA's eye, but whatever it was, I feel lucky." He was fortunate to have had the chance to get to know all of the seven Mercury astronauts and to work with those to whose missions he was assigned to support. "I consider Pete Conrad to have been an important mentor," he says. "I am grateful for the years of our friendship and deeply saddened that he is not with us today."

When Jerry interacts with the public one of the questions most frequently asked is, "What are the things young persons should do to prepare themselves for a career as an astronaut?" He first of all suggests that the question might be generalized by substituting the word "astronaut" with "successful." Then the formula consists of setting goals, plus enthusiasm, plus hard work, plus keeping one's options open in order to seize opportunity as it passes by.

Epilogue:

Thoughts on Jerry Carr

Jerry had just the right blend of the consummate professional and empathetic friend to make an outstanding Leader and Commander.

Jerry the professional was very determined and also fun to watch. After launch and rendezvous with the Skylab Space Station, it came time to physically linkup. The docking port to our new home stared us in the face. With great precision and care, as Jerry always exhibited, he maneuvered us in for the capture. We felt a gentle thud, then saw ourselves floating away in the direction from which we had just come. Jerry had input too much care and not enough velocity to affect a capture. Beads of sweat, hemispheres in zero gravity, soon dotted his determined face. Not able to resist it, I softly muttered the words "wimps" and "US Marines" in the same sentence. Our next approach was considerably swifter and the contact much "firmer," so much so that I'm thought it would've rolled down our socks if our legs were aligned in the right direction. We did not bounce off and the Station knew, without question, that we had arrived.

Jerry the professional, was also highly mission focused, even if his definition of "mission" did not match the accepted norm. It's no secret that early in our mission we found ourselves always falling behind the rigid, detailed timelines that the good folks on the ground had carefully labored over each night. We couldn't get ahead, we couldn't break even, and we couldn't quit. It was depressing. Exhibiting no creativity whatsoever, I stoically kept running, day after day, week-after-week, as fast as I could on what I pictured as the first rodent treadmill in space. I probably would have run right up to the day we came home if Jerry had not been both smart and courageous enough to intercede. He recognized that one of the larger objectives of our mission was to understand and put into practice how long-duration missions should be conducted. He reasoned correctly that people down here don't work against a minute-by-minute timeline, hour-after-hour, day-after-day. In contrast to anyone else on the ground or in Skylab, his vision dictated that, unlike the launch and re-entry on either end of our mission, the middle 82 days should be conducted in the same loose but efficient fashion that we humans have evolved to over thousands of years of effort. He had the courage to follow his vision and, after a heart-to-heart talk with the ground, we abruptly changed our mode of operation. The net result of Jerry's vision and courage was that all of us in Mission Control and Skylab were much more efficient, smarter, and happier.

Was it Jerry's vision or his empathy for his friends, Bill and I, which gave him the courage to go against the so many other professionals? It was probably a good measure of both. But it is almost exclusively his empathy that continuously makes him so supportive of those around him. Bill and I each had our specialties on the mission, and Jerry was always highly supportive of us getting what we needed in training and in flight to perform well. He never hogged the spotlight but made sure we each had our say. In the many times I have encountered or worked with Jerry after our flight, his magnanimous nature has remained unchanged. Always highly supportive of his family, he has literally gone the extra mile, quite a few extra miles in fact, in supporting his wife Pat's career as an artist. She creates the designs and plans for her sculptures; Jerry, always with a smile on his face, does the heavy lifting, even if it entails moving hundreds of pounds of sculpture over hill and dale to the most meaningful locations. He remains just as happy in the back seat as he does driving.

In addition to his vision, courage, and empathy, Jerry consistently demonstrates an unyielding and hard altruistic core. He shows respect and, in turn, earns it. And, of course, he is competent, as anyone who is a Commander among the best of the best has to be. All these qualities make him a great leader and loyal friend.

The world could use a whole bunch more Jerry Carr's!

Edward G. Gibson (PhD)
Former NASA Astronaut - Science Pilot SL4

Appendix 1

A Career Timeline Gerald (Jerry) Paul Carr

8 Oct 1909	*Thomas E. Carr (Jerry's father) is born*
12 Apr 1911	*Freda Wright (Jerry's mother) is born, 50 years to the day before Yuri Gagarin becomes the first man in space.*
14 Sep 1926	*Pat Tapscott (later Musick) who became Jerry second wife is born in Los Angeles California.*
23 Jan 1930	*William R. Pogue (Jerry's Pilot on Skylab 4) is born in Okemah, Oklahoma*
22 Aug 1932	Born in St Anthony Hospital, Denver, Colorado, USA.
20 Sep 1933	*JoAnn Petrie (Jerry's first wife) is born in Pomona, California*
8 Nov 1936	*Edward G. Gibson (science pilot Skylab 4) is born in Buffalo New York*
Sep 1938- Jun 1940	Attended first and second grade at Mountain View School, Denver.
Jun 1940	Family moved from California,
Sep 1940-Jun 1941	Attended Chapman Avenue School in Gardena
1941	Family moved to Santa Ana California which Jerry adopted as his hometown
Sep 1941-Jun 1944	Attended Wilson Elementary, Santa Ana, California
Sep1944-Jun 1947	Attended Willard Junior High School
Sep 1947-Jun 1949	Attended Santa Ana High School
14 Oct 1947	*Chuck Yeager breaks the sound barrier in the X-1*
23 Oct 1949	Enlisted in the USN Reserve
1949 - 1950	Airman Apprentice USNR
17 Sep 1950	Honorably discharged USN Reserve
18 Sep 1950	Commenced studies at the University of Southern California, Commissioned Midshipman USN, ROTC.
Summer 1951	Naval cruise aboard *USS Steinaker* (DD-863) for one month
Summer 1952	Two weeks spent at naval aviation pilot training, NAS Pensacola, Florida, followed by two weeks amphibious training in Little Creek, Virginia
Summer 1953	Third naval summer cruise aboard *USS Worcester*, visiting Europe and the Mediterranean
4 Jun 1954	Graduated UofSC, with BS degree in Mechanical Engineering;
4 Jun 1954	Commissioned 2nd Lieutenant USMC
20 Jun 1954	Married JoAnn Petrie in Santa Ana

1 Aug 1954	Commenced USMC Office Basic School, Quantico, Virginia
Feb 1955 - May 1956	Flight training, USMC
4 Dec 1955	1st Lieutenant USMC
25 May 1956	Designated Naval Aviator awarded Pilot Wings
May 1956- Aug 1959	Assigned Marine All-Weather Fighter Squadron 114, Cherry Point, North Carolina
31 Jul 1955	First daughter, Jennifer Ann born
4 Oct 1957	*Soviet Union launches sputnik opening the space age.*
31 Jan 1958	*United States orbits its first satellite Explorer 1*
1 Oct 1958	Captain USMC
3 July 1958	First twins, daughter Jamee Adele and son Jeffery Ernest born
9 Apr 1959	*First group of American astronauts selected for Project Mercury. Group includes Marine pilot John Glenn*
Aug 1959 - 1961	Student USN Postgraduate School, Monterey California; graduated with BS degree in aeronautical engineering
1961-1962	Student, Princeton University; graduated with MS in aeronautical engineering
12 Apr 1961	*Soviet cosmonaut Yuri Gagarin becomes the first human to fly into space. Competed 1 orbit in 108 minutes*
5 May 1961	*Astronaut Alan Shepard becomes the first American in space during a 15 minute sub-orbital flight*
20 Feb 1962	*Astronaut John Glenn, becomes the first American to orbit the Earth. Completes 3 orbits in 4.5 hours*
Aug 1962	*Soviet cosmonaut Gherman Titov spends one day in orbit*
4 Apr 1962	2nd son John Christian born
Aug 1962	*Soviet cosmonaut Andrian Nikolayev spend three days in space*
Aug 1962-Mar 1965	Assigned Marine All-Weather Fighter Squadron 122, Beaufort, and California
Sep 1962	*NASA selects a second group of astronauts - all test pilots*
Oct 1962	*Astronaut Wally Schirra completes six orbits in almost 8 hours*
May 1963	*Astronaut Gordon Cooper spends one day in space*
Jun 1963	*Soviet cosmonaut Valeri Bykovsky spends five days in space*
Oct 1963	*NASA selects a third group of pilot astronauts. Group includes a Marine named CC Williams*
12 Mar 1964	Second twins born. Third daughter Jessica Louise and third son Joshua Lee
18 Mar 1965	*Soviet cosmonaut Alexei Leonov completes the world's first spacewalk - EVA*

Mar 1965- May 1966	Assigned Test Directorate, Marine Air Control Squadron Three, Santa Ana, California
Jun 1965:	*NASA selects a fourth group of astronauts. This time they are scientists rather than pilots. Included is a physicist named Ed Gibson.*
Jun 1965	*Gemini 4 astronaut Jim McDivitt and Ed White spend four days in orbit. Ed White becomes the first American to walk in space*
22 Sep 1965	Major USMC
Dec 1965	*Frank Borman and Jim Lovell complete two weeks in space aboard Gemini 7*
1 Apr 1966	Informed by Alan Shepard of his selection as one of nineteen new NASA astronauts
4 Apr 1966	Officially named as one of 19 Group 5 Pilot Astronauts by NASA. Bill Pogue is selected with him.
2 May 1966	Reported to NASA Manned Spacecraft Center, Houston, Texas for astronaut training
6 Oct 1966	Assigned to the CB technical support team for the development and testing of the Apollo Lunar module
22 Dec 1966	Named Support crew member (LM specialist) Apollo 3
Jan 1967	Crews stood down due to loss of three astronauts in Apollo 1 pad fire
24 Jul 1967	Jerry's father dies, aged just 57
21-27 Dec 1968	Shift CAPCOM Apollo 8, first manned lunar orbital mission
10 Apr 1969	Named Support crew member Apollo 12 (LM specialist)
20 Jul 1969	*Neil Armstrong and Buzz Aldrin step on to the moon during Apollo 11. Mike Collins orbits in CSM*
14-24 Nov 1969	Shift CAPCOM Apollo 12
1970	Early CB representative (with Joe Engle and Jack Lousma) for development of Lunar Roving Vehicle.
Summer 1970	Preliminary work for assignment as BUp LMP Apollo 16, leading to assignment as LMP Apollo 19. Commander would be Fred Haise, CMP Bill Pogue
Jun 1970	*Soyuz 9 cosmonauts Andrian Nikolayev and Vitally Sevastyanov complete an 18-day space marathon*
1 Jul 1970	Lt Colonel USMC
1970 Sep	Apollo 19 is cancelled, Jerry loses the Moon and the chance to become the 16th person to walk on the surface. Reassigned with Bill to the Skylab space station program as a potential crew Commander.
Early 1971	Received assignment as Commander of third Skylab crew.
Apr 1971	*Soviets launch Salyut 1, the world's first space station.*
19 Jan 1972	Officially named Commander Third Skylab mission (Skylab 4)

Apr 1973	*Military-orientated Salyut 2 is launched but suffers early in-flight problems and is not manned. Cosmos 557, an intended Soviet civilian space station, suffers a similar fate in May*
May 1973	*Unmanned Skylab 1 orbital workshop is launched*
Jun 1973	*Skylab 2 crew set record with 28-day mission*
Sep 1973	*Skylab 3 crew set a new 59-day record*
16 Nov 1973-8 Feb 1974	Commander Skylab 4, third manned mission; set record of 84 days in space
25 Dec 1973	Jerry performs his first EVA with Bill Pogue, the second of the mission - 7 hours 01 minute
29 Dec 1973	Jerry's second EVA this time with Ed Gibson, 3 hours 39 minutes. Mission's third excursion outside
3 Feb 1974	Jerry and Ed perform the fourth (Jerry's 3rd) EVA of the flight. 5 hours 30 minutes
8 Feb 1974	Colonel USMC
Nov 1974	*Ed Gibson resigns from NASA*
29 Mar 1974	Jerry Carr Day in Santa Ana, California.
May 1974	Named Head of a Support Group, Shuttle Branch Office, CB, working on cockpit design
Jun 1974	*Military Salyut 3 Soviet space station is launched*
1 Sep 1975	Retired USMC with rank of Colonel
Sep 1975	*Bill Pogue resigns from NASA*
Dec 1975	*Civilian Salyut 4 space station launched*
Jun 1976	*Military Soyuz 5 space station launched*
25 Jun 1977	Retires from NASA and joins Bovay Engineers
Sep 1977	*2nd generation Salyut 6 space station launched*
Mar 1978	*Soyuz 26 cosmonauts break Skylab 4 record and set new one of 96 days.*
1978	Jerry and JoAnn divorce
July 1979	*Unmanned Skylab space station re-enters Earth's atmosphere and is destroyed*
Sep 1979	Jerry marries Pat Musick
Feb 1981	Joins Applied research Inc
12 Apr 1981	*STS-1 (Columbia) is launched, on 20th anniversary of Gagarin's flight*
Apr 1982	*Salyut 7 is launched*
1983-1985	Project Manager University of Texas McDonald Observatory projects

1985	Jerry and Pat form CAMUS, a family company working in both Aerospace and Art. Contracts include Space station definition work with Boeing, and studies for a return to the moon and manned exploration of Mars
1985	Jeff Carr begins work with Media Service at NASA JSC. He joins NASA PAO in 1987, eventually becoming the Acting Director for Public Affairs at JSC before joining United Space Alliance in 1996
Mar 1986	*First core module of Mir space complex is launched*
Dec 1986	Jerry and Pat move into their new property in Arkansas, which remains their home, studio, office and base for operations for the next 20 years.
Mar 1995	*American astronaut Norman Thagard surpasses Skylab 4 record with a mission of over 115 days*
1997	Jerry retires from CAMUS aerospace operations and stops flying as a private pilot.
Oct 1998	*77-year-old Senator John Glenn flies on STS-95 Shuttle mission*
Nov 1998	*first element of International Space Station is launched*
2000	Ozarks woodland Sculpture Garden is begun
2002	Stone Songs on the Trail of Tears work is created
2006 Sep	Jerry and Pat move from Arkansas to a new home in Vermont. Though the move of personal items was quite quick, it took almost 12 months to move all of Pat's art collection and sculpture pieces still in storage and stages of fabrication.
22 Aug 2007	Jerry Carr completes his 75th solar orbit to add to his 1214 Earth orbits accumulated on Skylab

Appendix 2

Awards and Honors

Youth, Scouting & College Years (1932-1954)

1941	Certificate of Perfect Attendance 4th Grade Wilson School
1942	Santa Ana Public Library Junior Department Vocational Reading Club Certificate for reading 10 good books during summer vacation
26 Sep 1943	Certificate of Achievement, Young Men's Christian Association (YMCA) for passing Junior Commando Tests of Santa Ana YMCA Santa Ana Public Library Junior Department Vocational Reading Club Certificates for reading 25 good books during summer vacation.
Mar/Apr 1945	General Eisenhower War Service Medal for Extraordinary Patriotic Achievement in collecting 1000 lbs of waste paper, Boy Scout Troop 24
10 Jun 1946	Certificate of Honor for achieving 29 words per minute typing skill at the H.M. Rowe Company
19 Nov 1946	Elected degree of membership to the International Order of DeMolay for Boys (Masonic Temple)
6 Jun 1947	Willard Echo Staff of Frances E. Willard High School Distinguished Service to the School and Community Award
9 Jun 1947	Junior High School Promotion Certificate Santa Ana

Scout awards

10 Mar 1945	1st Class Scout
Mar 1945-Mar 1948	Merit badges earned: Rabbit raising, safety, physical development, reading, handicraft, path finding, scholarship, mechanical drawing, first aid, personal health, public health, cooking, bird study, life saving, swimming, civics, camping, pioneering, woodwork, wood carving, wood turning, journalism, firemanship, dog care, carpentry, public speaking, business, masonry, metal work, gardening, poultry keeping, aerodynamics, airplane structures, aerodynamics, leather craft
April 20 1945	Star Scout
May 24 1946	Life Scout
15 May 1947	Eagle Scout
Jun 1947	American Legion Distinguished Achievement Award, Francis Willard Junior High School.
Summer 1948	Order of the Arrow, Lodge 298, National Brotherhood of Boy Scouts Honor Group
18-25 Jun 1949	California Boys State American Legion Certificate of Merit.
20 Jun 1949	Elected to State Bar of California (Honorary) - entitled to "practice" law at Boys State.

10 Jan 1950	Elected Student Body Commissioner of Finance
22 May 1950	University of California, Los Angeles (UCLA) Alumni Association scholarship for entering students
6 Jun 1950	Certificate of Life Membership Seal of Chapter 13 California Scholarship Federation for consistent and superior scholarship and service.
9 Jun 1950	Santa Ana High School Student Government Certificate of Merit as Commissioner of Finance for 1949/1950
Jun 1950	University of California (UC), Alumni scholarship for entering students
15 Jun 1950	Graduated Santa Ana High School

College years

18 Sep 1950	Entered the University of Southern California as an NROTC Midshipman
3 May 1951	Elected to Phi Eta Sigma engineering scholastic Fraternity; Honorary member Phi Eta Sigma Gold Legion
17 Jun 1951	Inducted into Tau Kappa Epsilon Fraternity, Beta Sigma Chapter
17 May 1952	Designated NROTC Midshipman Battalion Executive Officer Certificate of Merit for highest scholastic marks in Naval Science, Basic Naval ROTC Course University of Southern California, LA 17th District American Legion Department of California, Aqueduct Post 343
9 Dec 1952	Inducted into Blue Key National honor fraternity, USC, Membership No. 30B567
Spring 1953	Inducted into Skull and Dagger, USC men's honorary.
1953 – 1954	University of southern California scroll of honor for meritorious achievements in distinguished contributions to student advances
1953 – 1954	Prytanis (President) of the Beta Sigma Chapter of TKE
Spring 1954	Awarded the Howard Jones Memorial Award at USC for athletics leadership and scholarship.
19 Apr 1954	Appointed Battalion Commander, NROTC, USC
12 Jun 1954	Graduated from USC with a BS degree (Mechanical Engineering) and the rank of Second Lieutenant, USMC
1954	Top Teke of the year: Beta Sigma Chapter, Tau Kappa Epsilon Fraternity
1954	Member, University of Southern California Alumni Association

USMC Years (1954-1966)

Nov 1954	USMC Pistol and Rifle Expert medals
18 Dec 1954	Graduated USMC 3rd basic school (from 19 Jul)
29 Jan 1955	Graduated USMC basic school post graduate administration course, awarded certificate.
25 May 1956	Designated Naval Aviator

1958	National Defense Service Medal
3 Mar 1959	Letter of Commendation by Commander Carrier Division Two
6-9 Jun 1961	Completed course on Nuclear Weapon Orientation, Training command US pacific Fleet, Nuclear weapons training center, Pacific
Jun 1961	BS Aeronautical Engineering USN Postgraduate school
31 May 1962	Qualified as an air cushion vehicle pilot, US army, "On this day flew to an altitude of 6 inches without oxygen," at Princeton University.
Jun 1962	MS Aeronautical Engineering degree Princeton University
16 Jul 1962	Completed NAMTRADET #1007 F8U Phase 1 pilot familiarization, Naval Air Technical training command, Naval Air Mobile Training Command.
1962	US Marine Corps Expeditionary Medal and Armed Forces Expeditionary Medal for Service in Cuba
Jan 1963	Awarded MIM Verification Anatomist certificate as an expert "anatomist in dissecting and in osculating" an F-8E Crusader aircraft upon completion of his tour at the Chance Vought facility, Dallas, TX where the airplane was dismantled and re-built to verify the newly published Maintenance Instruction Manual (MIM).
1 Jul 1963	Top Gun Award for Air to Air Sidewinder Missile Firing, VMF122 2nd Marine Air Wing, FY63 COMPEX (Competitive Evaluation Exercises) for attaining highest score of any pilot.
13 Aug 1963	Achievement Award from the Commander General 2nd Marine Air Wing

USMC Dates of Promotion:

Airman Apprentice, USN Reserve 23 Oct 1949, Honorably Discharged USN Reserve 17 Sep 1950, Midshipman (NROTC) 18 Sep 1950, 2nd Lieutenant. 4 Jun 1954, 1st Lieutenant. 4 Dec 1955, Captain 1 Oct 1958, Major 22 Sep 1965, Lieutenant Colonel. 1 Jul 1970, Colonel. 8 Feb 1974, Retired Colonel USMC 8 Feb 1974.

NASA Years (1966-1977)

1966	NASA Monogrammed lunch box designated "Carr Moon Kit" containing: Horseshoe Nail Game Pocket telescope Compass Dre's Space Tube Candy – Moon Cherry flavor. Ye Olde Space Labbe National Hero Pills: 'To be taken before blastoffs, Moon landings, Tickertape Motorcades and Presidential Meetings' This was presented to Jerry by his old high school friends just before his departure from Santa Ana to Houston
1966	NASA astronaut Silver Lapel Pin
14 Apr 1967	Designated a qualified SCUBA diver at Key West, FL
Sep 1968	Designated helicopter qualified at Pensacola, FL
28 Oct 1968	Boy Scout award of appreciation for a presentation at the Eagle Scout recognition dinner, Orange Empire area.
1969	Presented Apollo 12 Tether by flight crew in appreciation for duty as CAPCOM (mounted on wall plaque)

1 Oct 1971	NASA group achievement award LRV Team (Apollo 15)
3 Mar 1972	USC certificate of appreciation for serving as president of the Houston USC Alumni Chapter 1970 – 1972
2 Apr 1973	NASA group achievement award, Lunar Landing Team (Apollo)
1973	Member of the Marine Corps Aviation Association
8 Feb 1974	Recovery ship shield, USS New Orleans LP11-11 Helio Support Squadron One
8 Feb 1974	Navy Astronaut Wings
20 Mar 1974	NASA distinguished service medal (Skylab 4)
29 Mar 1974	Santa Ana High School award for outstanding Alumni
15 Apr 1974	NASA group achievement award, Earth Resources Experiment Team (Skylab)
May 1974	Man of the Year award, headliners banquet Orange County Press Club, California
4 Jun 1974	Robert J. Collier Trophy for 1973
27 Jun 1974	Distinguished service medal, USN
8 Aug 1974	Membership to the Society of Experimental Test Pilots
1974	NASA Astronaut gold Pin
1974	City of Chicago Gold Medal
1974	City of New York Gold Medal
1974	USC Alumni Merit Award
1974	Boy Scouts of America Distinguished Eagle Scout Award
1974	AIAA Citation for contributions to astronautics
1974	Award of merit, USC Alumni Association
1974	Received rotational hand controller from SL4 CM
1974	Marine Corps Aviation Association Exceptional Achievement Award
1974	Federation Aeronautique Internationale (FAI) Gold Space Medal
1974	The De La Vaulx Medal
1974	V.M. Komarov Diploma
1974	Aviation Week magazine presentation of 6 laminated front covers featuring scenes from Skylab
Jul 1975	Portrait of Carr in CM presented by Office of Naval Research in appreciation of his work on Skylab ONR experiment
1975	National space club certificate achievement (all Skylab astronauts) presented at Dr. R. H. Goddard memorial dinner, Washington DC
1975	Robert H. Goddard Memorial Trophy (for 1975)

27 Aug 1975	Elected a Fellow of the American Astronautical Society (AAS)
7 Oct 1976	American Astronautical Society Flight achievement award (for 1975)
1976	Honorary D. Sc. Aeronautical Engineering, Parks College, St. Louis University
1977	Founding member, along with Wernher von Braun, of the National Space Forum, later known as the National Space Society
February 26 1977	AAS ANF award Banquet goblet

CAMUS Years (1977-Date)

1977	Joined Eldorado Bank, Tustin, California advisory board
1978	Registered Professional Engineer (Texas)
1978	Texas Society of Professional Engineers
1978	National Society of Professional Engineers
25 Jan 1979	Enrolled as a Registered professional engineer (No. 44480), the State of Texas
1982	Orange County Sports Hall of Fame, Ralph B. Clark Distinguished citizen award
	Member of Order of Daedelions flight #5098
1983	Member Aircraft Owners and Pilots Association
Jan 1985	Boeing certificate of recognition awarded to CAMUS as small business supplier of the month
Jul 1987	Founding director, Space Dermatology Foundation
1986-1992	NASA Headquarters OAST Space Systems Technology Advisory Committee (SSTAC). Served on this committee as human factors engineer and astronaut representative. In 1990 was Chairman SSTAC Ad Hoc Committee on Human Performance.
11 May 1988	Awarded piece of Skylab debris by NASA on 15th anniversary of launch of OWS "Skylab revisit" plaque
1988-1989	NASA Headquarters OSSA Life Science Strategic Planning Study Committee (LSSPSC). Served for two years as an astronaut representative on this committee.
Nov 1991	Human factors achievement award, IDEEA ONE
1991-1993	Arkansas Aerospace Education Center Construction Project Manager
4 Oct 1997	Inducted into the US Astronaut Hall of Fame
1982	Membership Association of Space Explorers (ASE)
1999	Inducted as a Chevalier du Tastevin, Nuit St. George, FRANCE
1999-2004	Trustee of the University of the Ozarks, Clarksville, AR

JACK R. LOUSMA

THOMAS K. MATTINGLY

BRUCE McCANDLESS II

EDGAR D. MITCHELL

WILLIAM R. POGUE

STUART A. ROOSA

JOHN L. SWIGERT, JR.

PAUL J. WEITZ

ALFRED M. WORDEN

Appendix 3

The Original Nineteen

As a NASA astronaut, Jerry Carr was selected to train with 18 other pilots (all male) to support early Apollo missions and as potential crewmembers for later missions to the moon and space stations using Apollo hardware.

Though in general this was achieved between 1966 and 1975, many of the missions they were selected for were cancelled long before reaching the launch manifest. Some members of the selection stayed at NASA to support early Shuttle development and remained active long enough to fly on the Space Shuttle between 1981 and 1990.

Skylab 4 Crewmembers

Jerry flew in space on only one mission, with fellow Group 5 astronaut Bill Pogue and Ed Gibson of the first scientist astronaut selection (1965, Group 4). Like Jerry, neither Pogue nor Gibson made a second trip into space.

Ed Gibson left NASA in November 1974 to conduct research in solar physics with the Aerospace Corporation in Los Angeles. From March 1976, he completed a year's assignment as a consultant to ERNO Raumfarhttechnik GmBH in West Germany before returning to NASA and the Astronaut Office at JSC in March 1977 as a Senior Scientist Astronaut (Mission Specialist). Later assigned as the launch CAPCOM for STS-1, but with little prospect of making another spaceflight, he left the space agency for a second time in October 1980. Subsequently, he worked in the aerospace industry with TRW and Booz, Allen and Hamilton for some years before setting up his own consultancy business and authoring two science fiction novels.

Bill Pogue remained at NASA until September 1, 1975, leaving to become vice president of the High Flight Foundation in Colorado Springs, Colorado. High Flight was a religious organization founded by fellow Group 5 astronaut Jim Irwin. Bill returned to NASA in 1976 as a consultant on programs to study Earth from space. He left for the second time in 1977 to become a privately employed consultant to aerospace and energy corporations. Bill also worked with Jerry on several CAMUS aerospace projects. Bill has also written science fiction novels and two popular non-fiction books on the questions most frequently asked of an astronaut.

The Unflown Members

The career paths of some of those selected in 1966 were as varied and challenging as Jerry's. Of the 19, only Ed Givens (killed in an automobile accident in June 1967) and John Bull never made it to space, although many had long years of waiting for their ride into orbit. John Bull was forced to withdraw from the astronaut program and active naval service in July 1968 when he was medically grounded by the discovery of a rare pulmonary illness. He had been working on the thermal vacuum testing of the Lunar Module with Jim Irwin and was assigned to the support crew of Apollo 8 when he withdrew from the program. Like Jerry, he would have very likely been assigned to a lunar landing mission had the flights continued. After leaving the astronaut program, Bull attended Stanford University Graduate School to study Aeronautical Engineering, receiving his MS in 1971 and his PhD in 1973. He then joined NASA Ames Research Center at Moffett Field, California. From 1973 until 1985, Bull conducted simulation and flight test research in advanced flight systems for both helicopters and fixed wing aircraft. He worked as Chief of the Aircraft Systems Branch and then as a research scientist in the Aircraft Guidance and Navigation Branch. From 1986 until his retirement from NASA in 1989, he managed NASA's wide range of programs in autonomous systems technology for space applications. Since leaving NASA (after 23 years of service) John Bull has provided consultancy services for aerospace research and technology programs.

The Golden Era

The remaining 17 pilots (including Jerry and Bill) selected in April 1966 all made it to space. Of these, nine (Duke, Evans, Haise, Irwin, Mattingly, Mitchell, Roosa, Swigert and Worden) flew as crewmembers on the final five Apollo lunar missions between 1970 and 1972 (Apollo 13 through 17). Mattingly and Haise were scheduled to fly with Jim Lovell on Apollo 13, but Mattingly was replaced by Swigert just days before the mission. The resulting explosion onboard the SM cancelled the planned lunar landing but Mitchell (Apollo 14), Irwin (Apollo 15) and Duke (Apollo 16) made it to the surface as Lunar module pilots. Joe Engle was bumped from Apollo 17 to allow geologist Jack Schmitt (Group 4) to walk on the moon at the end of the program. The CM Pilots in the group were Roosa (Apollo 14), Worden (Apollo 15), Mattingly (Apollo 16) and Evans (Apollo 17) with the latter three each conducting a deep space EVA on the way home.

During 1973, Paul Weitz and Jack Lousma flew as pilots and completed EVAs on the first two manned Skylab missions of 28 and 59 days respectively, before Jerry and Bill flew on the final Skylab mission with Ed Gibson. Vance Brand made the final Apollo-era flight, as CMP for Apollo 18 in July 1975, the first international manned spaceflight, in which an American Apollo docked with a Soviet Soyuz spacecraft.

Shuttling to space

All the remaining Group 5 astronauts worked on support, CAPCOM or back up assignments between 1966 and 1975, covering the Apollo, Skylab and Apollo-Soyuz programs. Don Lind and Bruce McCandless, despite working in support of Apollo and Skylab missions would, like Joe Engle, have a long wait before reaching space aboard the Space Shuttle.

As they were pilots, they were all technically eligible for command of Shuttle missions, though in practice it did not work out like that. Of the Apollo veterans, only Ken Mattingly flew as a Shuttle commander, aboard STS-4 in 1982 (the final Orbital Flight Test of the program) and STS 51-C in 1985 (the first classified DoD Shuttle mission). After Skylab, Jack Lousma later flew as Commander of STS-3 (1982) the third OFT and Paul Weitz as Commander of STS-6 (1983) the maiden flight of Challenger. Vance Brand completed three Shuttle missions as Commander, the first operational mission, STS-5 (1982), the tenth mission, STS-41B (1984), and the 39th mission, STS-35 (1990).

Joe Engle had received his USAF Astronaut Pilot wings in 1965 by flying the X-15 research aircraft higher than 50 miles. After losing his seat on Apollo, he was assigned to early atmospheric tests of the Shuttle in 1977 before finally making it to orbit as Commander of STS-2 in 1981. He returned to orbit as commander of STS 51-I four years later.

Bruce McCandless worked for years in support of Apollo and Skylab missions before support assignments on Shuttle EVA operations. He finally made it to space in 1984, not as a pilot on the Shuttle but as a mission specialist aboard STS-41B (commanded by his colleague Vance Brand). McCandless made history as the first person to fly an untethered MMU on that flight, something he had been working on since the Skylab days. He made a second flight (STS-31) to deploy the Hubble space telescope in 1990. Don Lind had the longest wait of the group. After losing an Apollo seat (probably Apollo 21) and serving as a back up to Skylab, he flew as a Mission specialist on STS 51-B (Spacelab) in 1985, nineteen years after being selected.

Moving on

Having fulfilled their ambitions, the group gradually turned to other challenges in their lives.

The tragedy of losing Ed Givens in June 1967 and the departure of John Bull in July 1968 meant that only 17 of the 19 astronauts selected in April 1966 were active at the time Apollo started flying. They all remained active astronauts through most of the Apollo and Skylab programs.

First to leave the Astronaut Office were Jim Irwin in July 1972 and Al Worden two months later. Following Apollo 15, both had been assigned with their Commander Dave Scott to back up the final Apollo mission, Apollo 17, but were reassigned following a disciplinary action involving commemorative stamps carried to the moon. Irwin left to form the religious foundation High Flight and later authored his autobiography *To Rule the Night*. Sadly, Irwin later suffered two heart attacks after the mission and died of a heart attack on August 8, 1991, aged 61. Al Worden was reassigned to NASA Ames Research Center in Moffett Field, California, initially as a senior aerospace scientist then as chief of the systems division. He resigned from NASA in September 1975 and joined High Flight with his fellow Apollo 15 crewmember Jim Irwin. He later became involved with energy management companies and created his own consultancy business. He also became an author, writing a book of poems entitled *Hello Earth, Greetings from Endeavour,* and a children's book called *A Flight to the Moon,* both published in 1974.

Ed Mitchell completed a dead end assignment as back up LMP for Apollo 16 in 1972 before retiring from NASA in October that year. He subsequently founded the Institute for Noetic Sciences, conducting research into the powers of the mind, as well as forming several consultancy companies. He co-authored a book, *Psychic exploration: A Challenge for science (1974)*, as well as *The Way of an Explorer* in 1996. Jack Swigert trained for the joint ASTP mission after Apollo 13, but when he was not named to the crew he left NASA in April 1973 to become executive director of the committee on Science and Technology of the US House of Representatives, until 31 August 1977. Eligible to return to the Astronaut Office as a Shuttle pilot (and probably the command of an early mission), he decided against it and resigned, choosing instead to run for US Senate. His 1978 campaign was unsuccessful. From 1979 until 1982, he entered the world of business management. He ran for the US House of Representatives in 1982 and was elected Republican Congressman for Colorado's Sixth Congressional District in November that year. Sadly during his campaign he learned that he was suffering from bone cancer. He died on December 27, 1982, aged 51, just a week before he should have taken up his seat in Washington. Bill Pogue left NASA for the first time in September 1975, followed three months later by Charles Duke. Duke had worked on the Shuttle program following Apollo 16. He became a distributor for Coors Beer in San Antonio until March 1978 and subsequently became a Christian lay minister, as well as running his own investment company. In 1990 his autobiography was published, co-written with his wife Dottie, and entitled *Moonwalker*.

Stu Roosa resigned from NASA on February 1, 1976. After Apollo 14, he had worked as a back up CMP for Apollo 16 and 17 and worked on early Shuttle development issues. He moved to Athens in Greece as vice president for international affairs for the US Industrial Middle East company. He returned to the United States in 1977 to become President of Jet Industries in Texas, then President and owner of Gulf Coast Coors in Gulfport, Mississippi in 1981. He died of pancreatic cancer on December 12, 1994, aged 61.

Jerry became the second member of his class to leave NASA during 1977. Three months earlier, in March of that year, Apollo 17 astronaut Ron Evans left the Astronaut Office to join Weston American Energy Corporation in Scottsdale, Arizona. After flying around the moon on Apollo 17, Evans was assigned as back up CMP for the 1975 ASTP mission. In 1978, Evans joined Sperry Flight Systems in Phoenix, rising to become director of space systems marketing before subsequently forming his own consultancy firm. He died of a heart attack on April 6, 1990, aged 56.

The final member of the group to leave NASA in the 1970s was Fred Haise, who should have flown to the moon with Jerry as Commander of Apollo 19, had it not been cancelled. Haise did go on to serve as back up Commander of Apollo 16 in 1972 before being assigned to Shuttle development from 1973. He served as commander of the first approach and Landing Test crew, evaluating the landing capabilities of a Shuttle Orbiter and flying Enterprise (OV-101) to several landings at Edwards AFB off the back of a converted Boeing 747. In 1978, Haise was named to command the third planned Orbital Flight Test, which at the time also included a rendezvous with the unmanned Skylab space station. However, several delays resulted in this objective being cancelled well before Skylab re-entered the atmosphere in 1979. Haise decide to leave the agency in June 1979 and joined Grumman Aerospace Corporation, working with that company for over 20 years until retirement.

By the early 1980s, the remaining Group 5 astronauts were all assigned to the emerging Shuttle program. Their chances of flying an early mission improved greatly, so the decision to stay until at lease one Shuttle flight had been completed was understandable. Once they had achieved their Shuttle flight, and with so many astronauts waiting in line for their first mission, many veteran astronauts decided to move on.

The first Shuttle Group 5 veteran to leave the program was Jack Lousma, who departed on October 1, 1983. He ran unsuccessfully in 1984 as a candidate for the US Senate from Michigan. Jack later worked on several projects with Jerry at CAMUS and has created his own high technology consultancy firm based in Ann Arbor, Michigan. Ken Mattingly was next to leave NASA. After flying on his second Shuttle mission early in 1985, he resigned five months later in June 1985 to Command the US Navy's Electronic Systems Command, the first Group 5 astronaut to return to active duty with his parent service. He also became Director of Space Systems for the Space and Naval Warfare Systems Command prior to retiring from the US Navy with the rank of Rear Admiral in 1990. He then joined Grumman (working with Fred Haise) at the company's Space Station Integration Division in Reston, Virginia, as Director of Utilization and Operations. He joined the Space Systems Division of General Dynamics in San Diego, California, in 1993, holding the position of Deputy Program Director for the Atlas launch vehicle. Subsequently, he served as Program Director for the X-33 reusable launch vehicle program for Lockheed Martin.

Prior to the loss of Challenger in January 1986, Don Lind had announced his intention to leave NASA.

In April 1986 he joined the faculty of Utah State University as Professor of Physics until his retirement. His autobiography *Don Lind: Mormon Astronaut* was published in 1985, shortly after completing his only spaceflight. Joe Engle retired from NASA in November 1986 and became an advisor to the Kansas National Guard and consultant to Rockwell International on the National Aerospace Plane. He also provided technical support to the Tom Stafford Task Force reviewing the Shuttle-Mir operations and ISS training issues. Paul Weitz served as Deputy Chief of the Astronaut Office following his STS-6 mission. In 1988 he assumed a NASA Management role as Deputy Director of NASA JSC, and in 1993 became Acting Director of the field center until his retirement from the agency in May 1994. Since then, he has pursued his favorite past time of fishing across the world as often as he can. Following his second space flight in April 1990, Bruce McCandless departed from the Astronaut Office on August 31 to become an aerospace consultant, working on various projects including the Shuttle Hubble Servicing Mission.

In 1992, Vance Brand became the final Group 5 astronaut to leave the Astronaut Office, after four spaceflights and 26 years of active astronaut service. He worked initially at Wright Patterson AFB in Ohio on the National Aerospace Plane, until transferring to NASA's Dryden Research Center when the NASP was cancelled. At Dryden, Brand served as Assistant Chief of Flight Operations, then Acting Chief Engineer, then Deputy Director for Aerospace Projects, and more recently as Acting Associate Center Director for Programs. Aged 76, Brand still worked at NASA over 41 years after his selection as an astronaut.

VANCE D. BRAND   JOHN S. BULL   GERALD P. CARR

CHARLES M. DUKE, JR.   JOE H. ENGLE   RONALD E. EVANS   EDWARD G. GIVENS, JR.

FRED W. HAISE, JR.   JAMES B. IRWIN   DON L. LIND

Appendix 4

Skylab 4

Mission Data

International designation 1973-90A
Launch date: November 16, 1973
Launch time: 09:01 EDT
Launch site: Pad 39B Kennedy Space Center Florida
Launch vehicle: Saturn 1B (SA-208A)
CSM: 118
Rescue vehicle: SA-209/CSM119 (backup)
Recovery: February 8, 1974
Recovery Time: 11:17 EDT
Recovery Site: Pacific Ocean

Third manned crew

Prime crew:	Commander	Gerald P Carr, 41, USMC
	Pilot	William R Pogue, 43 USAF
	Science pilot	Edward G Gibson, 37 Civilian
Backup Crew:	Commander	Vance D Brand
	Pilot	Don L Lind
	Science Pilot	William B Lenoir
Rescue Crew:	Commander	Vance D Brand
	Pilot	Don L Lind

Distance traveled: 70.5 million miles
Mission duration: 84 days 1 hour 15 minutes 37 seconds
Manned orbits: 1214

49th manned space flight
30th US manned space flight
14th Apollo CSM manned space flight
3rd manned Skylab crew
15th US and 17th world flight with EVA operations

Man-hour utilization: Skylab 4

Medical activities:	336.7 hours	6.1%
Solar Observations:	519.0 hours	8.5%
Earth resources:	274.5 hours	4.5%
Other experiments:	403.0 hours	6.7%
Sleep/rest and off duty:	1846.5 hours	30.5%
Pre/post sleep and eating:	1384.0 hours	23.0%
Housekeeping	298.9 hours	4.9%
Physical training/personal Hygiene	384.5 hours	6.4%
Other (EVA etc)	571.4 hours	9.4%
Totals:	6048.5 hours	100.0%

Experiment performance:

Solar astronomy	519.0 hours	33.2%
Earth Observations	274.5 hours	17.6%
Student experiments	14.8 hours	0.9%
Astrophysics	133.8 hours	8.5%
Man/systems	83.0 hours	5.3%

Material science	15.4 hours	1.0%
Life science	366.7 hours	23.5%
Comet Kohoutek	156.0 hours	10.0%
Totals	1563.2 hours	100.0%

Data returned

Solar observations	73,366 frames	
Earth observations (film)	19,400 frames	
Earth observations (magnetic tape)		100,000 feet

Extra Vehicular Activity

EVA1	22nd November 1973	6 hours 33 minutes	Pogue / Gibson
EVA2	25th December 1973	7 hours 1 minute	Carr / Pogue
EVA3	29th December 1973	3 hours 28 minutes	Carr / Gibson
EVA4	3rd February 1974	5 hours 19 minutes	Carr / Gibson
EVA Total for mission		22 hours 21 minutes	
EVA total for Carr		15 hours 51 minutes (3 EVAs)	
EVA total for Gibson		15 hours 20 minutes (3 EVAs)	
EVA total for Pogue		13 hours 34 minutes (2 EVAs)	

Mission achievements

Set world, mission and individual crew accumulated time in space (84 days 1 hours 15 minutes 37 seconds)

Held world endurance record for single/accumulated manned space flight for more than three years (until surpassed by Soyuz 26/Salyut 6, 96 days, 1977-78)

Held US space endurance/space station record for more than 21 years (until surpassed by Norman Thagard, Mir-21 crew March - July 1995)

Observed and photographed Comet Kohoutek

Increased length of time in space by approx. 50% (Skylab 3-59 days)

Jerry Carr

74th human (73rd male) to fly in space

47th American (47th male) to fly in space

13th of his selection (1966) to fly in space

10th of his selection to perform 1 EVA (7th for 2 EVAs; 3rd for 3 EVAs)

29th human (26th male) to perform 1 EVA

18th human (18th male) to perform 2 EVAs (also 18th American/male)

11th human (11th male) to perform 3 EVAs (also 11th American/male)

Bibliography

The writing of any book such as this requires a vast amount of references and source material. This project began in book form in 1994, though research by the author into the career of Jerry Carr and other members of the NASA astronaut team really commenced in 1969. Information assembled over these years came from the series of official NASA Astronaut Biographies released on each astronaut and updated periodically until shortly after their retirement from the NASA program. In addition, several periodicals provided useful news items over 1965-1977, including: *Aviation Week and Space Technology*, *Flight International*, the British Interplanetary Society magazine *Spaceflight*, NASA Activities and JSC Space News Round Up. The series of annual chronicles by NASA (Aeronautics and Astronautics) under the code SP-4000 (especially those for 1965-1977) and the extensive archive of Jerry's own collection were frequently referred to for both Jerry's biography and the larger Skylab story.

Taped and transcribed interviews:
A series of personal and telephone interviews were conducted in support of this project and subsequently transcribed.

> *Jerry Carr:* August 1988; August 1989; September 1994; December 1997; December 1998; July 2000; October 2002, September 2003; November 2004.
> *JoAnn Carr:* May 2002; October 2002
> *Jennifer Carr:* May 2002
> *Jeff Carr:* May 2002
> *John Carr:* May 2002
> *Jamee Carr:* May 2002
> *Jessica Carr:* May 2002
> *Josh Carr:* May 2002
> *Pat Musick (Carr):* 1999; October 2002
> *Bill Pogue:* September 1999; July 2000
> *Ed Gibson:* January 2004
> *Al Haraway:* November 2002

The following publications were of great help and are suggested for future reading:

Jerry Carr: (biographical references)
1966	Nineteen pilots will join the astronaut team. NASA News MSC 66-22
1966-77	NASA Biographical Data Sheets, Gerald P. Carr, April 1966; July 1966; November 1973 (Press Kit); May 1974; September 1975; June 1977
1977	Astronaut Carr Leaves NASA to Join Engineering firm, NASA JSC News 77-36, 22 June 1977
1983	Gerald P Carr entry in NASA Astronaut Biographical Data Record Book, Group 5 April 1966, David J Shayler, AIS Publications April 1983
2000	NASA JSC Oral History Project, Gerald Paul Carr, 25 October, transcript, Biography sheet 8 October 1999.

Boy Scouts of America movement (Chapter 2)
1942	Revised Handbook for Boys 1st Edition (35th printing), Boy Scouts of America, New York

United States Marine Corps (Chapter 3)
1955	3rd Basic Course, Able company, Marine Corps School, commemorative book
1960	FDR Mediterranean Cruise crew commemorative book
1962	Model Studies to Determine a Winged GEM Configuration for the Curtis-Wright air Car, by Captain Gerald P. Carr USMC and Captain John J Metzko, USMC, Dept of Aeronautical Engineering Princeton University Report No 608 May 1962
1962	Guide book for the Marines (8th revised edition, 1st printing, dated 1 June 1962), published by the Leatherneck Association Washington DC
1963	The Marine Officers guide by General G.C Thomas USMC Retired, Colonel R.D. Heinl Jr USMC and Rear Admiral A.A. Ageton USN Retired

NASA Astronaut Program (Chapter 1, 4 and 10)
1966   Project Apollo Flight Crew General Training Plan, prepared by Raymond G.
          Zedekar, Assistant Chief for Crew Training, Flight Crew Support Division, NASA
          MSC. NASA 66-CF-1
1973   Skylab 4 Press Kit, NASA
1983   NASA Astronaut Biographical Data Book Group 5 April 1983, David J. Shayler,
          AIS Publications
1994   Deke, the US Manned Space Program from Mercury to the Shuttle, Donald K.
          Slayton with Michael Cassutt, Forge Books
1999   Who's Who in Space The ISS Edition, Michael Cassutt. McMillan
2000   NASA JSC Oral History Project - Ed Gibson (April)
2000   NASA JSC Oral History Project - William Pogue (July)
2007   NASA Scientist Astronauts, David J. Shayler and Colin Burgess; Springer Praxis

Apollo Program (Chapter 4 and 5)
1978   The Apollo Spacecraft, a Chronology Volume IV 21 January 1966-13 July 1974,
          Ivan D. Ertel and Roland Newkirk, NASA SP-4009
1979   Chariots for Apollo: A History of Manned Lunar Spacecraft, Courtney Brooks,
          James Grimwood and Lloyd Swenson, NASA SP-4205
1989   Where No Man Has Gone Before, a History of Apollo Lunar Exploration Missions,
          William David Compton, NASA SP-4212
1998   Genesis, the Story of Apollo 8, Robert Zimmerman, Four Walls Eight Windows
1999   Exploring the Moon: The Apollo Expeditions, David M. Harland, Springer-Praxis
1999   Apollo 8, the NASA Mission Reports, Ed Robert Godwin Apogee Books
1999   Apollo 12, the NASA Mission Reports, Ed Robert Godwin Apogee Books
2001   Apollo 15, the NASA Mission Reports, Ed Robert Godwin Apogee Books
2002   Apollo: The Lost and Forgotten Missions, David J. Shayler, Springer -Praxis
2006   Apollo the Definitive Source Book, Richard W. Orloff and David M. Harland,
          Springer-Praxis

Skylab Program (Chapter 6 through 10)
1973   Skylab, a guide book. Leland F. Belew & Ernst Stuhlinger, NASA SP-107
1973/4   Skylab 1 / 4 Technical Air to Ground Voice Transcripts (A Tapes) NASA Test
          Division Program Operations Office, JSC JSC-08652
1974   Skylab 4 Technical Crew Debrief, February 22 1974 JSC-088099, NASA JSC.
1974   Skylab 1 / 4 Onboard Voice Transcription, CM and AM recorders (B Tapes)
          JSC Test Division Program Operations Office, March 1974
1974   MSFC Skylab Mission Report - Saturn Workshop, NASA Technical Memorandum
          NASA TMX-64814
1974   Skylab: The Three month Vigil, David Baker. *Spaceflight*, 16, nos. 11, and 12,
          17 No 1 (1975) British Interplanetary Society
1977   Skylab a Chronology, Roland Newkirk, Ivan Ertel with Courtney Brooks, NASA
          SP-4001
1977   Biomedical Results from Skylab Ed Richard S. Johnston and Lawrence F. Dietlien,
          NASA SP-377
1977:   Skylab: Our First Space Station, Ed. Leland Belew, NASA SP-400
1977   Skylab Explores the Earth, NASA SP-380
1977   Skylab: Classroom in space, Ed Lee B. Summerlin, NASA SP-401
1978   Skylab EREP Investigations Summary, NASA SP-399
1979   A New Sun: the Solar Results from Skylab, John A. Eddy, NASA SP-402
1979   Skylab's Astronomy and Space Sciences Ed Charles A. Lundquist, NASA SP-404
1983   Living and Working in Space: A history of Skylab, W. David Compton and
          Charles D. Benson., NASA SP-4208
1985   Astronaut Primer, William Pogue
2001   Skylab: America's Space Station, David J. Shayler, Springer-Praxis

Space Shuttle Program (Chapter 10)
2001   Space Shuttle: the history of the National Space Transportation System - The
2004   The Story of the space Shuttle, David M Harland, Springer-Praxis

Space Stations (Chapter 11)

1984	Space Station Habitability Recommendations Based on a Systematic Analysis of Analogous Conditions, Jack W. Schuster, ANACAPA Sciences Inc
1984	Aerospace Crew Station Design Proceedings of a Course given at the International Center for Transportation Studies (ICTS) Amalfi, Italy October 19-22,1983, Amsterdam, The Netherlands, Elsevier Science Publishers B.V.
2000	The History of Mir 1986-2000, Ed Rex Hall, British Interplanetary Society; also Mir the Final Year supplement , Ed Rex Hall BIS 2001
2002	From Imagination to Reality, the ISS, Ed Rex Hall, BIS
2002	Creating the ISS, David M Harland and John E. Catchpole
2002	From Imagination to Reality, the ISDS Volume 2 Ed Rex Hall, BIS
2005	The Story of Space Station Mir, David M. Harland, Springer-Praxis

Manned Spaceflight History (General)

1980	History of Manned Spaceflight, David Baker, New Cavendish
1997	Walking to Olympus: an EVA Chronology, David S.F. Portree and Robert C. Treviño, NASA Monograph #7
2000	Challenges of Human Space Exploration, Marsha Freeman, Springer-Praxis
2004	Walking in space, David J. Shayler, Springer-Praxis
2006	Praxis Manned spaceflight log 1961-2006, Tim Furniss David J. Shayler with Michael D Shayler, Springer-Praxis

CAMUS (Chapter 11)

1988	EVA at Geosynchronous Earth Orbit, Gerald P. Carr, Nicholas Shields Jr., Arthur E. Schulze and William R. Pogue, NASA Technical Reports JSC
1988	Advanced EVA Systems Requirements Definition Study Phase 2 EVA at a Lunar Base, Gerald P Carr, Valerie Neal, Nicholas Shields Jr, William R Pogue, Harrison H. Schmitt and Arthur E. Schulze, NASA Technical Report JSC
1989	EVA in Mars Surface Exploration Final Report, prepared for NASA Advanced EVA Systems Requirement Definition Study, NAS9-17779, Phase III, (a CAMUS/Boeing sub contract to NASA)
1990	Advanced EVA Requirements in Support of the Manned Mars Mission, Gerald P. Carr, William R. Pogue and Nicholas Shields Jr. NASA Technical Reports JSC
1995	Italy and Space Habitat Modules, Ed by Giovanni Caprara, McGraw Hill (CAMUS consultancy)

Pat Musick (Chapter 11 and 12)

1992	Epilogue Series Catalogue
1994	Artists of the Spirit by Mary Carroll Nelson
1997	Rural Arkansas
1999	The Texas Tour Catalogue
2005	Stone Songs on the Trial of Tears, Journey of an Installation, with Jerry Carr and Bill Woodiel, University of Arkansas Press

INDEX

BONUS DVD-VIDEO

Included on the accompanying bonus DVD-Video

"Skylab - Space Station 1" - a documentary about the entire Skylab program including interviews with all three crews

"Four Rooms - Earth View" - a documentary about the Skylab program including all nine crew members

"Jerry Carr Training Footage"
      Including the following rare footage narrated by Jerry Carr:

      Jerry Carr training with the Lunar Module for Apollo 12, MESA deployment,
      Apollo 12 Back-up crew training with ALSEP familiarization and deployment,
      Skylab food preparation and taste-testing,
      A tour of the Orbital Work Shop
      Martin-Marietta M509 MMU Training

Jerry Carr training footage also features rare clips of Pete Conrad, Alan Bean, Rusty Schweickart, Dick Gordon, Vance Brand, Deke Slayton, Joe Kerwin and others.